Table of Contents

AUTHOR'S MINI-BIOGRAPHY

Edwin Wise develops CAD/CAM software during the day and explores the edges of mad science at night through his R&D company Simulated Reality Systems. He has written game software for Broderbund and Dynamix, and manufacturing software for Point Control, Building Blocks, and now as a partner of Wittlock Engineering LLC. He has written two books on robotics (*Applied Robotics*, and the sequel to it that you now hold) as well as *Animatronics: A Guide to Animated Holiday Displays*. Edwin feels that Halloween should be celebrated 365 days a year.

INTRODUCTION

Before you rush home with this book and start building from its pages, ask yourself, why are you building a robot? What do you want your robot to do? What are you trying to accomplish by building a robot? Are you looking for a practical appliance that will vacuum your house or mow your lawn? Do you see robotics as the wave of the future and want to get a leg up on the techniques involved, perhaps to get some educational (and hard-won) experience under your belt? Or maybe you just think robots are fun, and you are hoping for some new ideas to try out? Whatever your goals are, I hope this book can help you.

Note that this book is the second of a series. *Applied Robotics* introduces the reader to robotics projects. Most of the projects in that first book are simple both mechanically and electronically. This second book does not so much *extend* the work in *Applied Robotics* as *kick it up a notch* (if I may borrow this culinary phrase). The projects are bigger, more complicated, and for the experimenter who has a few desktop robots under his or her belt already, more interesting.

This book is not an end but only a single step in the journey of robot creation. And hopefully not a *first* step; this is a challenging book, with challenging projects. You'll note, for example, that once you pass Chapter 8, the projects become more and more abstract until, by the last chapter, they are little more than directions and guidelines. This is because each chapter in the later parts of the book stands in for many books' worth of information! You need to be devoted to the robotic cause to work through them all, and you must be willing to do some extra research to fill in the gaps. The Internet, where you can find most of the research papers that paved my way, extends this book. Be sure to check out the References list in Appendix B at the back of this book for complete information on those papers, as well as details on a number of other resources.

The Internet is used to expand this book in a more direct way as well. To keep this book a reasonable (and hence *affordable*) size, we have put some of the seriously technical details on the Internet at www.simreal.com, making this book something of a cyborg itself; part of it is electronic!

Instead of using this Introduction to provide my traditional listing of the tools and materials you will need to create the projects in this book (answer, "lots, including the Internet"), I will instead give you a whirlwind tour of the contents of the book, to help set the stage, and perhaps convince the idle bookshop reader that, yes, you want to buy this book!

It all begins in Chapter 1, which talks about motors. Not just little motors, but bigger motors—motors that are big enough that you start to care about their electromagnetic interference (EMI) characteristics.

Chapter 2 is about batteries and their use, including power supplies that run off of them. (You can learn even more about batteries in Appendix A.) My favorite part of Chapter 2 is where we build a switching power supply—these had always been mysterious to me before, and now I'm talking about how to make one. Very exciting.

Once we have some motors on the table and the power system of the robot has been laid out, we are ready for Chapter 3, the rolling platform itself. True to form, I make this from welded steel, though the design is not so complex that you can't bolt one together just as easily.

Chapter 4 covers your motor control options, providing details about using off-the-shelf R/C car products (don't laugh, these are both easy to use and pack quite a punch), and introducing the concept of home-built speed controllers. If that last one catches your fancy, there's a project online just for you.

Chapter 5 fills out the rolling platform with a fairly simplistic upper body, suitable for holding the laptop "brain," introduced later, plus a simple and, admittedly, somewhat wobbly, pan-tilt camera "head."

Rounding out the mechanical work are Chapters 6 and 7, where we create an arm and hand. Though I have not achieved perfection in those projects, they provide an excellent and unique base for a complex and effective manipulator.

Chapter 8 begins the control electronics with the requisite reflex microcontroller project. This book uses the Cygnal 8051-based MCU chip, though to be honest this MCU could be substituted by any microcontroller of your choice.

Chapters 9 and 10 use the MCU systems built in Chapter 8 to provide motor control signals and sensory input to your nascent robot. Chapter 11 tops it off by adding a laptop computer onto the reflex control network, to do the really heavy thinking. The final project in Chapter 11 captures images from the robot's onboard camera.

Rolling into Chapter 12 the book becomes more abstract as we discuss the creation of an interesting brain, and begin discussing various technologies you could use for it. Chapter 13 heats up the discussion with some detailed projects in neurons and neural networks—stuff you should be able to sink your teeth into.

The book coasts to an end through Chapters 14 and 15, with a variety of essays on learning, spatial mapping, vision, and language. These chapters, for the most part, build on the technologies explored in Chapter 13.

This is my robot book. I hope you enjoy it.

ACKNOWLEDGMENTS

I have to give lots of credit to Melissa Wise and Marla Wise, for reasons they are both fully aware of. I had a nice line to write here about each of them but, carelessly, I lost it, so—just thanks. Of course, I have to give a nod to Debbie Abshier and her gang at Abshier House Publishing for putting up with my endless changes in plan (by the way, Debbie, I rearranged the final chapters again). Of course, this book would not be half as good as it is without the diligent efforts of that gang—Joell Smith-Borne, Kim Heusel, Michael Woodward, Phil Velikan, and probably a stack of other people I'm not aware of. Interestingly enough, I've never met any of these people, who are spread out across the country.

And finally, I acknowledge you, the reader—without your support, comments, and interest in the first book, this second volume would never have been created.

CHAPTER 1

AND SO IT BEGINS...

Figure 1-1 shows our ultimate goal of the first half of this book—a robot! The next set of chapters goes through the various steps needed to create this machine in the flesh, as it were. Then, from Chapters 11 through the end of the book, we explore various ways of making this machine come alive.

The most noticeable element of any electromechanical device (e.g. robot) are the motors...if for no other reason than these are the obvious source of that machine's motion.

Figure 1-1 The robot

ROBOT MOTORS

We are only going to use one particular type of motor in these projects—the brushed, permanent magnet, direct current (PMDC) motor. There are other choices out there, too, like AC motors, stepper motors, brushless motors, and so forth.

AC motors use cyclic power, like what comes out of your wall socket, and are not suitable for battery-powered applications. This removes a whole area of motor technology from consideration.

DC motors come in a variety of styles. The basic apply-voltage-and-it-runs motor you find in most projects is the PMDC motor named above. You can also find field-energized motors (where an electromagnet replaces the permanent magnet on the stator), which are found in series-wound or shunt-wound flavors. There is also the brushless DC motor, where special control logic replaces the brushed commutator.

This list just touches the surface of the many options available, but there is plenty to talk about for just the PMDC motor, so let's get on with that.

HOW PMDC MOTORS WORK

To understand how PMDC motors work, you need to know how the pieces are organized inside the motor, and how the electric flow and magnetic fields behave.

PMDC motors consist of three basic parts. The stator provides the nonmoving (static) magnetic field. The rotor (or armature) is the part that turns and carries the electric wires that provide the motive force in the motor. The third part is the commutator, and it switches (commutes) the incoming electricity.

You can see the outer can of a small PMDC motor in the left of **Figure 1-2**. Inside this can are the two permanent magnets that make up the stator of the motor. The rotor and armature are shown in the center, including the contacts at the left end that are the moving part of the commutator. You can see the brushes for the commutator are shown to the right.

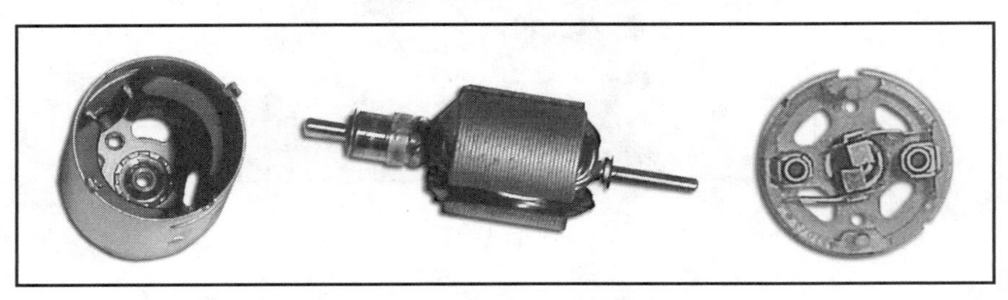

Figure 1-2 Stator, rotor, and commutator

With the two stator poles and three armature poles, the rotor creates a complex dance of magnetism and electricity as it turns. Let's start the discussion with the barest minimums—the simplified (almost cartoon) version of a PMDC motor.

CHARGE MOTION IN A MAGNETIC FIELD

A pair of magnets in a stator create a fixed magnetic field, from north to south (see **Figure 1-3**). This is known as the B-Field, or B-Vector. *B* is a vector, so it has both a direction (from north to south) and a magnitude (measured in *webers*).

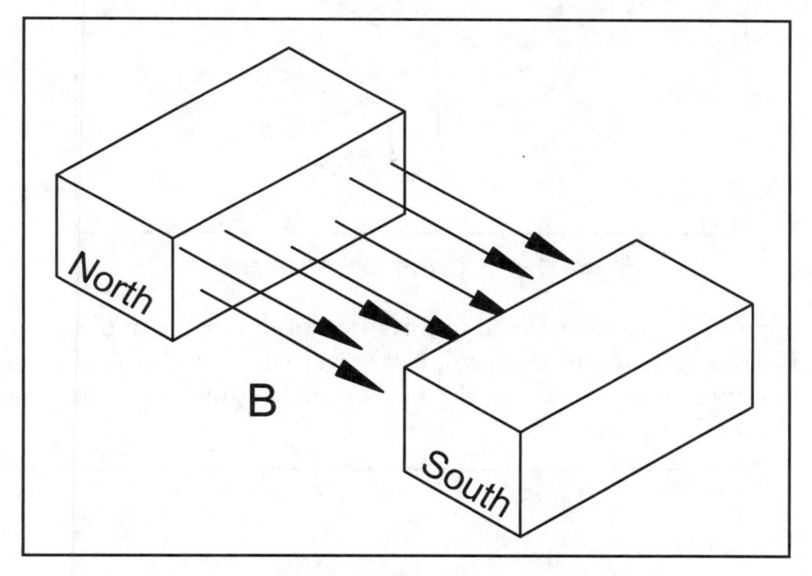

Figure 1-3 Magnets

The weber is a measure of magnetic flux and it represents the continuous lines of force that connect the magnetic poles. One weber is equal to 10^8 *maxwells* (another measure of magnetic force). Additionally, the *tesla* is a unit of magnetic flux density and is equal to 1 weber per square meter, or 10^4 *gauss*. Finally, 1 gauss is equal to 1 maxwell per square centimeter, or 10^{-4} webers per square meter. But today, we'll simply call it *B*.

1 weber = 10^8 maxwell

1 gauss = 1 maxwell/sq. cm.

1 tesla = 10^4 gauss

1 tesla = 1 weber/sq. m.

Adding a wire to the magnetic field, we get **Figure 1-4**. Negatively charged particles move through the wire; we'll call their motion the vector *I*.

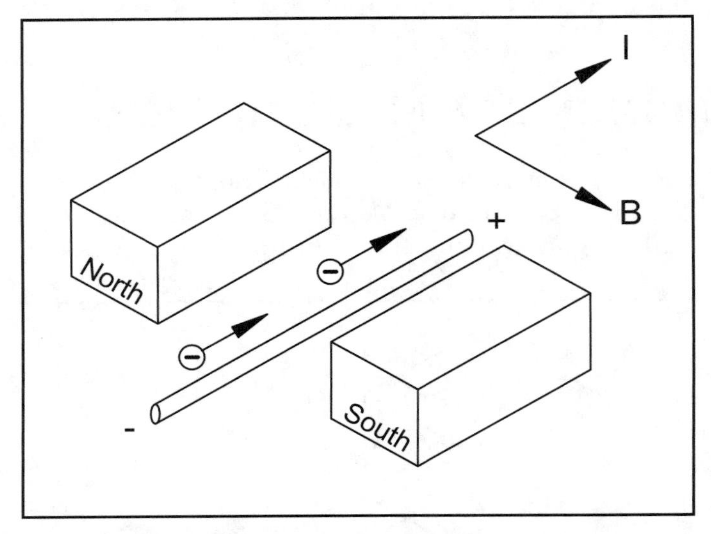

Figure 1-4 Charge motion in a wire

Traditionally, positive charges are used in this discussion, moving in the other direction. Since both the direction and the charge sign are reversed, the ultimate effect is the same; $(-)(-)I = I$. Our applications focus on the motion of electrons, so we use the negative-charge form.

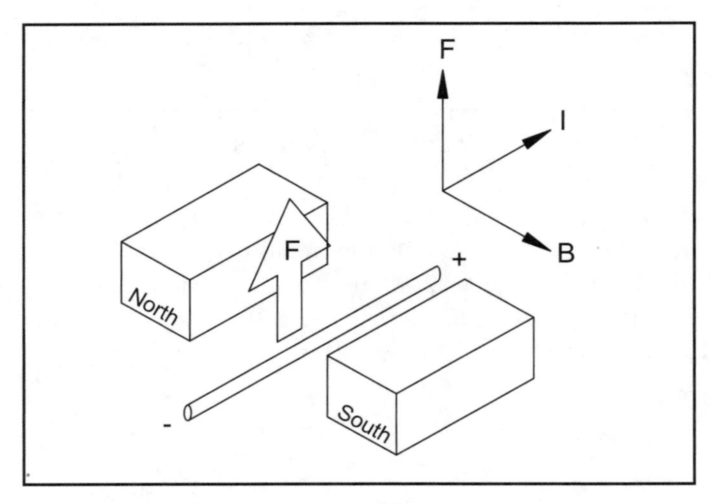

Figure 1-5 Induced force

The charge I moving at right angles to the magnetic flux B creates a physical force F in the wire, as you can see in **Figure 1-5**. The magnitude of this force corresponds to the angle of vector I to vector B as well as to the amount of wire exposed to the flux, represented by the wire's length L:

$$F = I * L \times B$$

(Note: L x B is the L vector cross the B vector, multiplied by the I magnitude; look to a math text for a discussion of cross products).

Conversely, physically moving the wire through the magnetic field creates a reverse motion v in the charged particles, as shown in Figure 1-6. This is how generators work, and it is also the source of the back electromotive force (back EMF) in a running motor.

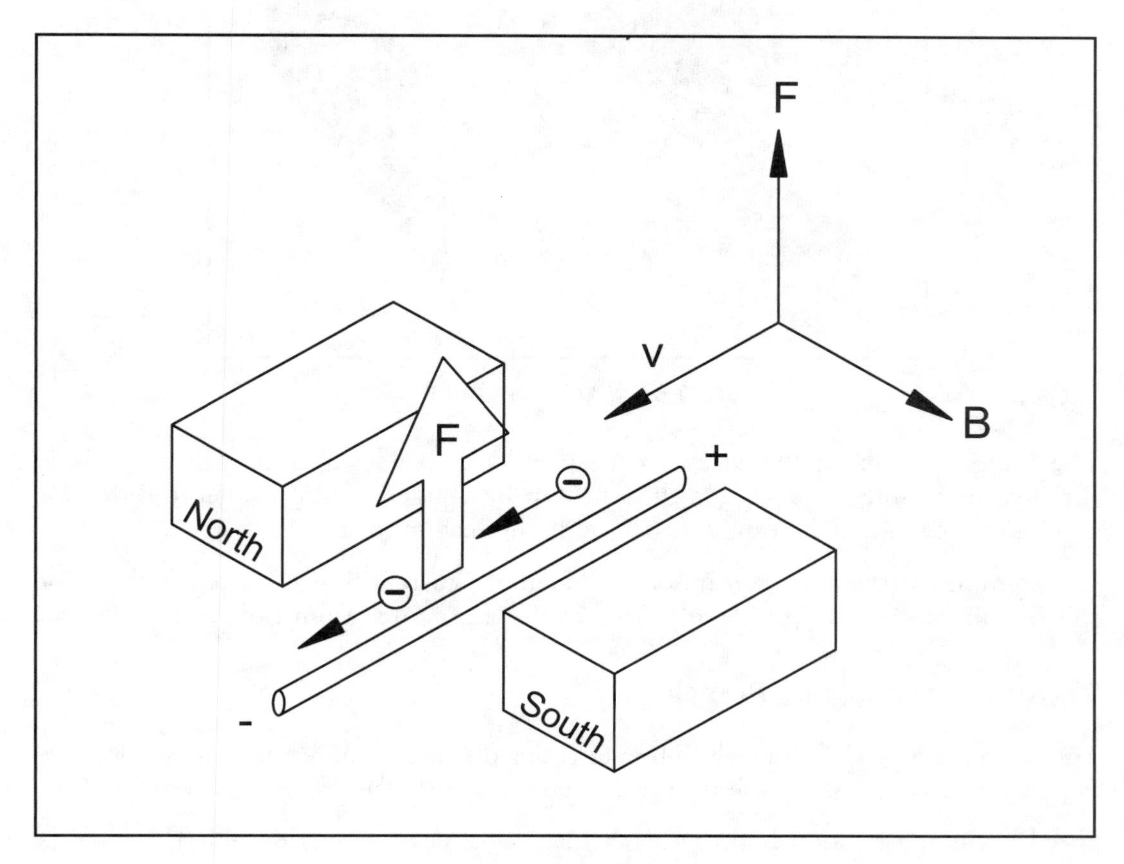

Figure 1-6 Back EMF

Moving from the single-wire case back to something that resembles an actual motor, you can see the effect of looping the wire back through the magnetic field in **Figure 1-7**. The forward charge motion creates an upward force on the left half of the rotor, while the reverse charge motion creates a downward force on the right half. Actual motors will typically have more than one turn, to maximize the effective length L of the wire in the magnetic field (while also trying to keep the resistance low, to maximize current flow *I*).

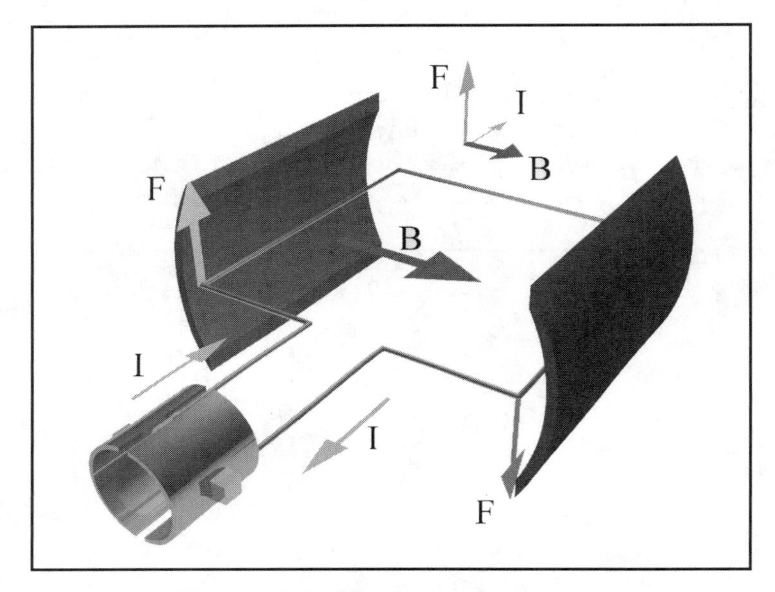

Figure 1-7 Simple armature loop

The charge motion on the wire that lies parallel to B does nothing, since the force varies with the sine of its angle with respect to B (charge motion parallel to B doesn't add to the force, and motion across B adds the maximum force).

The commutator switches continuously change the current flow direction as the rotor spins, maintaining an optimal angle to get the most force from the system.

TORQUE, SPEED, AND POWER

Torque, τ, is a *twisting* force, defined by a force at a distance from the center of rotation (see **Figure 1-8**). Torque is typically specifed as pound-feet (lb/ft) or ounce-inches (oz/in).

Speed, ω, is the rotational speed (or angular velocity) of a rotating object. Speed is typically specified as rotations per minute (RPM).

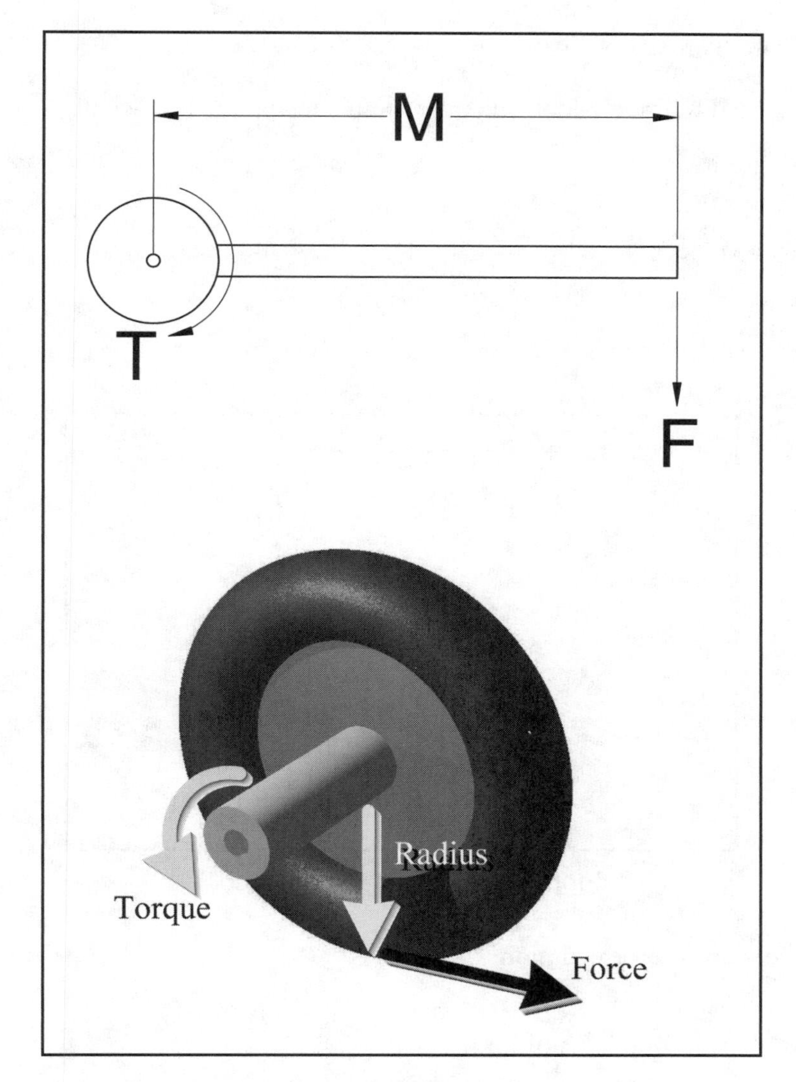

Figure 1-8 Torque, speed, and power

Power specifies the amount of work the motor is able to perform, as defined by a force applied over time. Power may be specified as watts (W), Newton-meters per second (Nm/s), pound-feet per second (lb-ft/s), or horsepower (HP).

If you are interested, the precise definitions (mathematical relationships and other background information) can be found at http://www.simreal.com.

MOTOR CONSTANTS

Electrically, a PMDC motor looks like the circuit in Figure 1-9, where:

E = Applied Voltage

I = Motor Current

L = Winding Inductance

R_T = Terminal Resistance

V = Back EMF Voltage

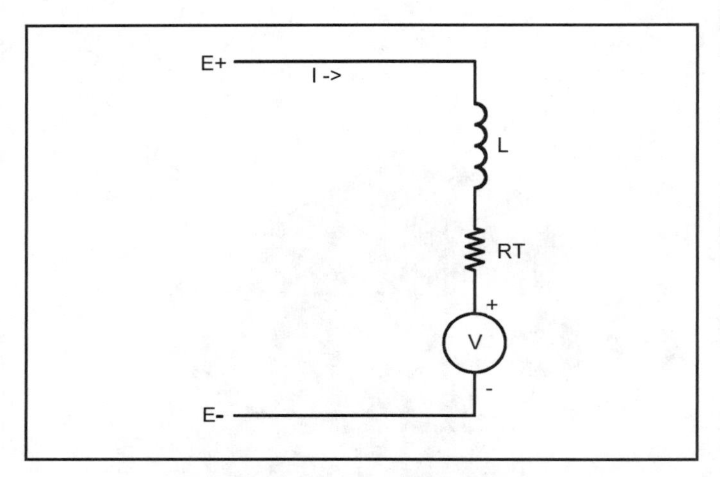

Figure 1-9 Basic Motor Circuit

This gives the basic motor equation:

$$E = I * R_T + V + L * \frac{\Delta I}{\Delta t}$$

If the motor is driven by an electronic circuit (as opposed to a relay or a switch with negligible resistance), R_T should include the entire circuit resistance.

Of course, look into a motor catalog, such as Globe Motors (www.globemotors.com) or Pittman (www.pittman.com) and you won't see the motors defined in terms of τ, ω, or the other values explored so far. Sure, there may be the stall torque or no-load RPM, but there are also the values K_T, K_E, and so on. What do these mean? Looking in my Pittman catalog, let's take a tour of some motor constants.

I_{NL} and I_P are the no-load current and peak (stall) current. T_C and T_{PK} are the continuous and peak (stall) torques, and S_0 is the no-load speed. R_T and L are the terminal resistance and inductance of the windings. E is the reference voltage.

Ignoring a bunch of parameters, we cut to the key constants.

The back-EMF constant K_E is the "voltage per unit speed" of the motor, where the speed may be defined in radians per second (rad/s) or per thousand RPM (KRPM). The sign for voltage V depends on the direction of rotation.

$$V = K_E * \omega$$

$$K_E = \frac{V}{\omega}$$

The torque constant K_T is the "current per unit torque" of the motor, where the torque may be defined as inch-ounces per amp (in-oz/A), newton-meters per amp (nM/A), or other torque per amp measurement. The torque applied to the motor is composed of two components, the internal torque losses τ_F or τ_M (or T_F or T_M) due to friction, eddy current, etc., and the external load on the motor τ_L (or T_L).

$$A = \frac{\tau_F + \tau_L}{K_T}$$

$$K_T = \frac{A}{\tau_F + \tau_L}$$

The motor constant K_M is a measure of the motor's "size," and is the stall torque at one watt input power. When you are comparing the K_M for different motors, be sure that they are measured in the same units.

$$K_M = \frac{K_T}{\sqrt{R_T}}$$

With these motor constants, we can recast the basic motor equation to:

$$E = \left(\frac{\tau_F + \tau_L}{K_T} * R_T \right) + (K_E * \omega)$$

SELECTING MOTORS

Given the onslaught of values, formulas, and ratings above, how do you begin to select an actual motor?

The first two—and for the hobbyist, the most important—criteria are price and availability.

The motors for the projects in this book are described below. There are four places where motors are used, and each placement has a different motor evaluation.

WHAT VOLTAGE?

The power supply for this robot is going to be a single 12-volt battery (Chapter 2), so it would be handy if all of the motors ran at 12 volts. They won't, of course...that would be too easy.

If you have a choice of voltages for your system, higher voltages are more efficient for motors than lower voltages.

Electrical power is defined in watts:

$$watts = volts * amps$$

Double the voltage and, to get the same power output, you can halve the amps. 10 volts at 10 amps is 100 watts of power; 20 volts at 5 amps is also 100 watts.

Though you are supplying the same power regardless of the voltage, the efficiency comes into the picture when you consider the resistance of the circuit (e.g. motor, wiring, and control electronics). Current flowing through a resistance creates heat, which is wasted power. The wasted heat is the current *squared* times the resistance ($H = I^2 R$). Reducing the current by 1/2 reduces the heating to 1/4. So for a given power, it is better to use a higher voltage at a lower current.

For a given motor, doubling the voltage will double the no-load speed. While the current draw will remain the same, since it only depends on the applied torque, the power output by the motor also doubles. For example, calculating the power output P_t for two no-load speeds (1,000 RPM and 2,000 RPM) on a t_{stall} =5 lb-ft motor loaded at a torque of t=2.5 lb-ft gives:

$$P_\tau = \omega_{nl}\tau - \left(\frac{\omega_{nl}}{\tau_{stall}}\right)\tau^2$$

$$P_\tau = 1{,}000 * 2.5 - \left(\frac{1{,}000}{5}\right) * 2.5^2$$

$$P_\tau = 2{,}500 - (200 * 6.25)$$

$$P_\tau = 1{,}250$$

...and:

$$P_\tau = 2{,}000 * 2.5 - \left(\frac{2{,}000}{5}\right) * 2.5^2$$
$$P_\tau = 5{,}000 - (400 * 6.25)$$
$$P_\tau = 2{,}500$$

Halving the motor's stall torque τ_{stall} to 2.5 and the applied torque τ to 1.25 (and, hence, the current draw) for a second example gives:

$$P_\tau = 2{,}000 * 1.25 - \left(\frac{2{,}000}{2.5}\right) * 1.25^2$$
$$P_\tau = 1{,}500 - (800 * 1.5625)$$
$$P_\tau = 1{,}250$$

To compensate for the halved torque at doubled speed, gearing can be applied so the final output is again at the original speed and torque.

When you drive a motor at over-voltage, you should be careful of how the motor is loaded. The stall current increases with the voltage, since it is dependent on the motor's resistance and the voltage (Ohm's Law, $I = E/R$). When you double the voltage, you are at risk of quadrupling the heat generated in the motor and circuit, leading to the early demise of your system. One way to avoid this situation is to use a current-limited power supply, capping the system's available power.

DRIVE MOTORS

The drive motors for the robot need to be strong, slow, and efficient. Of course, they also need to be affordable and available for purchase. The need for this last criterion is not as far-fetched as you might think. Just because you can find a component in a catalog or on a website doesn't mean you can actually *buy* it in single quantities. Some things are available only to those customers who need a thousand or more, though you can often get samples, if you sound convincing.

I was in luck when I went looking for a drive motor for this robot. Glenn Currie, at my local robot club (the Robot Group, in Austin, Texas, www.robotgroup.org), had a stack of old wheelchair motors he could sell me cheap (**Figure 1-10**). Given an opportunity like this, I couldn't resist. Even though they are wormgear motors (worm gearing tends to be

Figure 1-10 Drive motor

less efficient than spur gearing), and well used, the price was right and they were designed to carry heavy loads.

When you are looking for motors, be sure to check out wheelchair repair shops, as well as the surplus catalogs and stores in your area. New motors can cost $200 to $600 each, regardless of size, so start with surplus.

When you are looking for used motors, two choices will pop up immediately: car starters and electric window motors from old vehicles. Unfortunately, though they are cheap and readily available, use them with care and trepidation.

Electric window motors can actually make very nice robot drive motors, if they are used carefully. These motors are designed for intermittent use, so if you run them at full power for long periods of time, they may overheat.

It can be difficult to figure out what a surplus motor can do because surplus catalogs don't often give detailed specifications—you are lucky to get the RPM and current draw specifications.

Car starters are not at all useful for most robots. They are designed for huge current draws, such as 100 to 200 amps and more, over short periods of time, just a few seconds.

You may also find that your motor will run at a different speed in forward and reverse. Putting a motor on each side of the robot, each one facing a different direction, can make the robot turn in circles when you want it to go forward.

Motors have a *timing* setting in their commutator. This is done for performance--a timed motor will run more efficiently in one direction, holding the current in the windings at just the right angle with respect to the stator, for that motor's designed load and speed. A motor designed for equal forward and reverse performance will be less efficient.

As you can see in Figure 1-10, my drive motors have a shaft on each side of the gearbox. The motors point the same direction and should run at the same speed, regardless of whether they are timed for symmetrical reverse operation or not. In practice, the motors are still mismatched and tend to pull the robot into circles. Later, we will create a control system that can make them work together smoothly.

You may also want to gear your motor down, so the robot runs at a stately pace. If you are running this large, heavy robot indoors and around people, you don't necessarily want it to go fast. You can calculate the top speed of the motor using the equation found on www.simreal.com.

My wheelchair motors are a wild card in the otherwise orderly progression of knowledge in this book. I don't know anything about them at all—except that, as geared and controlled at 12 volts on this robot, they make it move forward at a slow walking speed. We'll talk about these motors, their gearing, and their control in later chapters.

Speed, Acceleration, and Climb

If you want to calculate the power of your robot's drive motor, you need to take three things into consideration: the robot's intended top speed, how fast you want it to accelerate to that speed, and what kind of hills you want it to be able to climb.

In general, you can determine how much energy it takes to push the robot using your long forgotten high-school physics and some rough estimates of the robot's weight.

Of course, what you are most *likely* to do is to use whatever motors are convenient and affordable and then adjust the drive gearing until it works....

If you are one of the few who want to perform the calculations, point your Web browser to www.simreal.com to find the math behind the magic.

Arm Motors

The design goal for the arm is for it to be able to pick up a full bottle of soda and hold it at arm's length. That's about a two-pound weight, plus the weight of the arm itself. Whether we can achieve this lofty goal remains to be seen, but it helps to start with a target.

Looking at some rough sketches, I make a wild guess that the motor may need to put out about seven pound-feet of torque at the shoulder. Though that doesn't sound like much, it's 1,344 ounce-inches...and when you look for a motor that can do this, you'll see how hard it is to find one.

One of the things you do as a robotics hobbyist is collect catalogs. All sorts of catalogs. As many catalogs as you can get.

Looking through the Pittman DC motor catalog, there are a few nice off-the-shelf devices with continuous torque ratings of 500 oz/in (more if you abuse them).

Instead of driving the joint directly (the 59 RPM of one sample motor is a bit fast for an arm; that's 1/4 rotation in 250 mS), I plan on gearing the final motor down 1:3 or 1:4. This Pittman motor would be great for the arm! The catch, of course, is the cost: almost $200 per gearmotor. I only paid $35 for each drive motor, so I'm not about to pay $200 per arm motor.

Looking deeper, I find that almost all of the gearmotors in all of my catalogs cost at least $100, and sometimes they require another $80 or so for a gearbox. This will never do.

The other thing that a robotics hobbyist must do is to look for ways to subvert ordinary mass-market objects into a more sublime duty.

Running off to my local home improvement superstore (you know the one), I checked out the electric drills and screwdrivers. For the most part, none of them give torque, current, or RPM ratings for their product. They wouldn't want to confuse the customer with extraneous facts, I suppose, and instead try to sell their tools with sheer style and intimidation.

Then I find one, a Black & Decker electric screwdriver, which is rated at 40 lb/in of torque (though a disturbing report on the Internet rates it at about half that). The more expensive version of this screwdriver (the Pivot Driver) even has the motor in its own little detachable module (see **Figure 1-11**). Included in the cost is a nice NiCd battery and a charger, for future projects.

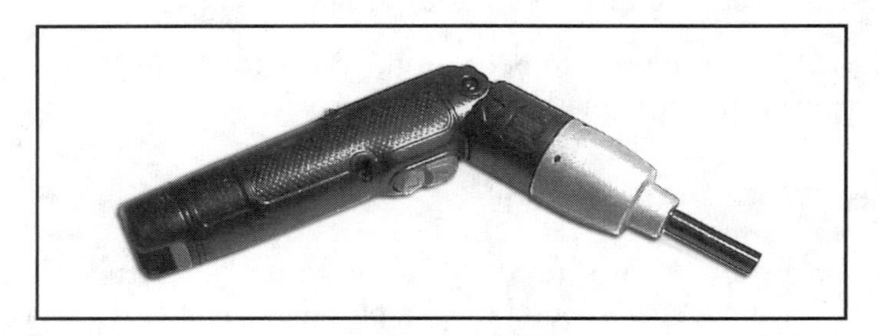

Figure 1-11 Black & Decker electric screwdriver

A brief moment with the Phillips screwdriver and I have the switch and battery pod in one hand, and the gearmotor in the other. Very nice, and the motor pod even has a rectangular protrusion at the back so it can easily be keyed into place.

Another test of wills with a pointy object and some pliers, and the entire motor pod comes apart into its component pieces (**Figure 1-12**). I don't recommend you do this carelessly, however, since the anti-backdrive pins (not pictured) are hard to get back into place, and easily lost (hence, not in this picture).

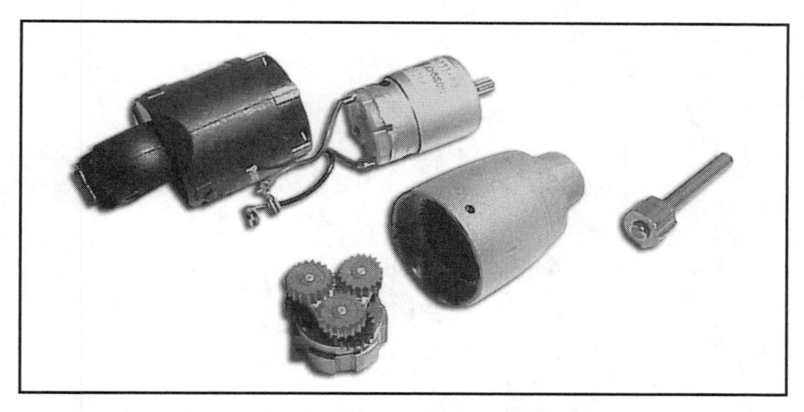

Figure 1-12 Screwdriver components

An aside—this electric screwdriver has an anti-backdrive mechanism. This consists of a bunch of little rolling pins that jam the head of the gearbox when you try to turn the shaft by hand. While designed to allow the originally intended end user to drive screws by hand with the motor off, it allows us to hold the robot's arm in position without expending any electricity. In practice, the motor can be backdriven, but only if you really work at it.

Looking at Figure 1-12, you can see the component pieces of this motor module. Not much, really. There is the plastic base at the top left, a small motor (which is pictured in pieces in Figure 1-2), the planetary gearbox with two rows of gears and the backlash fitting to the left of the cast gearbox, and the output shaft. Though the gears are plastic, the whole thing seems durable enough.

Back to the motor's ratings. Forty lb/in translates to 640 oz/in. That is probably the stall torque and not the continuous operating torque. However, with an additional 4:1 speed reduction planned in the arm's mechanism, and running at half the stall torque value, there should be about 1,280 oz/in of continuous torque available, with peaks at 2,560 oz/in. This is 6 lb/ft continuous torque, with 13 lb/ft stall torque. A quick bench test with an ammeter and a firm grip on the motor shaft gives an estimate of 6 amps current draw at near-stall (and there is a fallacy in this test, to be disclosed in a later chapter. See if you can figure out why this estimate is useless before then). This all fits the guesstimate for the arm.

The only catch is that these are 3-volt motors; running them at a higher voltage could be risky. Anyway, I bought a half-dozen of them. This is going to be fun!

HAND MOTORS

The arm is designed for large motions, and it has large motors with large mechanisms to drive it. The hand, or more accurately, the fingers, is designed for small motions and it has smaller power requirements.

With the hand design in this book, which is based on a 1990 NASA patent (US Patent #4,921,293), R/C servos with pull-cables seem to be the perfect solution (see **Figure 1-13**).

I have a box of small, cheap, and, hence, weak R/C servos in a box in my project closet. If these show promise, I can easily upgrade to the more expensive and stronger servos, such as the Airtronics 94358Z High Performance servo with 200 oz/in torque (at over $100 each, this is a painful upgrade though).

NECK MOTORS

The camera sensor at the head of the robot needs a pan-tilt mechanism, so it can act as a head. Two more servos should suffice as the neck.

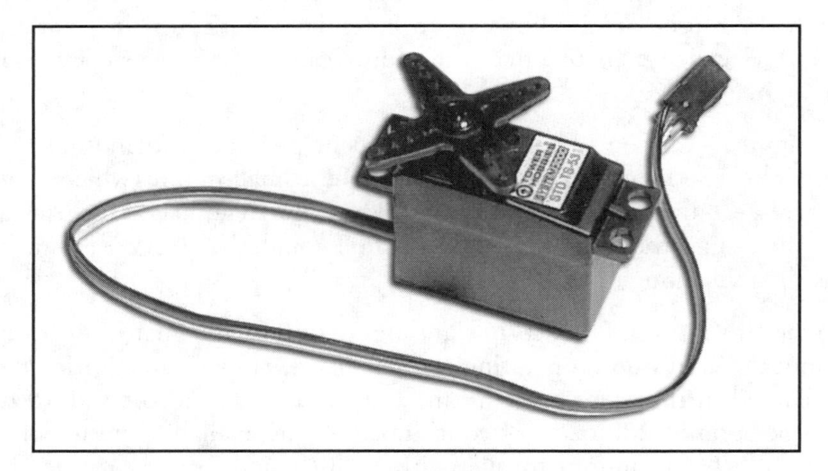

Figure 1-13 R/C servo

OTHER NOTES

BREAKING IN YOUR MOTOR

A new motor, right off the shelf, will typically have a round commutator sitting under flat brushes. This gives a minimal contact area for the current to flow into the motor, so under high power it will tend to spark and pit the brush surface, reducing the motor's performance.

To prevent this, run the motor at about 1/2 voltage and without any loading, so that it is free-running, for an hour or two. This will wear the brush faces into a better fitting curve. This procedure can actually improve the performance of your motor by 10% or more.

Some expensive motors don't need this break-in period—but they will certainly see fit to mention this feature in their advertising. If you are unsure, it's best to break the motor in first if you want the maximum performance.

INTERFERENCE SUPPRESSION

Due to the mechanical switches in the motor (the brush/commutator interface), the motor can and will spark and generate some electromagnetic interference (EMI). One remedy for this problem, if you are experiencing it, is to connect a 0.01 µf capacitor across the motor leads. A more difficult remedy is to attach a 0.01 µf capacitor across the brushes (or each pair of brushes, if there are more than two) inside the motor.

You don't necessarily want to connect a capacitor from the leads to the motor's case, unless you can match it to the frequency of the EMI. Otherwise, you can cause more problems than you solve, coupling the EMI back into the power supply. This seems to be a topic of some dispute, since other sources do recommend connecting the capacitors to the motor case. This may be an area best served by trial and error.

If they are sufficiently long, you should also twist all motor and battery supply wires. Also, keep all high-power lines away from sensitive circuits in the robot.

For R/C servo motors, there can be an additional source of control problems. On R/C servos, the power and control signal are typically carried on a common 24 gauge 3-wire harness. This wire has a resistance and when the servo starts up it can cause a power drop at its end of the harness. This power drop affects the control signal as well. You could put a big capacitor at the servo end of the harness, but a better solution would be to make a new connector with 18 gauge or larger wire for the power feed. Also, keep the cable as short as practical, and use a good quality connector.

CHAPTER 2

BATTERIES

While it is interesting to learn about the fundamentals of batteries—their chemistry, construction, and so forth—when you are trying to find a battery to use for a project, you are more interested in the practical questions: How much power (voltage and current) will this battery give me? How long will it last in my application? How large and heavy is it? How can it be recharged? And, in some applications, what happens when it is tipped upside down or punctured?

The inner details of the battery are, to this admittedly geeky author, fascinating stuff. For a tour through the insides of your battery, visit www.simreal.com

You can find more information about the chemistry of batteries and how they work in more detail from the battery manufacturers. Most manufacturers have nice Web sites on the subject, which you can find using any decent Web search engine.

TYPES OF BATTERIES

The first dividing line between battery types is whether they are one-use, so-called "primary" batteries, or rechargeable, also called "secondary" batteries. The rechargeable batteries use a reversible chemical reaction to create their electricity. When power is applied to the battery, as opposed to being extracted from it, the battery "resets" its internal chemicals in preparation for a new discharge cycle.

Once you get past the primary/secondary division, you run into the many different battery chemistries. There are a few very popular battery types; the carbon/zinc and alkaline batteries on the primary side, and the lead/acid and NiCd on the secondary side. There are also many exotic battery chemistries that you are not likely to find at the local Radio Shack, such as sodium/sulfur, vanadium, and the supercapacitors.

There are many considerations in selecting the right battery, not the least of which is cost. Due to their popularity, availability, and relatively low cost, most projects use an alkaline, NiCd, or lead/acid battery.

There are a number of important factors in battery selection:

- Cost. A cheap battery isn't always the most cost effective. You need to take into account the entire life-cost of a battery. How long will it last? How much energy can it deliver? Can it be recharged? How many *times* can it be recharged? Only the rich can afford cheap batteries.

- Voltage (and voltage curve). Different chemistries will have slightly different per-cell voltages, and will follow a different curve from their fully charged state, to their discharged state.

- Capacity. How much power will it provide before it is discharged? The capacity of smaller batteries is typically measured in milli-amp Hours (mAh, which is 1/1000 of an amp Hour), while larger cells may be measured in amp Hours (Ah). Capacity is not just determined by the chemistry and physical size of the battery, but by the mechanical structure as well. In batteries, it really does matter who the manufacturer is, so if it's important to you, study the discharge graphs for the batteries you are considering.

- Drain rate. Some batteries are only useful for a slow, steady current trickle, and others can be tapped like a fire hydrant. Lithium and NiMH are better at low loads taken over a long time, while lead/acid and NiCd are better for high loads.

- Discharge profile. Different chemistries lose their charge in different patterns; some have a level voltage until they suddenly drop to nothing. Others will have a steady decline in voltage until they are no longer useful.

- Self-discharge rate. Batteries lose charge by just sitting on the shelf. Different batteries, however, lose their charge at different rates. Alkaline cells can take five years to discharge to 80% of their rated voltage. Lithium cells take twice this long. Secondary batteries lose charge much faster; about six months for lead/acid batteries, a few months for Li-ion, 30 days for NiCd, and just a few weeks for NiMH. High temperatures will decrease the shelf life of any battery.

- Charge rate. If you are using a rechargeable battery, you might care about how quickly it can be charged. New technologies are improving the recharge rate of batteries all the time.

- Cycle life. How many times can your rechargeable battery be recharged? Depending on how they are used, a lead/acid battery can be recharged up to thousands of times, as can NiCd batteries. NiMH and Li-ion only give several hundred recharge cycles.

- Size and Weight. Different chemistries have different energy densities, so the same amount of power may fit into different size packages. Given your power needs, will the necessary batteries fit into your project?

- Temperature. Some chemistries do better at low temperatures than others. High temperatures can degrade some batteries' life spans. In general, batteries do well in the 0° F (-20° C) to 140° F (+60° C) range, which is more than enough for most applications.

A note on the specifications given for the batteries: the discharge graphs are provided for convenient reference only; any given battery may behave somewhat differently. Since all batteries come in a variety of sizes, the graphs are given for a characteristic cell size. Refer to manufacturer's technical data for additional sizes and information.

A variety of operating parameters are listed for each cell type.

- The nominal cell voltage is the voltage at full charge, while the discharged voltage is the level where the battery is considered to be "dead." Note that most cells are considered to be "good" well below the 1.0 volt point.

- The average capacity of the battery is then listed for several representative sizes (AA, C, and D cells).

- The internal resistance determines the maximum discharge rate the battery is capable of. The lower the resistance, the more instantaneous current is available. The figures shown here are rough estimates, as the actual values can vary wildly from manufacturer to manufacturer.

- The energy per weight (in Watt-hours per Kilogram) of the battery gives a normalized measurement you can use to compare battery technologies.

- The recommended discharge load helps determine what type of application the battery is optimal for.

- The shelf life.

SOME PRIMARY BATTERIES

Batteries were invented in the early 1800s when experiments by Alessandro Volta created an electrical current from the chemical reactions between two dissimilar metals. The original battery (or "voltaic pile," after its function and structure) used stacks of zinc and silver disks for the electrodes, separated by a porous separator soaked in a sea water electrolyte. Though this makes for a large and bulky battery, it was the only practical source of electricity in the early 19th century.

CARBON/ZINC

The carbon/zinc battery has a limited shelf life, and is susceptible to leaking its corrosive electrolyte. You don't see them in use much these days.

ALKALINE

The alkaline cell (or, more specifically, alkaline manganese dioxide cell) is similar to the carbon/zinc cell but with an improved electrolyte, a lower internal resistance, and a higher discharge rate and power density. Since they are not rechargeable, however, they are not often used for utilitarian robots (though alkalines do seem to be a popular choice for combat robots).

SOME SECONDARY BATTERIES

The chemical principles behind rechargeable batteries were demonstrated in 1802 by Johann Ritter, but it took another sixty years before rechargeable batteries became practical. Among other practical limitations, there wasn't a convenient source of electricity available to recharge them until the advent of the steam-driven generator.

LEAD/ACID

The largest problem associated with the lead/acid battery is the damage caused by leaking acid. German researchers addressed this problem in the early 1960s by developing a gelled electrolyte. Working from another direction, other researchers developed a way to completely seal the battery, preventing leaks. Either way, the sealed lead/acid (SLA) battery needs little or no maintenance, which, while costing more, can be an advantage in some situations.

The low internal resistance of these batteries means they can dump a lot of power into your circuit quickly. Shorting out the battery is a sure way to reduce its life by boiling the electrolyte and warping the electrodes. While the low energy density makes this a heavy power source, this is an easily available battery that comes in many large sizes. The discharge curve is also fairly friendly, with a steady voltage and then smooth dropoff that gives plenty of warning. Discharge characteristics for a 12-volt battery are shown in **Figure 2-1**.

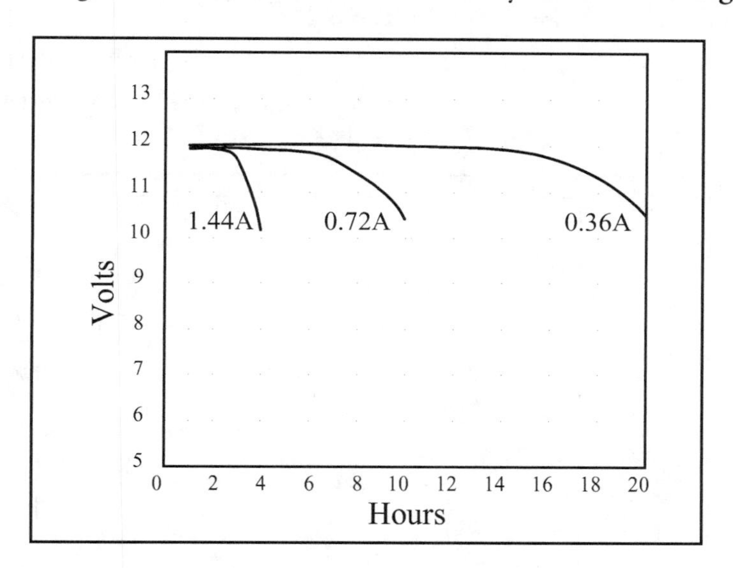

Figure 2-1 Discharge characteristics for a 12-volt (six-cell) 7.2Ah battery at different loads

The SLA is used in many of the larger robots. If you decide to use an SLA battery in your project, be sure to get a deep-cycle marine or wheelchair battery, as opposed to a regular car or motorcycle battery. SLA battery characteristics are shown in **Table 2-1**.

Cell Voltage (nominal)	2.0
Cell Voltage (discharged)	0.8
Internal Resistance (Ri in Ohms)	.02
Energy by Weight (Wh/Kg)	30
Cycle Life (recharge cycles)	200-500
Shelf Life (to 80% capacity)	6 months

Table 2-1 SLA battery characteristics

NICKEL CADMIUM (NICD)

The original version of the NiCd battery used a vented, unsealed cell that required regular maintenance. In the 1940s they perfected the sealed NiCd cell (though they do retain a need to breathe a bit), which is maintenance free. Today's NiCd batteries can take a lot of abuse, both mechanical and electrical, and are cheaper than other batteries in cost per hour of use. Characteristics are shown in **Table 2-2**.

This battery has a surprisingly high capacity for current delivery. The AA battery shown in **Figure 2-2** has a recommended maximum continuous current draw of 9 amps, with 18 amp pulses allowed.

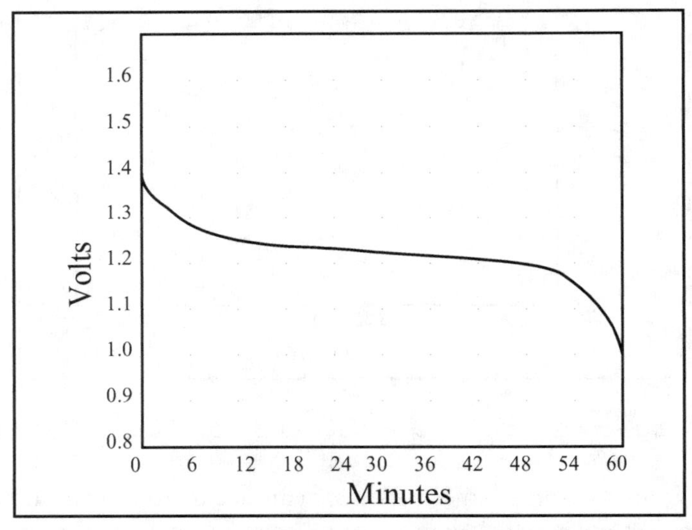

Figure 2-2 AA battery, 800mA load

Cell Voltage (nominal)	1.2
Cell Voltage (discharged)	0.9
Internal Resistance (Ri in Ohms)	0.04 – 0.10
Energy by Weight (Wh/Kg)	50
Cycle Life (recharge cycles)	500-5000
Shelf Life (to 80% capacity)	6-8 weeks

Table 2-2 NiCd battery characteristics

Detailed information for all of these batteries (and more) can be found at www.simreal.com.

SELECTING A BATTERY

What battery should you actually use in your robot? For the projects in this book, I'm using a 12-volt sealed lead/acid battery. Specifically, the Sears DieHard 27141 AGM Wheelchair battery.

This is about a 19Ah battery (it specifies a reserve capacity of "32") that measures 5" wide by 7.5" long and 6.25" high. The two battery terminals stick up another 3/4" or so. This package is filled with lead and fluid, so it is also rather heavy, weighing in at about 20 pounds.

Though this seems large and heavy, it doesn't provide as much power as I would like. One of the biggest problems in mobile robots is the power source; it takes a lot of power to drive an electric robot for any length of time (e.g. several hours).

Why did I choose this particular battery? I actually wanted to get a gel cell, but they were out of stock and the sealed battery seemed just as good. So, first, because it was readily available.

Though more modern battery technologies offer higher (and in some cases, *much* higher) energy densities, they are more complicated to charge. The SLA battery is easily trickle-charged through a simple interface—no charging stations to drop the cells into and no worrying about cell reversal in a six-cell battery pack. So, second, the stability, simplicity, and robustness of the SLA technology.

Your choice may lean more toward NiCd or Li-ion, and that's okay. Just be sure to match your charging system to your choice. Even within the lead/acid technologies there are different charging requirements—open cell batteries charge differently from sealed and gel.

VOLTAGE REGULATION

Though I want only one battery on the robot (for simplicity in charging, and because of available space) I know I will have several different voltage requirements from that one battery.

The drive motors will take whatever the battery has to offer—12 volts on down to the point where the motor controllers stop working, which I'm guessing is around 9 volts, though it could be six or less.

The arm motors are rated at 3.6 volts, though I know they can be driven at 6 volts. The hand servos want to run at 6 volts as well.

The microcontroller electronics all want 5 volts and, if you wanted to build with low-power components, you would need to provide 3.3 volts for them as well.

Now for the clincher: the main brain for this robot will be a Compaq Presario laptop that I got for a good price at an auction. While I could run this computer off of its internal batteries for an hour or so, I would prefer to run everything off of one power source. Looking at the wall adapter for the laptop, I see that it claims to be providing 19 volts at 3 amps.

The power control system for this robot is going to have to work overtime to support all of these devices. To complicate things, the motors will be feeding noise into their power supplies as they operate, and heavy intermittent motor loads may cause large power droops. All of this as the battery fades from 12 volts at its peak down to five or six volts at its fully discharged nadir.

Looking at this laundry list of problems, it looks like it might have been easier to go with multiple battery packs; perhaps a bunch of NiMH or Li-ion "D" cells in a set of packs, with power taps at 3, 6, and 12 volts (or so). Looking at the cost of these batteries, plus the need to buy and customize a specialized charger for them, the SLA battery with all of its voltage issues doesn't look so bad after all.

LESS FROM MORE

There are several ways to reduce a battery's voltage for a project. In this section we explore two of them—the venerable linear power supply, and the switching regulator.

PROJECT 2-1: LINEAR POWER SUPPLY

The 12-volt main supply will fry the robot's electronics, so it is necessary to reduce it to a steady 5 volts. The classic way to do this is with a linear regulator, as shown in **Figure 2-3**.

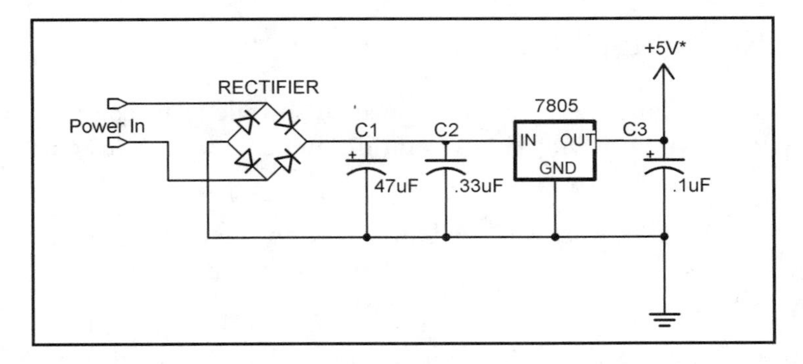

Figure 2-3 7805 linear regulator

I normally like to use a bridge rectifier (such as International Rectifier's DF02M) on the input to my power supply as a safety measure. I've plugged things in backward before and destroyed perfectly good chips that way. The penalty is about $0.50 cost, a bit of board space, and (more critically) about two volts lost in the diodes. You can also use a single diode, at a lower cost, with less voltage drop in the system, to block rather than rectify a reversed power input.

C1 and C2 act as input filters. C3 is an output filter. Their values are not critical. Together, they help stabilize the voltage from the power supply.

The working part is the 7805 linear regulator. It operates by adjusting the flow through a transistor to limit the available voltage to the load. A series-type regulator has a variable impedance element in series with the load. A shunt regulator will put a variable impedance element in parallel with the load, shunting power to ground. A series regulator is typically a three-terminal device (such as the 7805), and the shunt regulator is two-terminal (such as a zener diode).

Though the 7805 regulator itself only consumes about 4 mA, it is an inefficient way to regulate electricity. The efficiency of the series regulator depends on the ratio of input voltage to output voltage:

$$\eta = \frac{V_{out}}{V_{in}}$$

The linear regulator is fairly efficient when it is performing a minimum amount of regulation, but when it is converting a 12-volt input to a 5-volt output, it is only about 41% efficient. That is a lot of wasted power.

Say the input to this circuit is 12 volts from a fresh battery. The client circuit is drawing, for example, 200 mA at 5 volts, or 1W. Internally, we need to shunt 7 volts at 200 mA, effectively converting it to heat. That's 1.4W wasted and 1.0W used, out of 2.4W supplied.

Of course, that is just one example and the old, inefficient 7805 regulators are actually fairly hard to find now. They are being replaced by more efficient Low Drop-Out (LDO) linear regulators (such as the LM330), which have a voltage drop out of about 0.4V (as opposed to 2V). But these still act like a variable resistor. The benefit of LDO regulators is that they can operate with a lower input voltage, raising their efficiency when used on the appropriate power source.

The other benefit of the linear regulator is that it is extremely easy to use and inexpensive. But for our application, it isn't the best solution.

SWITCHING POWER SUPPLIES

To avoid the inefficiencies of the linear voltage regulator, it is necessary to look to the switching voltage regulators. Of these, there are two common types: the buck converter and the flyback converter. Buck converters (or "down" converters) are built around an inductor, while flyback converters are built around a transformer. Of the two, the buck converter's characteristics are matched closer to the needs of our application and it avoids a bulky transformer.

Flyback converters are good for low to moderate loads, and are also used when you need to isolate the load from the power source. Buck converters can handle the larger load demands we will place on it.

Other types of switching power supplies include the boost (or "up") converter, which is very similar to the buck converter but configured to increase voltage. There is also the buck/boost converter, which can go either direction, and the isolated forward converter which is has a similar configuration to the flyback converter but with up/down conversion capability. An excellent discussion of all forms of power electronics can be found in the great book *The Essence of Power Electronics* by J.N. Ross.

Switching regulators are more particular about their environment than linear regulators. They also generate a stack of electromagnetic interference that is best counteracted with a nice, large ground-plane on a printed circuit board, where most of the bottom of your board is covered with copper and connected to ground.

This project looks at a series of related buck regulators from National Semiconductor. Others exist, clearly, but it simplifies our task to focus on a specific implementation.

National Semiconductor has an interesting (and amazing, to me, since I've never seen this before from a manufacturer) design, simulation, and kit-ordering system with two parts called WeBench (www.national.com/appinfo/power/webench/) and BuildIt. Using this system, you can specify a power supply to your satisfaction and then order a kit with all of the necessary parts at a very reasonable cost. This is, in fact, probably the best way to purchase the parts for this project since the inductor can be hard to find. At the time of this writing, all of the parts plus a circuit board cost $17 plus $11 shipping.

The schematics in **Figure 2-4a and Figure 2-4b** show a power supply suitable for a low-power project such as a 500mA regulator using the LM2674, or a 1A regulator using the LM2672. It also shows a high-power switching regulator using the more powerful chips LM2673 or LM2679, which provide more features as well (not used here).

The recommended component values are detailed in the **Table 2-3**, **Table 2-4**, **Table 2-5**, and **Table 2-6** for the different chips and for different recommended power

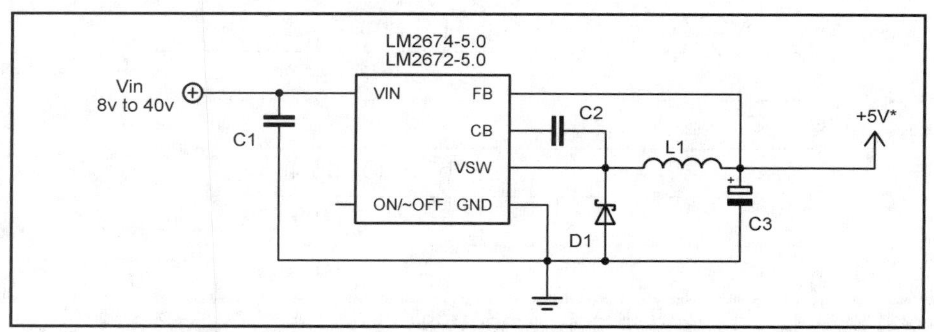

Figure 2-4a Low-power switching regulator

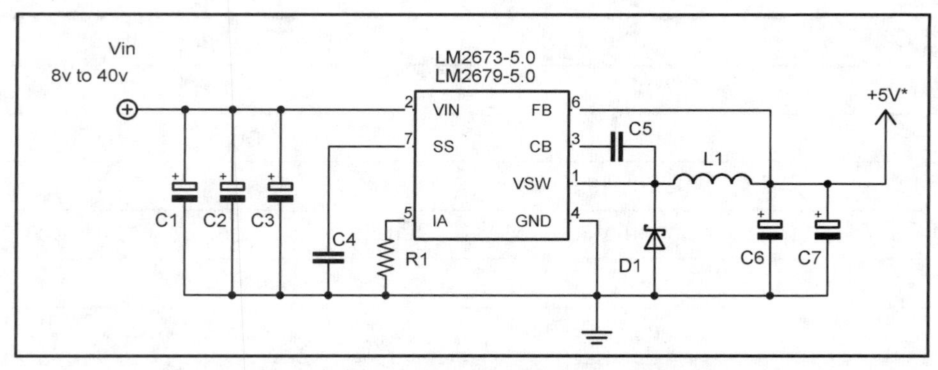

Figure 2-4b High-power switching regulator

loads. These values were taken from National's WeBench online calculator. The input voltage was specified for the battery's range of 8.0 volts to 13.0 volts, with a fixed output voltage of 5.0 volts.

LM2674	0.5 amp
C1	47µF (0.34Ω)
C2	10nF
C3	47µF (0.11Ω)
D1	Schottky (40V, 2A, 0.5Vf)
L1	68µH (0.153Ω)

Table 2-3 Recommended component values for the LM2674 at 0.5 amp

LM2672	1.0 amp
C1	120µF (0.14Ω)
C2	10nF
C3	68µF (0.095Ω)
D1	Schottky (40V, 2A, 0.5Vf)
L1	33µH (0.079Ω)

Table 2-4 Recommended component values for the LM2672 at 1 amp

LM2673	2.0 amp	3.0 amp
C1 (C2, C3)	2@ 10µF (0.3Ω)	3@ 10µF (0.3Ω)
C4	2.7 nF	2.7 nF
C5	10nF	10nF
C6 (C7)	1@ 150µF 0.08Ω)	1@ 150mF 0.08Ω)
D1	Schottky (30V, 4A, 0.42Vf)	Schottky (30V, 4A, 0.42Vf)
R1	7.15KΩ	7.15KΩ
L1	22µH (0.085Ω)	22µH (0.036Ω)

Table 2-5 Recommended component values for the LM2673 at 2 amps and at 3 amps

LM2679	4.0 amp	5.0 amp
C1 (C2, C3)	3@ 15µF (0.22Ω)	3@ 33µF (0.2Ω)
C4	2.7 nF	2.7 nF
C5	10nF	10nF
C6 (C7)	2@ 120µF 0.085Ω)	2@ 120µF 0.085Ω)
D1	Schottky (30V, 4A, 0.42Vf)	Schottky (30V, 7A, 0.45Vf)
R1	4.99KΩ	4.99KΩ
L1	15µH (0.027Ω)	15µH (0.02Ω)

Table 2-6 Recommended component values for the LM2679 at 4 amps and at 5 amps

Thanks to the WeBench package, finding the part values was easy. But finding the actual parts for sale (even with the manufacturer's part numbers and detailed specifications) can be an interesting challenge. Part substitutions are allowed, as long as you are careful to match the various characteristics.

Which of these designs will be the most useful for this robot? The electronics packages will only draw an amp or so—each microcontroller board taking about 100mA. However, there is no real cost detriment to building the largest possible power supply—and it might come in handy for later projects. To simplify this chapter, we will explore and build only one of these power supplies, the 5 amp LM2679.

Project 2-2: Switching Regulator

The power supply shown in **Figure 2-5** is essentially the same as in Figure 2-4, with the addition of D2, a 5.1 volt Zener clamping diode, to further smooth any glitches in the output. There are also two additional capacitors, C8 and C9 at 0.47µF each, mounted close to the chip to prevent ringing (oscillations in the signal between the chip and the capacitors). Assembling the project is simple enough, with a few caveats.

For any switching power supply, it is recommended that you use a printed circuit board with a broad ground plane, to minimize signal coupling through the circuit, and to reduce electromagnetic interference (EMI).

The input capacitors should be placed very close to the input pin of the LM2679 chip. The series inductance (resistance to change) of the input wire or printed circuit board (PCB) trace under this high-current load could create ringing signals in the input. It may also be necessary to place a 0.1µF to 0.47µF capacitor in parallel with the input capacitors to prevent this ringing (as shown at C8 in Figure 2-5).

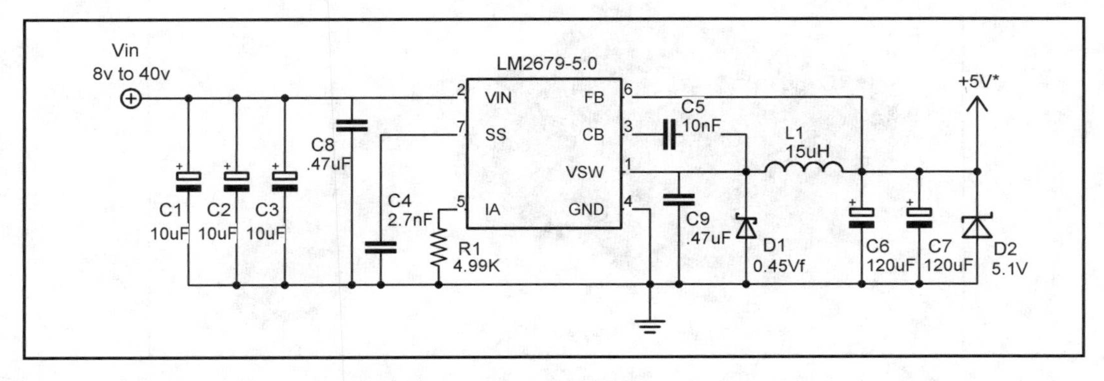

Figure 2-5 5-amp switching regulator

While the details of operation and component selection for the switching power supply are interesting, they are also detailed and complicated, and beyond the scope of this chapter. Refer to the LM2697 data sheet if you want to know more about the whys and wherefores of this versatile chip.

Though we only need one or two of these power supplies, it might be worthwhile to create a printed circuit board through a prototyping service. There are several manufacturers out there (such as Alberta Printed Circuits, or ExpressPCB) who will create custom circuit boards. When you create your own layout and make several copies of the board, they can be surprisingly economical. Or you can use the WeBench kit. Either way, using a circuit board will improve the quality of the power supply and reduce the switching noise.

More from Less

For light-duty voltage step-up, a flyback converter (such as the LM2585 or LM2588) would work nicely. These regulators also work in boost mode with an inductor instead of a transformer.

However, the laptop computer in this project needs 19 volts at 3 amps (or so says its wall transformer), and this is beyond the power of these converters. Using the principle of maximum simplicity (and mass-produced cost reduction), I looked for an off-the-shelf solution and found a PROWatt125 power inverter on sale for $40. This little box turns 12 volts of car battery power into 110 volts, and it can handle 125 watts of power (see **Figure 2-6**). Designed specifically to run laptops from car batteries, it was perfect for the job. And for $40, it was cheaper than I could build using a prototype printed circuit board.

Figure 2-6 12V to 110V power inverter

Of course, the laptop converter then steps this 110 volts back down to 19 and the power supply inside the computer generates the usual array of internal voltages. Using this series of converters is going about this the hard way. Until I can find a direct 12-volt adapter for the laptop, though, I'll live with the inefficiency.

VOLTAGE MONITORING

All of these power supplies (and the devices they are supporting) will take a heavy toll on the single 32Ah battery that drives the robot. It is important, then, that the robot be able to carefully monitor the current battery voltage, waiting for that all-important drop from 12 volts to 8 volts as it reaches the end of its charge.

The microcontroller boards have analog inputs, so it should be a simple matter to watch the battery voltage directly. Unfortunately, the 12+ volts present at full charge will quickly destroy the 5-volt analog-to-digital converter. The simplest solution is to make a resistor bridge to divide the battery voltage down to a range the microcontroller can safely monitor (see **Figure** 2-7).

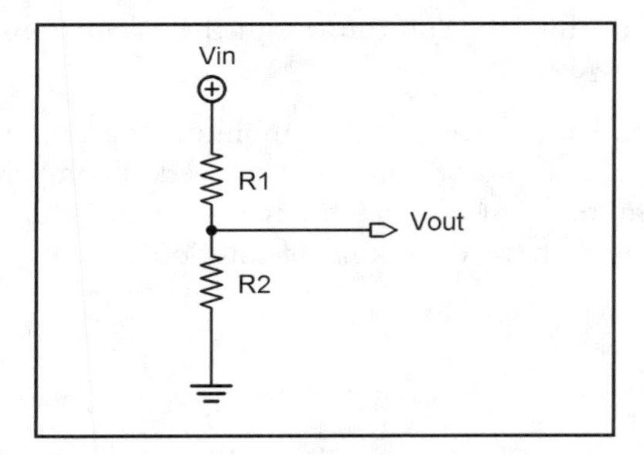

Figure 2-7 Voltage divider

The voltage V_{OUT} is determined by V_{IN} and the ratio of R1 and R2:

$$V_{OUT} = V_{IN} \times \frac{R1}{R1 + R2}$$

Selecting R1 and R2 to reduce V_{IN} by a factor of three would bring the 12 volts (plus extra for safety) into an acceptable voltage range. The resistors should also be sized so that they provide enough current for the analog input to work with and to minimize waste. We can arbitrarily choose values for R1 and R2 with these constraints in mind:

$$R1 = 2{,}000$$

$$R2 = 4{,}000$$

$$V_{OUT} = \frac{V_{IN}}{3}$$

Assuming a full range analog to digital conversion range of 0 through 255 (for zero to 5 volts), a full charge of 12 volts would read:

$$\frac{255}{5} \times \frac{12}{3} = 204$$

A low charge of 8 volts would read:

$$\frac{255}{5} \times \frac{8}{3} = 136$$

This only makes use of about 1/4 of the converter's dynamic range, so it is inefficient—yet it is also sufficient for this job. You can re-run the calculations for the exact resistor values available on your workbench.

For a more expensive (and also fancier) battery monitor, you can look to the chip makers for a solution. Of course, at $6 to $8, this may be overkill compared to a pair of 2-cent resistors. A quick perusal of DigiKey shows "Battery Gauge" chips from Microchip and TI, and Maxim is bound to have some kind of solution as well.

POWER SYSTEM

Schematically, the complete power system looks like **Figure 2-8**. This abstraction ignores such details as where to put everything in the robot, how the power rail is set up, and so forth.

The battery, and its associated emergency stop switch (E-Stop), is shown at the left. If you want to include an E-Stop switch, you may have to search a while to find a good high-current switch. Industrial suppliers may have something, as might marine suppliers (if you don't mind paying extra).

The next column shows the various power converters. There are the 5-amp buck converters for the microcontrollers, and the 110 volt inverter for the laptop.

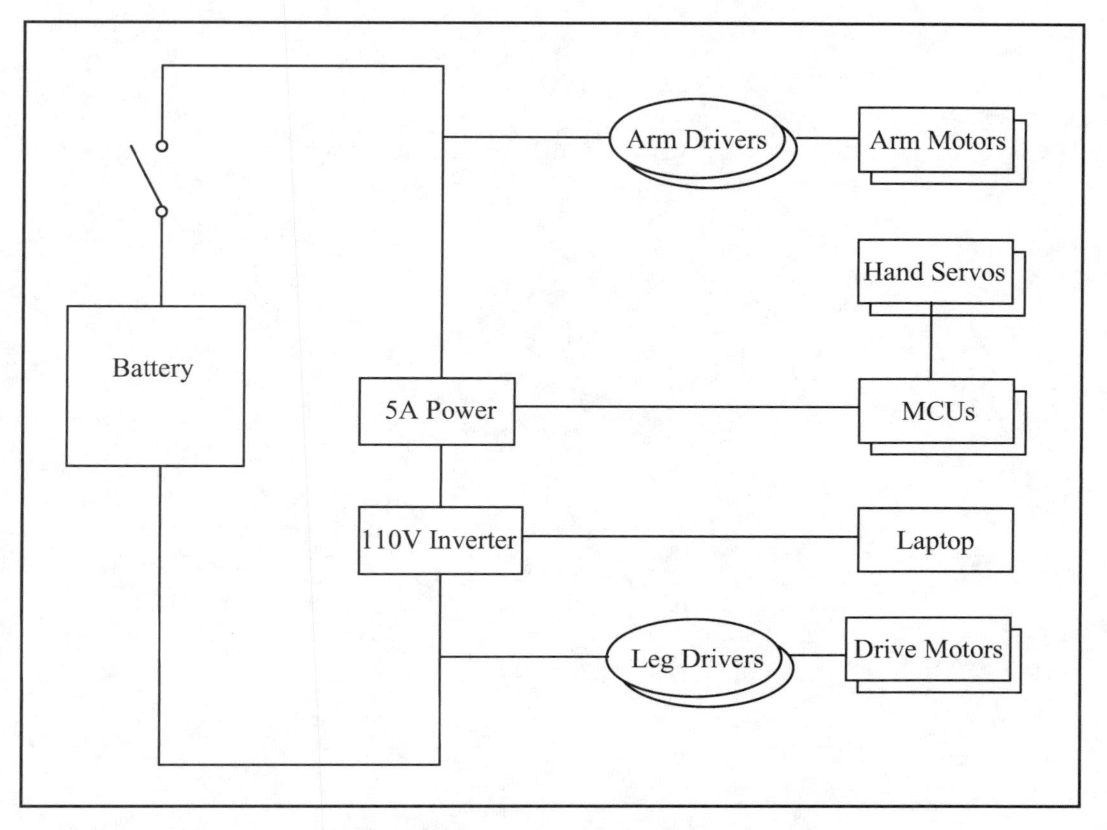

Figure 2-8 Complete power system

The third column shows the power drivers for the arm and main drive motors. Note that the hand servos stack off of the microcontrollers—which could be a problem if the power slump from their starting current draw is too severe. For high-power servos, you will probably need to supply drive current from a separate supply.

Finally, there are the end devices, the motors and computers themselves.

The construction and organization of the various pieces of this puzzle are explored throughout the rest of this book.

CHAPTER 3

ROLLING PLATFORM

If you are working your way through the book from the beginning, you should have a pile of motors in one hand and a bunch of power supply stuff in the other (figuratively speaking). Looking ahead in the book, there are more circuit boards, reflex control systems, laptop computers, arms, heads, and endless arrays of sensors to come. All of these things have one thing in common—they need to be attached to a common supporting framework.

At the very core, in the very minimum configuration, a wheeled robot consists of a sturdy framework, wheels, motors, a battery, and a control system. All together, this constitutes the rolling platform. With the exception of the control system, which is discussed in the next chapter, this chapter works you through the creation of a stable, sturdy, and compact rolling platform.

A rendering of the very basic elements of the rolling platform is shown in **Figure 3-1**. A photographic tour of this same platform as physically realized is given in **Figure 3-2a** and

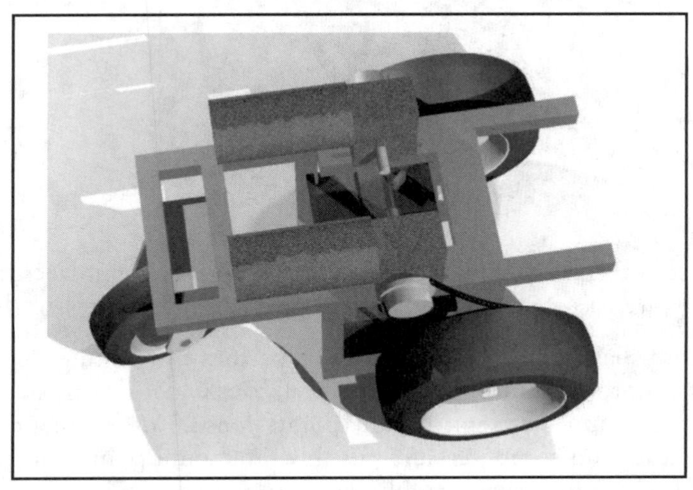

Figure 3-1 Rolling platform

Figure 3-2a Rolling platform top view

Figure 3-2b Rear caster

Figure 3-2b. Most of this chapter will rely on computer-generated images for construction photos, since they are easier to manipulate and give a good likeness of the project.

The rolling platform has several different tasks to perform: mounting point for the drive wheels and drive motor; mount for the caster wheel, battery compartment, and power bus; hard points for the bumper system; mount points for the upper platform, which will include the arm, head, and computer; and possibly the mount points for some of the low-level electronics, such as the power supplies.

DESIGN DECISIONS

There are a number of conflicting goals involved in designing a robot, as there are in designing any complex device. These involve trade-offs in size, weight, and functionality.

SIZE

I want the robot to be able to operate comfortably in my office and home, as well as being comfortable in convention centers, parking lots, and even unpaved outdoor areas. To operate in human areas the robot needs to conform to human space standards, give or take a bit. To operate in robust environments, such as parking lots and outdoors, it needs a stable platform with large wheels.

People are comfortable navigating spaces that are 18" wide. Most inside doors are 24" to 32", and outside doors are 32" to 36". Hallways also fall in the 32" to 36" range for homes, though public buildings have much larger halls and spaces. Indoors, you won't find many bumps larger than 3/4" high, excluding stairways—and we won't be driving up or down stairways with this robot, at least not on purpose.

Inside of this footprint we need to place a battery that is large enough for an extended run (1/2 hour or more), wheels suitable for outdoor travel, and all of the other electronic devices that will make the robot more than an expensive radio-controlled toy.

Most of the more complex components—the computer, head, and other electronics—can live in layers above the basic rolling platform. The platform itself is heavy enough and stable enough to carry a fairly tall payload without much threat of tipping over.

The arm, however, needs space in the front of the robot. The design criteria of the arm requires that it, among other things, be able to reach the floor, and be able to tuck out of the way inside the robot's circle of operation.

Figure 3-3 is a diagram of the rolling platform's footprint. The large outside circle shows the 31" space the platform fills when it spins around its center. The smaller circle at the left of the platform shows the caster wheel's area of influence. Though the caster is shown outside the major turning space, it tucks into that circle when the turn begins. Note also that the caster dominates a large space in the robot's interior.

Where the caster ends, the battery tray begins. The 20-pound lead/acid battery is slung as low as possible inside the robot, as close as possible to the spinning caster wheel. You can see that the battery tray is just to the left of the robot's turning center, which is directly between the two drive wheels. This weight behind the drive wheels helps keep the robot stable; any sudden stops or downhill slopes would have to lift the battery around this axis before the robot would tip over.

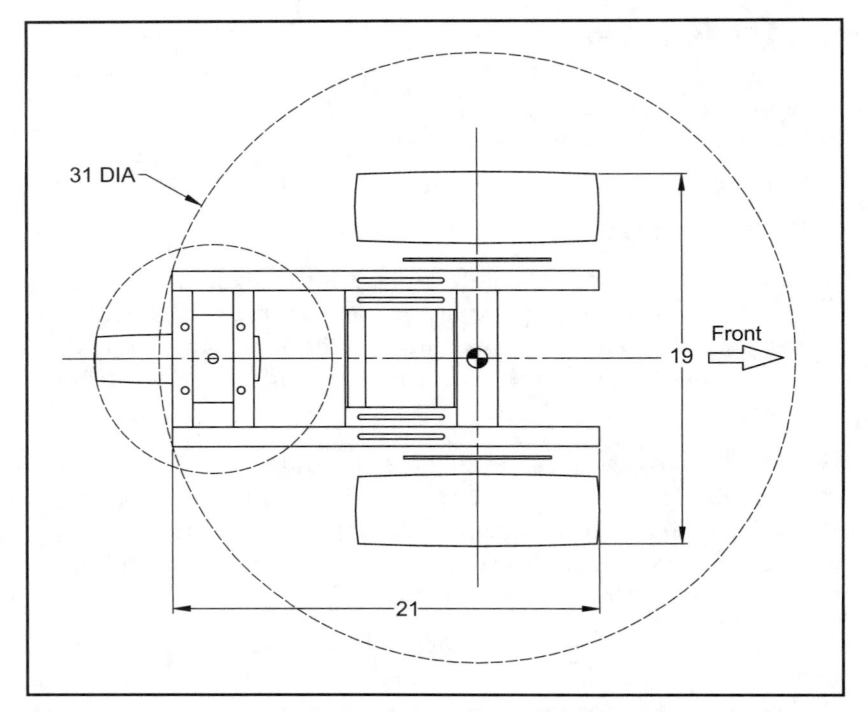

Figure 3-3 Rolling platform's footprint

The drive motors mount into the slotted bars to either side of the battery compartment. The drive motors sit above the bars, while the battery is slung underneath them.

The battery can't be placed directly over the caster, though this might appear to be the best place for it. In terms of space management, a battery high over the caster would interfere with the drive motors (although these motors *could* be relocated). More importantly, too much weight placed directly over the spinning caster wheel causes it to behave badly. The caster, while it can physically carry the battery's weight, tracks the robot's motion better when it is lightly loaded.

At the front end of the robot is an empty space that the arm's mechanism will fill.

CONFIGURATION

The drive wheels in the rolling platform are placed forward of the platform's center. This moves the turning center closer to the front of the robot (away from the geometric center), and increases the size of the circle the platform requires to turn in. However, it also increases the stability of the robot.

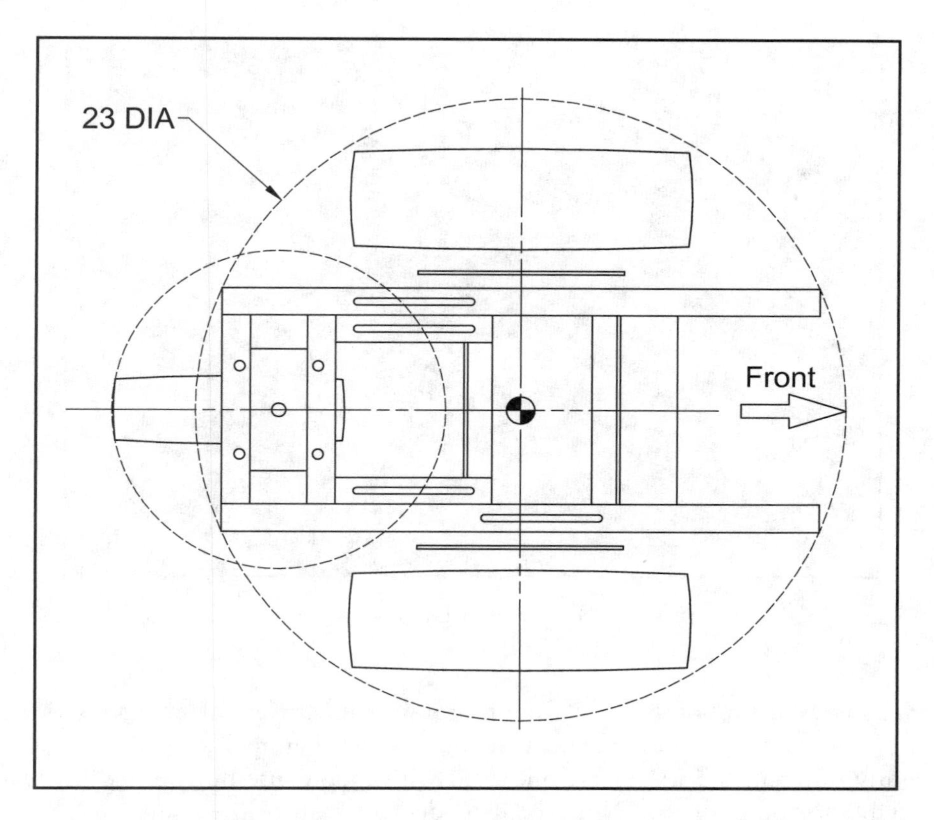

Figure 3-4 Symmetrical platform option

Figure 3-4 shows the robot platform with the drive wheels centered. This gives a 24" turning circle, which is a nice tight turn. But it also puts the battery forward of the wheels' axis. With the center of gravity forward of the robot's turn center, any quick stop or unfortunate behavior from the arm, can tip the robot forward.

The addition of a forward caster would fix this, but this fourth wheel would also consume valuable space where the arm should go.

Smaller casters would ease this problem, but could also sacrifice much of the robot's ability to travel over bumps.

Similar indoor/outdoor issues plague the selection of drive wheels. The broad drive wheel used here is excellent for outdoor use; it provides good traction, fair bump resistance, and a low-pressure footprint. But for indoor use they scrub when the robot turns and this can damage carpets or mar hard floors. If the robot was to be used exclusively indoors, narrow wheels like those used for wheelchairs would be better.

Figure 3-5 Kevin Derichs's robot

Figure 3-5 shows one robot, built by Robot Group member Kevin Derichs, that was built with essentially the same components as the platform in Figure 3-2. Kevin's design goals were slightly different; he wasn't as concerned with indoor use but wanted his platform to be suitable for outdoor use. Note the extended position of the caster.

Figure 3-6 shows the rolling platform of Tom Davidson's Eyebot. Tom, another Robot Group member, again used the same basic pieces as the other platforms. Tom, however, wanted his robot to run in as small a footprint as possible, and was aiming at indoor use only. He put his drive wheels across the center of the robot, and put small, spring-loaded casters at the front and back. This is the Robot Group's "standard configuration" as pioneered by Glenn Currie and others in the mid-1990s.

The springs on the casters give them some compliance, for a smoother drive as they go over bumps. For the outdoor robots, the pneumatic wheel on their caster fills this shock-absorbing role.

MATERIALS

Looking at the last three robot platforms, you will see that they are all made of welded steel tubing. Though fairly heavy, steel tubing is strong and, as long as you have a welding rig, easy to work with. Metal is also durable.

Figure 3-6 Tom Davidson's EyeBot (left), and a close-up of the EyeBot platform

Of course, a steel robot base doesn't have to be welded. You can use straight, angled, and bent plates to bolt it all together. Some of the later projects use bolted metals instead of welds, for several reasons. A bolt is easy to take off. And a bolt doesn't care if you are fastening steel, aluminum, brass, or some combination of materials; just as long as there is a hole for it to pass through.

Although I've stuck with steel for the projects in this book, your robot platform doesn't have to be made from metal—plywood and plastic will work just fine.

DRIVE PLATFORM

Don't weld or assemble any elements of the drive platform until you have read through this entire chapter and have your purchased components in hand. The different components, such as the wheel, swivel caster, and battery, determine the dimensions of frame. Any change in component dimensions needs to be accounted for in the various frame elements.

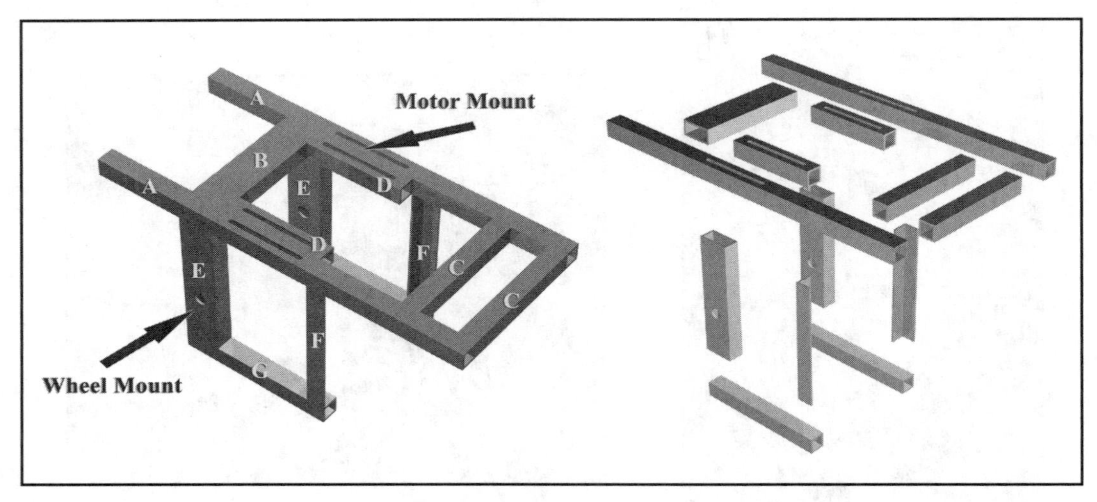

Figure 3-7 Drive platform

Figure 3-7 shows the bare bones of the drive platform. The slotted bars at the top (labeled Motor Mounts) are, clearly, for the motors to be mounted on. Each motor has four tapped holes in the frame for mounting. Bolts are run through the platform's grooves and into these holes in the motor. The grooves themselves allow the motors to slide forward and backward along the frame, to tension the drive chain.

Do not be tempted to mount your motor into single holes! Your drive chain needs to have an adjustable tension. If you cannot cut the appropriate grooves, you will have to include an additional spring-loaded chain tensioner. You can buy chain tensioners from industrial supply companies (such as MSC or Grainger), or you could build one yourself.

The drive wheels mount onto the large holes marked Wheel Mount in the picture. Additional holes not shown in Figure 3-7 will be explored as we cut and assemble the platform.

The parts illustrated in Figure 3-7 are dimensioned as shown in **Figure 3-8**.

All of the pieces are made of hollow steel tubing. The narrow pieces are 1" by 1" tubing, and the wide pieces are 2" by 1" tubing.

Pieces A, B, and C make up the main body of the frame.

A is 21" long and has one of the motor mount grooves along it. Exactly what diameter and spacing you make your grooves will depend on the motor you plan on mounting on the frame. In the case of my motor, the groove is 3.5" long and starts 8" from the front of the robot. The groove is also off-center along the bar.

The holes drilled in A are 1/4"-28 threaded holes centered in the piece. The front two are mounting points for the upper framework. The hole on the outside of **A** is for the front bumper. The two holes at the back of A are for additional bumper switches for the side and back of the robot.

B is the main cross support and it is 2" wide and 7" long. It is positioned 5" in from the front of the robot, directly above the wheel mounts.

C (both of them) have a dual function. They provide strength and structure at the rear of the robot, and they act as the mounting point for both the caster wheel and some of the bumper switches.

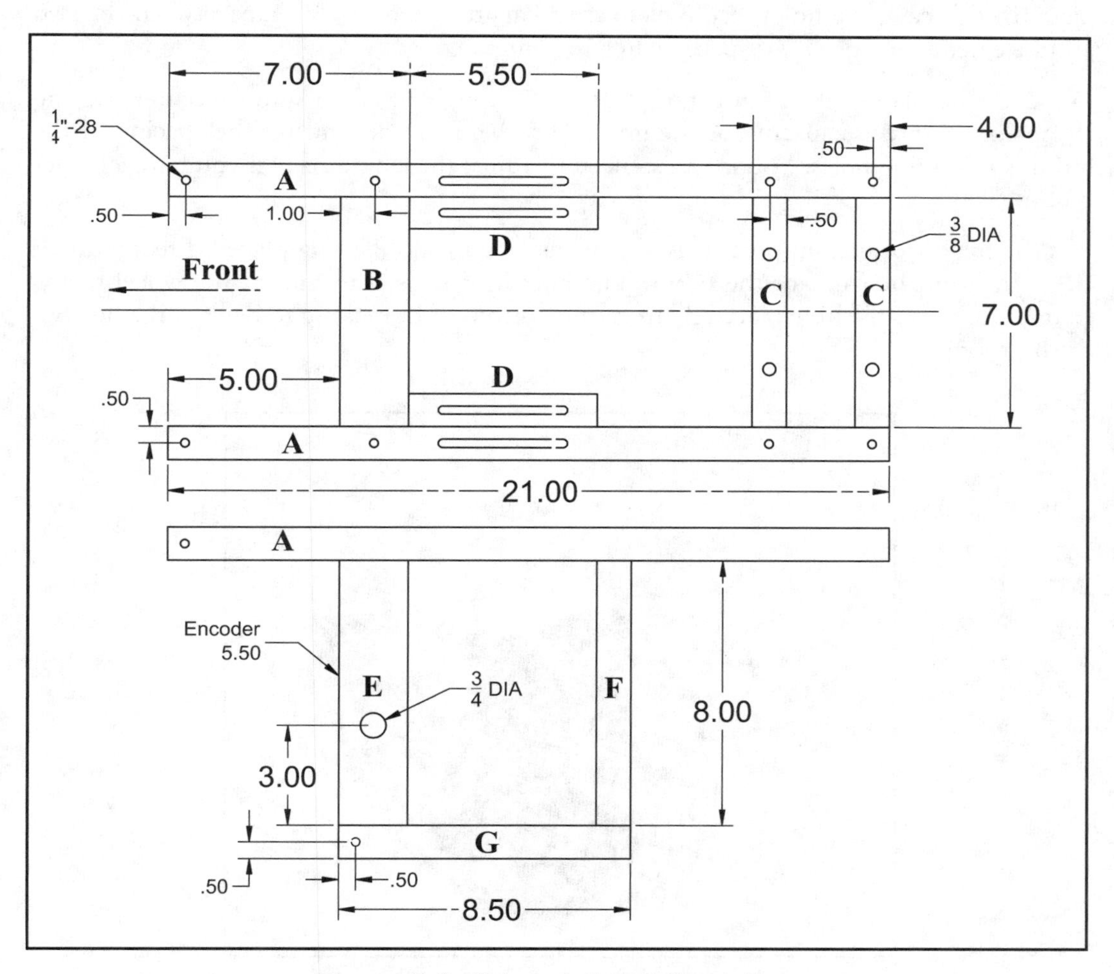

Figure 3-8 Drive platform dimensions

Figure 3-9 shows a representation of the swivel caster used on this robot. The actual caster came from Northern Tool and Equipment (www.northerntool.com). This caster stands about 10" high, and sweeps out about a 12" circle. While I used a black-rubber pneumatic caster, you can also get nonmarking gray casters in the same size range.

The height of the caster interacts with the ground clearance and tilt of the robot, so if you use a different sized caster be sure to adjust the other sizes in the platform. Take special care to lengthen A (and hence, position C further forward) if you use a larger caster. This frame is spaced to exactly accommodate this caster.

The spacing of the two C elements is determined by the width of the caster's mounting plate. In this case, the holes in the plate are 3" apart. There is a 2" gap between the two 1" C elements, giving 3" of space center to center.

D is the other half of the motor mount. It is 5.5" long with a 3.5" groove centered along it. Again, the precise spacing of the grooves in A and D depend on the spacing of the motor's mounting holes. You may also need to adjust the length of D if you have a longer motor.

E is the main upright support. It is a 2" wide tube that is 8" long placed directly under B, 5" in from the front of the robot. The hole in E is used to mount the wheel to the frame. This is a 3/4" hole placed 3" from the bottom of E, making it 4" from the bottom of the robot.

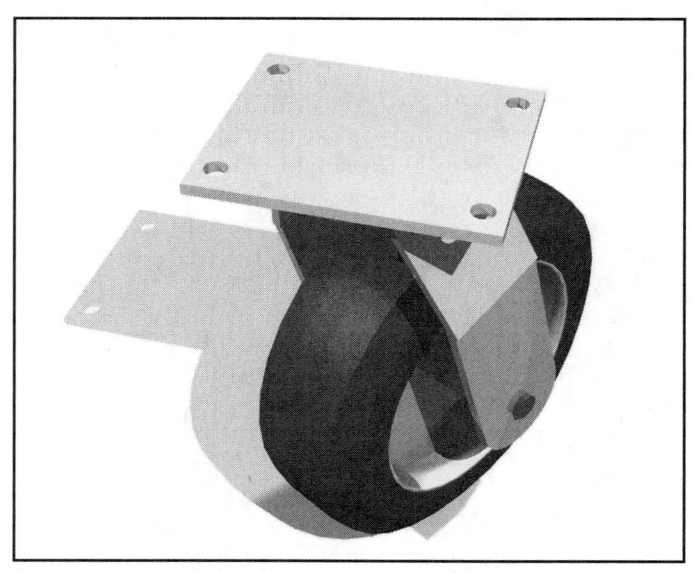

Figure 3-9 Swivel caster

Figure 3-10 Drive wheel

The drive wheel also came from Northern Tool (**Figure 3-10**). This is a 11.6" diameter wheel, with a large sprocket attached. The ball-bearing hub accepts a 3/4" bolt, which is used to fasten the wheel to the frame. A #35 chain ties the drive sprocket to the gearmotor mounted on pieces A and D. You will need two of these wheels, plus the bolts and nuts to mount them, and several washers per side to space the sprocket away from the frame.

The hole in E gives the frame almost 2" of clearance. With the 10" caster, there is about a 1/2" tilt from front to back. This tilt could be removed by adding spacers to the caster, or, at the expense of ground clearance, by moving the wheel mount holes up another 1/2". However, I like the tilt; it gives the robot a jaunty air.

F is a simple 8" support at the other end of G. This is shown as a 1" angle bracket. A simple 1" flat piece would also work. Whatever you use, you can't obstruct the inside corner of G, because the caster passes through this space as it swivels.

G is a spacer and brace for the drive platform, and makes up the side-rail of the battery tray. A 1/4"-28 threaded hole on the inside face of G serves as a mounting point for a front bumper switch.

BATTERY TRAY

The battery tray consists of two angle brackets H welded between the drive frame's G elements. The spacing between these brackets should be sufficient for your battery to sit on them (see **Figure 3-11**).

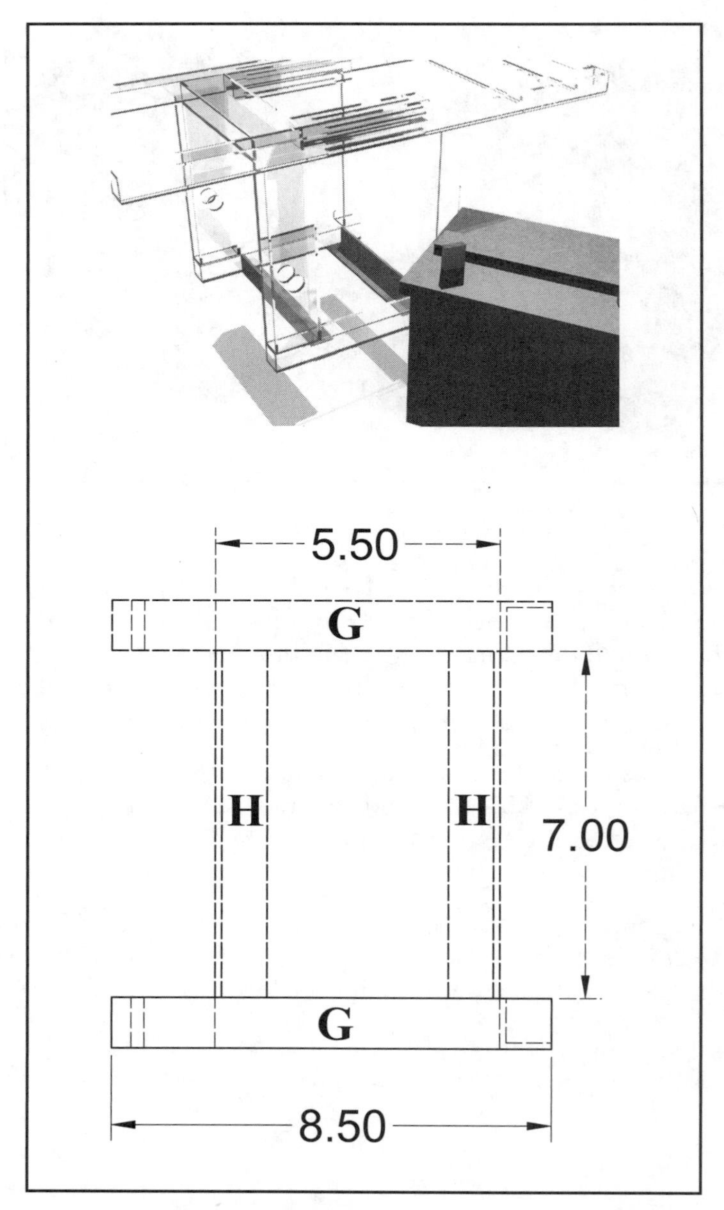

Figure 3-11 Battery tray (top) and dimensions

The battery supports G need to be spaced far enough back from C and the swivel caster so the caster's wheel doesn't jam against the battery.

The entire robot is built up around this battery, so it is important to have it in hand before cutting any metal. The height of the platform, as defined by E and F, needs to be sufficient to clear the battery terminals. The length of the battery tray as defined by G needs to be sufficient to hold the battery, and this length plus the size of the drive wheels and the area visited by the swivel caster as it turns determine the length of the robot platform A. The width of the robot's frame B and C is determined by the length of the battery.

So even though the battery tray is a very modest piece of the frame, the battery itself determines many of the frame's dimensions. The first frame I built didn't take this as seriously as it should have and the battery terminals shorted against the framework. I had to scrap that first frame due to insufficient planning.

BUMPERS

Desktop robots have a simple task for their bumper design. They typically have two simple wire whiskers spanning the front of the robot, attached to two lever switches (as seen in *Applied Robotics*, page 52).

A robot as large as this one, however, wants a little something more. In theory, if all of the other sensors and control logic are working correctly, the robot will never bump into anything. In practice, especially while the system is being trained, it is going to be ramming into things all over the place—possibly at high speed.

So not only do we want to sense when the robot has run into something, it would be nice if the bumpers were both durable and expendable. And it would be doubly nice if they could provide some cushioning as well.

Once the robot has struck an object it will take a moment before its momentum is lost as the motors slow or brake to a stop. Even if the motor could lock tight, the robot would shift and tilt from the sudden jolting stop. During this moment of time, as the robot stops it will continue moving in the direction of the contact, pressing the bumper into it. Then, as it attempts to free itself, the robot may actually snag or entangle the bumper with the sensed object—and then proceed to twist the bumper beyond all recognition. Any bumper design should provide a minimum of catch points. As mentioned earlier, it would also be nice if the bumpers were cushioning, durable, and easily (and cheaply) replaced.

SENSOR

There are many possible sensors available for a bumper system: a wire-and-microphone system, a ball-and-cage (or mercury) tilt switch, a pneumatic pressure sensor attached to rubber bladders, a force-sensitive resistor. I'm going to use my favorite device, the reliable and robust lever-actuated switch (see **Figure 3-12**). Using a switch with a fairly high triggering pressure eliminates many false signals.

The lever on the switch can be coupled with a spade connector providing a durable yet replaceable interface to the bumper. If you have problems with the bumper sliding off of the lever, you can either crimp the connector tighter or drill a small hole and wire (or bolt) the spade connector onto the lever.

Figure 3-12 Lever-actuated switch with lever and spade connector

The spade connector, of course, also fits onto a heavy wire that acts as the base of the bumper. My switches have a roughly 3/16" wide lever, so I purchased some 3/16" quick disconnects from Radio Shack. These, in turn, fit nicely on some 3/32" welding wire, used as the core of the bumpers' feelers (next section).

The electrical connections to the switches can also be made with the same 3/16" quick disconnects.

FEELER

I am a huge fan of welding wire. This wire is inexpensive, easily found (welding stores are everywhere), and easy to use. It is a steel wire plated over with copper. It comes in many sizes and, unlike piano wire, is easily bent into complex shapes with no threat of breaking. The copper coating makes it easy to solder, so it can be soldered into the quick disconnects instead of being crimped (**Figure 3-13**).

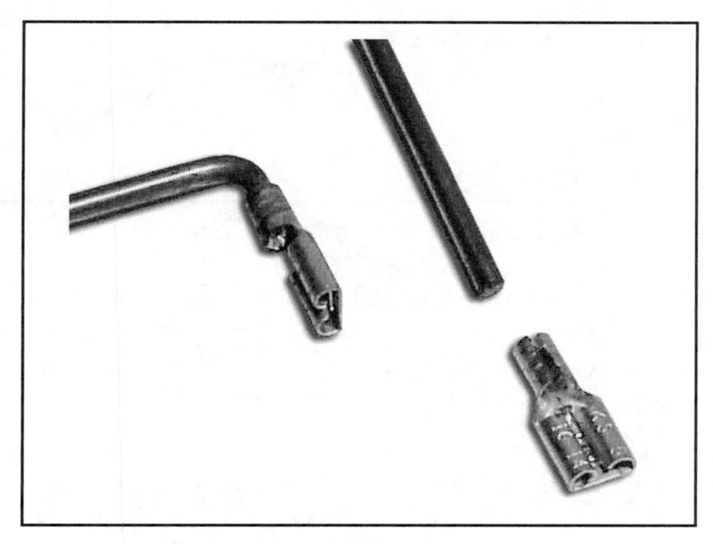

Figure 3-13 Wire and connector

Naked wire, however, doesn't make a satisfying bumper for this robot—it lacks the desired cushioning ability. It would probably work if you want a slimline bumper.

You can normally find dark gray foam cylinders used to insulate water pipes at your local hardware store. The smallest one I could find had a 3/4" diameter hole in the center, with a 1.5" outside diameter, but I suspect you can find smaller ones with a 1/2" inside diameter if you look hard enough.

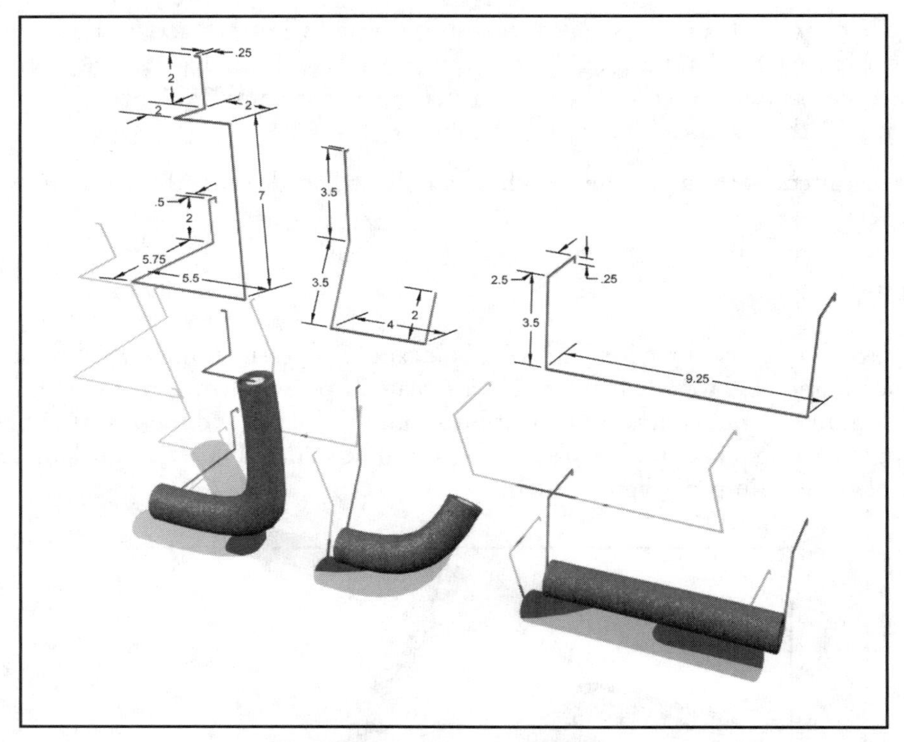

Figure 3-14 Bumper shapes

Taking a pile of wire and some of the insulation, I was able to quickly form a variety of interesting bumper shapes (**Figure 3-14**). Given the choice, select the insulating tubes that have sticky edges. These can be sealed into a continuous tube after they are fitted on the wire; otherwise they tend to bulge open at the corners.

The foam tubes don't want to stay on the wires without help. The traditional fixit-kit stuff could work here (e.g. zip-ties or duct tape), but that seems inelegant. I'm just now exploring the wonders of expanding insulating foam (e.g. Great Stuff), so I decided to lock the cylinders onto the wires with this stuff.

Great Stuff—I had forgotten how unspeakably nasty this foam is to use. It sticks to anything, forever. Once you get it on your hands, it doesn't have the common decency to harden but remains tacky for a very long time. While you are trying to get it off of your hands (note: acetone smells horrid and burns in any open scrapes and scratches you might have forgotten you had), it is oozing out of the holes in your project. So, pay attention to the stuff for ten minutes or so, and don't get it on your hands (or clothes, hair, porch, carpet, or pets).

The shape and size of the bumper wires is both a science and an art. The pieces that interface with the switches need to be carefully measured to provide a close fit without binding on the frame. However, the bulk of the bumper is artfully shaped to fit the contours of your robot. Feel free to make several sets of bumpers—the materials are inexpensive, and you'll probably need spares anyway.

MOUNTING BRACKETS

Now we have bumpers attached to wires. Wires attached to quick disconnect plugs. Quick disconnects snapped onto switch levers. But how to get the switches onto the robot?

Some quality time in the workshop gives me a set of switch brackets (**Figure 3-15**). These are short strips of 3/4" wide aluminum with three holes in them. The two small holes just fit the spacing of the holes in the switches. These are drilled and tapped to fit #4-40 bolts, to avoid needing the matching nuts behind the plate. The larger hole is a through-hole for a 1/4"-28 bolt. You could also use a 1/4"-20 bolt if you prefer the coarse thread bolts. For the thin walls of the frame, however, I chose 1/4"-28 so it would have more threads to grip on.

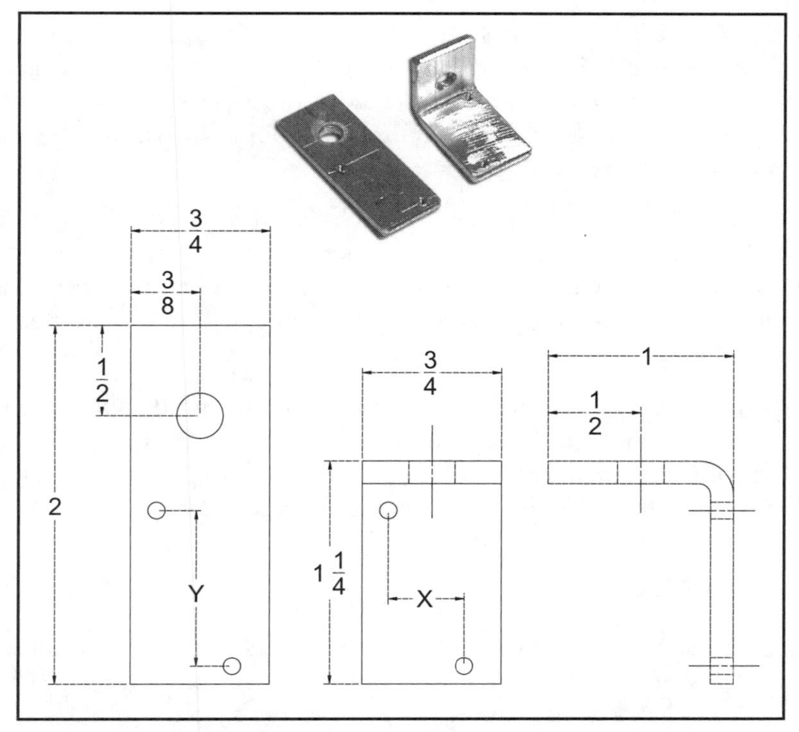

Figure 3-15 Switch brackets and dimensions

Figure 3-16 Switch mounts

Matching holes are drilled in the frame, and tapped to accept the 1/4" bolt. You can use nuts if you don't want to invest in the taps, but they are hard to work with in the tight spaces provided inside the frame's tubes.

Referring back to Figure 3-8, you will see five holes drilled (and tapped) into each element A. The one on the outer face of A holds one of the two front bumper switches, the other being on the inside face of element G. The two holes on the front half of A are for future use in mounting the torso and arm. The two in the back of A are used for the side bumper and the rear bumper switches.

You may notice that I am using a flat plate to attach the switch to the frame in some places (**Figure 3-16**). Though it might appear to be easier to tap and drill the two #4-40 holes directly into the frame, a few broken taps may convince you otherwise. It is far easier to tap a 1/4-28 hole than a #4-40 hole. The #4-40 holes tapped into aluminum is enough challenge for this builder.

After some experimenting, I found I could mount the bumper on the switch so that gravity would help hold it onto the lever (as shown in Figure 3-16 and **Figure 3-17**). Some of the bumpers are actually mounted backward from what you might expect. These use the rotation of the bumper to trigger the switch rather than a linear push.

WHEEL ENCODERS

Though they aren't attached until later, it is best to drill and tap the mounting holes for the wheel encoder sensors now. Once the robot is assembled this part of the frame becomes inaccessible. Marked in Figure 3-8, the encoder's hole is placed 5.5" from the bottom of the robot (4.5" from the end of element **E**), and 1/4" in from the outside edge.

Figure 3-17 Front (above, left), side (above, right), and back bumpers

The encoders used in this project are Fairchild Optoelectronics part #QRB1134 (**Figure 3-18**). These will be discussed in more detail later, with the encoder disk that will be mounted on the wheel. Be warned, however, that we will be putting an encoder image on the gear of the drive wheels later. You may want to skip ahead to Chapter 9, Sense and Control: Drive System, and do this now before you attach the wheels.

Figure 3-18 Wheel encoder

CIRCUIT MOUNTING

The final use for the lower frame is as a mount for the power supplies and reflex circuit boards.

There is precious little room left in the lower frame. The front area is still reserved for the arm. The battery takes the central cavity with no room to spare. The rear caster dominates the back. Above the frame, the front is reserved for the arm and head. The motors take up the central area above the battery. However, there is room above the caster to mount some more things.

PAINTING

Once all of the holes have been drilled, the frame assembled, and the grease and oxides cleaned off of your metal, you can paint it. Invest in a high-quality paint and primer. You have already spent a small fortune in time and money building this robot, and you have only just begun. Make it look nice.

The test robot for this book was assembled, painted, disassembled, drilled, tapped, oiled and cleaned, scratched, dented, folded, spindled, and mutilated. And still, the battered black paint job makes it look nicer than the traditional rusted steel (**Figure 3-19**).

Figure 3-19 Robot assembled so far

PUTTING IT ALL TOGETHER

Now that you have read the construction details, you can start building your frame. Of course, you may also want to wait until Chapter 5, Body and Head, and do the lower frame and upper frame at the same time.

First, be sure of your dimensions and placement. Figure 3-7, Figure 3-8, and Figure 3-11 give the basic idea, but your dimensions may be different depending on your battery and other possible differences in design requirements.

With plans in hand, cut all of the metal elements to length. Then drill and deburr all of the holes. Tap the holes that need tapping.

Weld or bolt the frame elements together, taking care to keep everything well placed and at right angles.

Grind your welds flat if you've welded things together, and clean the metal using the caustic, environmentally hazardous chemical of your choice.

Prime and paint everything. Let it sit for a couple of days, so the paint can dry and harden completely.

Bolt the drive wheels onto the frame, using several large washers to ensure the proper spacing between the frame and the wheels, so that they don't rub or bind. Bolt the rear caster wheel into place, as well.

Attach a sprocket to the drive motor, and loosely bolt the drive motors onto the frame. Position the motors in their slots near the wheel-side of the robot. Measure, cut, and assemble the drive chain to fit this position. Loop the chain over the wheel and motor sprockets. Snug the chain by moving the drive motors back and tighten the motors into place on the frame.

Using a lock washer (I used a toothed washer instead of the split washer) between the bracket and frame, fasten the switch brackets onto the frame. Then, fasten the switches onto the brackets and slip the bumpers onto the switch levers.

That's it!

CHAPTER 4

Stock Power Drivers

There are three types of motors used in this robot. The high-power drive motor (tested at 36 amps at dead stall, at 12 volts), the arm motors (estimated earlier at about 5 to 6 amps at near-stall), and the hand servos (current draw unknown).

The hand servos, in the mystic way of R/C servos everywhere, have the power drivers and PWM circuitry built into the package, along with the feedback potentiometer, motor, and gears. These are a miracle of modern surface-mount components and a joy to behold.

The drive and arm motors do not give us such an easy time.

Given a choice, you are almost certainly better off buying a motor driver and controller from a hobby store like Tower Hobbies (www.towerhobbies.com), an electric golf cart store like 4QD (www.4qd.co.uk) or Vantec (www.vantec.com), fellow robotics hobbyists like Team Delta (www.teamdelta.com), and so on. This chapter explores the R/C speed controller.

You can find plans and information to build homegrown speed controllers at www.simreal.com. They may not cost less or be easier to use, but it is educational to make your own controllers. And, with the knowledge of how to make an ESC of your own, you can customize the behavior or size to fit your precise needs; at least, if you don't mind suffering.

R/C ESC

The radio-control hobbyists provide a mass market for a variety of interesting and useful products. In this section, we focus on the Electronic Speed Controller (ESC) from the electric car segment of that industry.

There are two broad categories of ESC available: those that have a reverse direction and those that do not. For robot use, be sure to get a reversing ESC if you want to have any hope of backing out of corners.

In the category of reversing ESCs, there is a bewildering array of options and choices...and prices, ranging from $30 to $200, depending on your source. A good source of these supplies is Tower Hobbies, conveniently available online as noted on the previous page.

There are two key values to match on your speed controller: the voltage and current ratings. This robot operates on 12 volts, and the drive motors are in the 40 amp range.

15 Turns at 10 Cells?

ESCs come with different specifications than you might be familiar with as a robotics hobbyist. Some of the controllers do specify their abilities in familiar comfortable terms like "Rated Forward Current, 320 amps," but then they say things like "Motor Limit, 12 Turns at 6 Cells," or "BEC 6.0 volts / 3.0 amps."

Turns

The term *motor turns* specifies the number of times the wire has been wound around the armature of the motor. Fewer turns make for a faster, but lower torque, motor. Conversely, more turns make a slower and stronger motor. Low-turn motors are the "high-performance" motors in electric-car racing and are generally more expensive.

I'm not entirely sure how the number of turns relates to more familiar motor specifications, though I have to assume that a lower-turn motor will in general have a lower resistance and more current draw.

Cells

Input voltage to speed controllers is specified in cells, meaning the number of NiCd battery cells used to power the system. Each NiCd cell has a nominal fully charged voltage of 1.2 volts, with a discharged voltage of about 0.9 volts. So, a 10-cell controller should operate on a 9- to 12-volt power supply, and a 6-cell controller on 5.4 to 7.2 volts.

Most ESCs, and all of the inexpensive ones, seem to run in the 6-cell voltage range, almost making me wish I had started this project with two 6-volt batteries instead of a single 12-volt battery.

BEC

In plain English, this is the Battery Eliminator Circuit. It took a while before I could find a reference that spelled this out in words instead of initials. When I did, though, several anomalous behaviors of my speed controllers came clear—it all made sense!

Figure 4-1 shows the typical circuit for a radio receiver and servo, such as can be found on a model airplane. The receiver is hooked up to the battery, and feeds both power and the control signal to the servo. Inside the servo is the PWM control circuitry and power driver. Life is good.

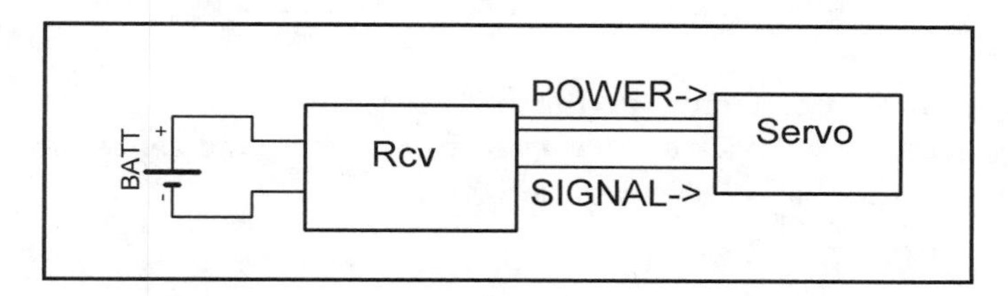

Figure 4-1 R/C receiver and servo

Moving over to electric cars, or, in our case, robots, you get a circuit more like that shown in **Figure 4-2**. The receiver still has its power supply, but it only needs to share a ground and signal with the ESC. The ESC, in turn, will be attached to a more powerful battery to drive the higher-current motors associated with racing. Now you have two sets of batteries to buy, charge, and maintain.

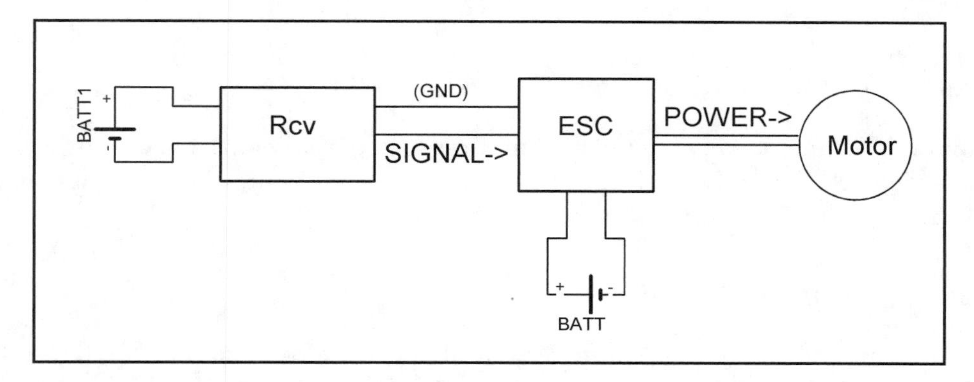

Figure 4-2 R/C receiver, ESC, and motor

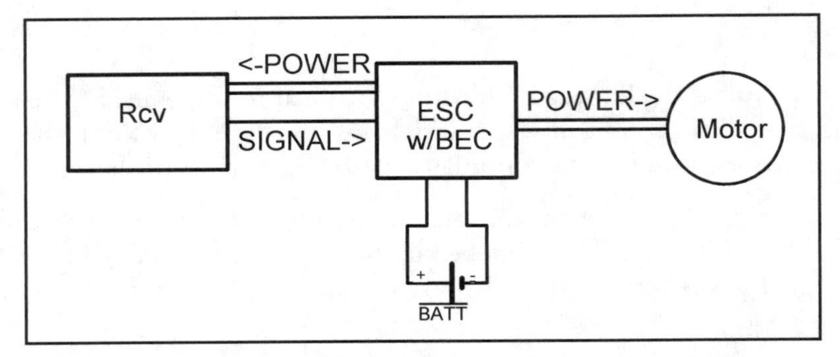

Figure 4-3 R/C receiver, ESC with BEC, and motor

Adding a battery eliminator circuit to the ESC lets you remove the receiver's battery. Instead, the ESC feeds power back up to the receiver through the control harness normally used to supply power to the servos (**Figure** 4-3).

Now, if they could just find a way to eliminate that final battery... .

OTHER OPTIONS

Here is a short grab bag of other ESC options to watch for.

ON-RESISTANCE

Like size, this is an attribute where smaller is better. A high-power ESC, like the Novak SuperRooster can have an "on" resistance of 2mΩ (0.002 Ω) or less, giving a very high efficiency.

REVERSE VOLTAGE PROTECTION

Without this feature, if you plug the ESC into the battery backwards, the circuits will fry. While a fried ESC is a bad thing, especially if you paid $200 for it, a little care in hooking to the battery can serve in lieu of this protection.

THERMAL PROTECTION

As you run current through the controller, it will heat up—slowly at 0.002Ω, but still, it will heat. If it heats up too much, things can burn out and stop working. However, thermal protection will turn the controller off first. Sure, your robot will stop moving, but once it cools down it will start up again.

Selecting the ESC

Given all this information on ESCs and the knowledge of what your motors require, it should be a fairly simple task to pick out an ESC that is well matched to your needs. Or you can do what I did, and get an ESC that is ridiculously overpowered for the task at hand (not that I had much choice among the 10-cell ESCs).

If you've read *Applied Robotics* you may recall my philosophy on tools, and, yes, an ESC can be considered a tool if you plan on reusing it: always buy the best you can afford. And at the time I bought the ESC for this project, I could afford the Novak Super Rooster reversible electronic speed controller.

This little (1.63" x 2.02" x 1.22") unit with the festive purple anodized heat sinks packs an amazing 320 amps of forward current, and 160 amps of reverse and braking current; almost ten times what I need for the drive motors. The on resistance is 0.002Ω, and it conveniently runs on 12 volts (6 to 10 cells). No reverse voltage hookup protection, but it is thermally protected. It works like a charm, too. One of the reasons I chose Novak is because it's the brand that some of the other Robot Group people used, so I was confident that it would work.

Hooking It Up

Your ESC may not come with connectors on the battery or motor wires, and if it does they are probably not the "right" type of connectors.

The best power connector that I've had the privilege of using is the Anderson Powerpole. I've had ample opportunities to wade through an impressive array of connecting technologies in my projects, and this is the very first connector that was actually a joy to use. It's even a crimp connector, and I really don't like crimp connectors; they are hard to use if you don't want to buy the amazingly expensive crimping tools, and even then I have my doubts. The connector used in this section is the 15/45 Powerpole, good with 15, 30, and 45 amp contacts. Tower Hobbies sells them as the DuraTrax Powerpole Connector.

You can buy Anderson connectors in sizes up to 350 amps, should you have that need, and they all seem to have the same friendly mechanical structure.

There are two necessary parts to a Powerpole connector: the plastic housing, and the crimp contact.

The plastic housing comes in a variety of colors, and inside of it there is a stainless-steel leaf spring that maintains constant contact pressure under even the most difficult environmental conditions. The housings are genderless and mate with themselves. They also slide together side-by-side and lock into multiple-connector rows or blocks.

The optional items include connector block housings and retaining pins to keep the blocks from coming apart, which they won't do unless you put them under a lot of stress; this is a tight fitting connector.

The connectors are plated with silver for a highly conductive, low-resistance contact. The connector for 12 to 16 gauge wire (the ESC has 14 gauge wire leads) can carry 30 amps of current. The only caveat is that the contacts are not designed for "hot" connects or disconnects, so turn off the power before unplugging things.

For this connector, you strip off just enough insulation for the wire to fit in the crimp tube (it has stops to keep the wire from protruding too far): in this case, 5/16". Then, with an appropriately sized crimping tool, crimp the contact onto the wire. I used an el-cheapo hand crimper, since my expensive ratcheting crimper didn't have the crimp dies for this size connector. Alternatively, you can solder it into place, taking care not to foul the contacts.

Finally, push the connector into the housing until it snaps into place under the spring with a solid click. If you want, pairs of housings can then be connected together into color-coded pairs.

The Super Rooster ESC comes with five sets of wires (see **Figure 4-4**). The 14 gauge red and black wires are for the battery connection, in the predictable red to plus, black to minus pattern. The other large wires, blue and yellow, go to the motor, yellow to plus, blue to minus. The motor connection polarity is not vital, since most radios can reverse channels as needed.

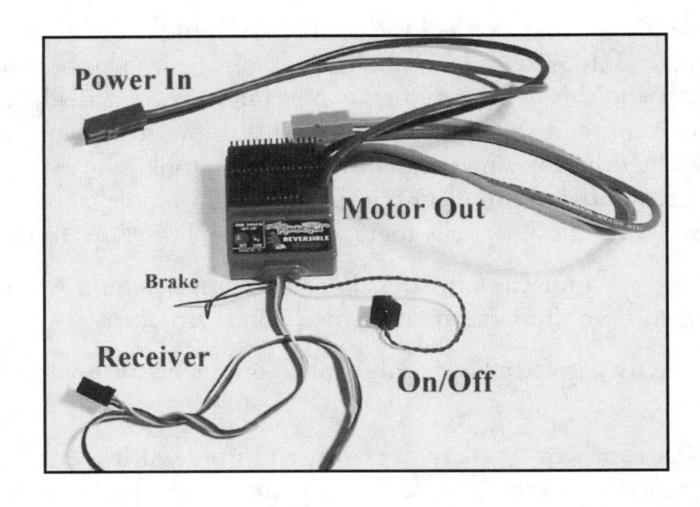

Figure 4-4 Novak Super Rooster ESC

The receiver wires are designed to plug into a standard R/C receiver, but see your ESCs instructions to determine *which* standard receiver, and how to convert to a different standard. The white wire carries the coded signal to the ESC, and the black wire allows the ESC to share a common ground with the receiver. Normally, the red wire would carry power from the receiver to a servo, but the Super Rooster has a BEC circuit so it supplies power up the red wire to the receiver. I'm not sure what two Super Roosters in parallel, each trying to supply 5 volts to the receiver, will do, but it seems to work in this platform without any visible problems.

The On/Off switch should probably be left in the "on" position and the power managed from a central switch on the robot. I actually cut off this switch and soldered the wires together.

Finally, there are the brake-light wires, which remain unused in this system.

The power and motor wires each received 30 amp Powerpole connectors, with color-matched connectors on the motor's leads and coming from the battery distribution block. This allows for easy removal of the ESC, as needed.

How the ESC gets attached to the robot is a question that will be better answered once more of the framework is in place. For now, it suffices to glom it all together with a variety of zip-ties, the electronics version of duct tape.

POWER DISTRIBUTION

We glossed over the topic of how to get electricity from the batteries and into the control systems in Chapter 2. We didn't have the need, or the tools, for it then. But now we need to hook up two powerful ESCs to the battery.

I thought about this problem for a long time. Originally, I just got a four-position terminal block from Radio Shack and screwed the wires into it. This was unlovely and needed replacing.

The final solution is simple and elegant. I went down to my local welding supply store and got a *very* nice 4 gauge welding cable, which I then cut in half. Welding cables are great—flexible, well protected, and you can get crimp-on terminals for them.

Using a bolt and a bench vise, I improvised a strong "crimp" to lock a terminal onto one end of the two power cables. These were then bolted onto the battery terminals.

The other end of each cable received a set of Powerpole connectors. Stripping off several inches of the cable insulation, it was conveniently organized inside as a set of seven stranded subcables. Each of these fit perfectly into a Powerpole terminal.

Before crimping on the terminal, I slipped a shrink tube over each strand, and a larger shrink tube over the whole cable. Once crimped, I shrank the little tubes. Once the terminals were put into the connectors I slid the large tube as far up as I could, until the spread of wires stopped its progress, and shrank it into place.

In the end, I had a row of seven ground connectors, and a row of seven power connectors. These two rows then snapped together into a nice 2x7 connector block! It was beautiful.

Testing the Drive System

If you are anything like me (and if you've worked your way through this book to this point, I'm afraid you may be), you will have a near-irresistible urge to see the platform move around. This is an important testing phase; you need to know if the controllers work, how fast the robot moves, whether you need to change the gearing, and so forth.

But so far, there is no brain to generate the control signals for the ESC. No signal, no motion.

Fortunately, radio-control transmitters and receivers have come down in price so that even a really fancy six-channel, computerized, custom-mixing enabled system is within the reach of most people. Even better, the hobby store will often buy back the servos that came with the radio that you don't need.

It also just happens that the ESC you installed on the robot (be it with zip-ties, double-sided tape, or chewing gum) is custom designed to operate on just the type of signal provided by a radio-control system.

So I ran out to my local hobby store, which had prices, amazingly enough, in the same range if not cheaper than my favorite online suppliers, but without the shipping charges, and bought a nice six-channel PPM transmitter and receiver set.

A note about transmitter types: AM is bad, FM is okay, and PCM or PPM is best, especially in the high-noise environment of a robot. You may notice that as you move from 2- and 3-channel transmitters to the 6-channel varieties, you go from the pistol-grip ground frequencies to the dual-joystick controllers of the air frequencies. It's a common practice for robots to operate on the air frequencies, though it is a better good-neighbor policy (and possibly the law, though I'm not qualified to comment on any legal issues) to see if your transmitter and receiver can be modified from air frequencies to ground frequencies.

Firing up the robot for the first time, I go through the ritual of setting the ESC controls to match the receiver. This done, I launch the robot for the first time.

It drives at top speed at a stately walking pace, just right for tight living-room travel. It turns nicely and controls well. I soon grow frustrated with the independent wheel controls and switch over to an "elevon" mixer setting on the transmitter—now one joystick axis is steering and the other speed/direction. This is even nicer!

Taking the robot to the club to show off, I find that the slow max speed is a bit low for outdoor use; I easily outpace the robot. It would benefit from a different gear ratio for outdoor use.

This done, I park the robot in a tidy location and prepare for the next steps in the project.

ESC Control Signal

Ultimately, and perhaps sooner, you will want to operate the ESC using a synthesized signal and not by way of the radio control transmitter. This is a robot, after all, and not a glorified R/C super truck. Apologies in advance to the Battle Bots, Robot Wars, and other combat robot builders and operators; you guys do great work and I wish some of you would write books of your own (hint, hint).

While a motor controller will send a pulse-width modulated signal to the motors to control their speed (more on this later), the R/C receiver sends a specially coded signal to the servo or ESC. **Figure 4-5** shows a representation of this signal.

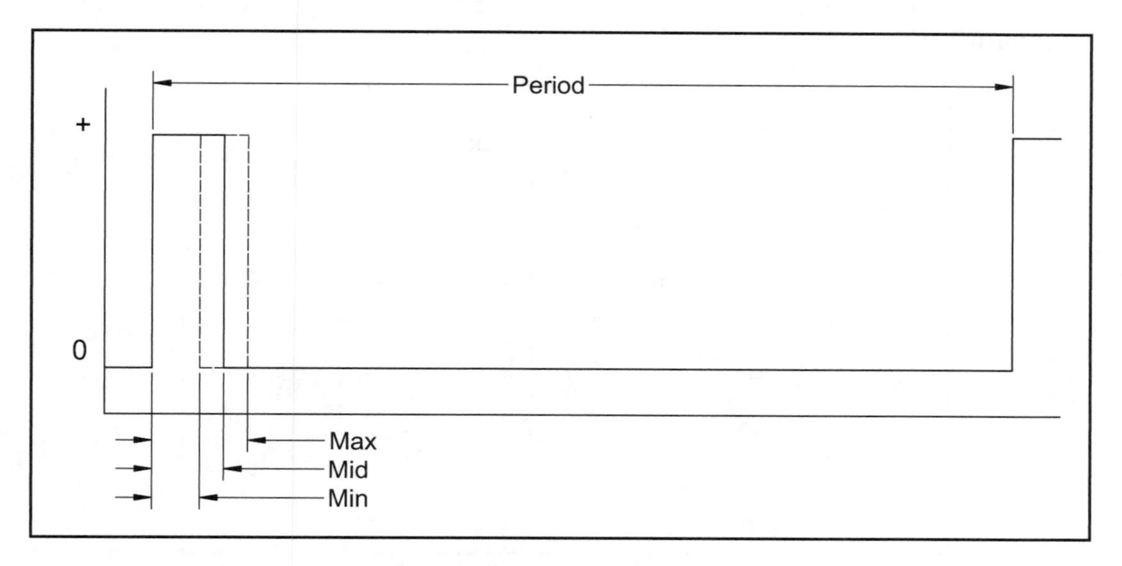

Figure 4-5 R/C control signal

Each channel of the receiver sends a pulse once every timer period, which is usually about once every 20 mS.

The pulse itself varies from about 1.0 mS to 2.0 mS. The minimum pulse width represents full left on a servo, or reverse on an ESC. The maximum pulse width is full right/forward. Between these extremes (roughly 1.5 mS) is the middle position on a servo, or stopped on an ESC.

Given a 1 mS change in pulse width (or 0.5 mS reverse pulse 0.5 mS forward), decoding the signal in 128 steps requires a timing sensitivity of 3.9µS. This is 62.5 cycles of a 16MHz clock...which may become significant later.

SYNTHESIS

A simple 555 timer circuit that can generate the R/C signal is shown in **Figure 4-6**. The switch wired to the reset pin of the timer turns the signal on and off. Resistor R4 controls the timing period, and R1 sets the minimum pulse width. Potentiometer R2 changes the pulse from a minimum to maximum range. Change the values of these three resistors to tune the circuit to your particular needs.

A nice system to generate R/C signals from computer control can be found at Scott Edward's Electronics (www.seetron.com).

We don't use this analog synthesis circuit anywhere in this robot, so we aren't going to dwell on it here.

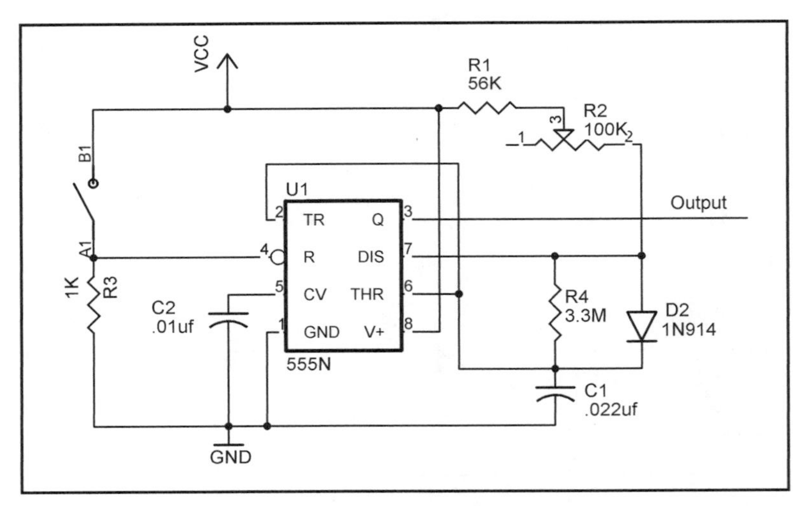

Figure 4-6 R/C signal generator

DECODING

If you wanted to decode the signal sent from the receiver using a microcontroller (and we will want to, later), the question also arises as to the timing of the pulses between channels (see **Figure** 4-7). Does each channel on the receiver start its pulse at the same time, are

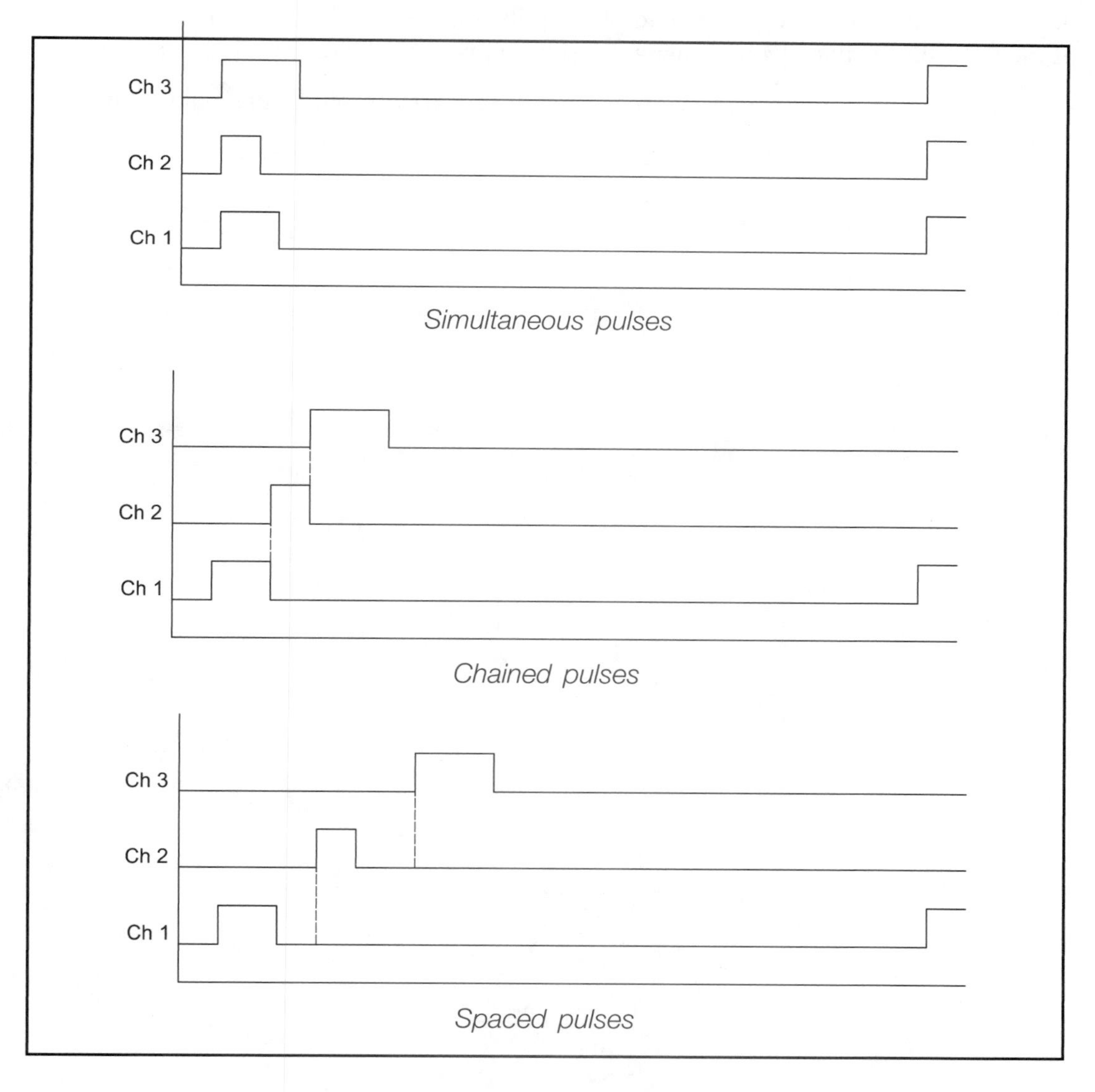

Figure 4-7 Pulse timing

the pulses chained hot on each other's heels, or are they evenly spaced out in the allotted 18 to 20 mS time period? On the Futaba receiver I tested, the pulses followed the chained pattern. I would be much happier if the pulses were evenly spaced, since I could logically-OR the pulses together and measure them using a single input channel.

Another question, then: as the control signals change width, does the period change or does the "dead time" between the end of the pulse on the last channel and the beginning of the pulse on the first channel change? Again, in my tests, the time period was invariant.

As in all things, your mileage may vary. Test your receiver yourself before making any rash assumptions.

CHAPTER 5

Body and Head

What does a robot need to be self-motivated? It needs a brain. The brain would then sit in the head. And the head needs an upper body to sit on, too. Okay, that's terribly cranial-centric, but let's go with it.

In this chapter, we lay the foundation for the robot's intelligence. We don't get to look at code yet—that is still several chapters away. But we are nearing the end of the construction. Soon, very soon, we will start to breathe life into it.

In the next chapter, we give it an arm, so it can interact with its world. (How would *you* like to have been born with just a torso and legs?)

Upper Frame

Biological analogies fail me here...the upper framework is like the robot's torso, yet the "brain" (e.g. the laptop computer) is mounted on it, head-like. Then there is the robot's eye—which will be much like the head—that sits in front of the laptop. Finally, there are a few corners of the framework that can serve as mounting points for additional low-level circuitry; the brain stem and nervous system, as it were.

Figure 5-1 shows a few views of the upper frame as it will sit on the robot. It is a removable component; it attaches by bolting onto the holes we prepared earlier in the lower frame.

An important part of the upper frame is the tray for the main computer; the skull for the brain, as it were. This must be sized to fit your laptop, or whatever you plan on using for the brains. In my skull, I am putting thin rubber pads along the metal channel to protect the dainty plastic shell of my brain (er, computer).

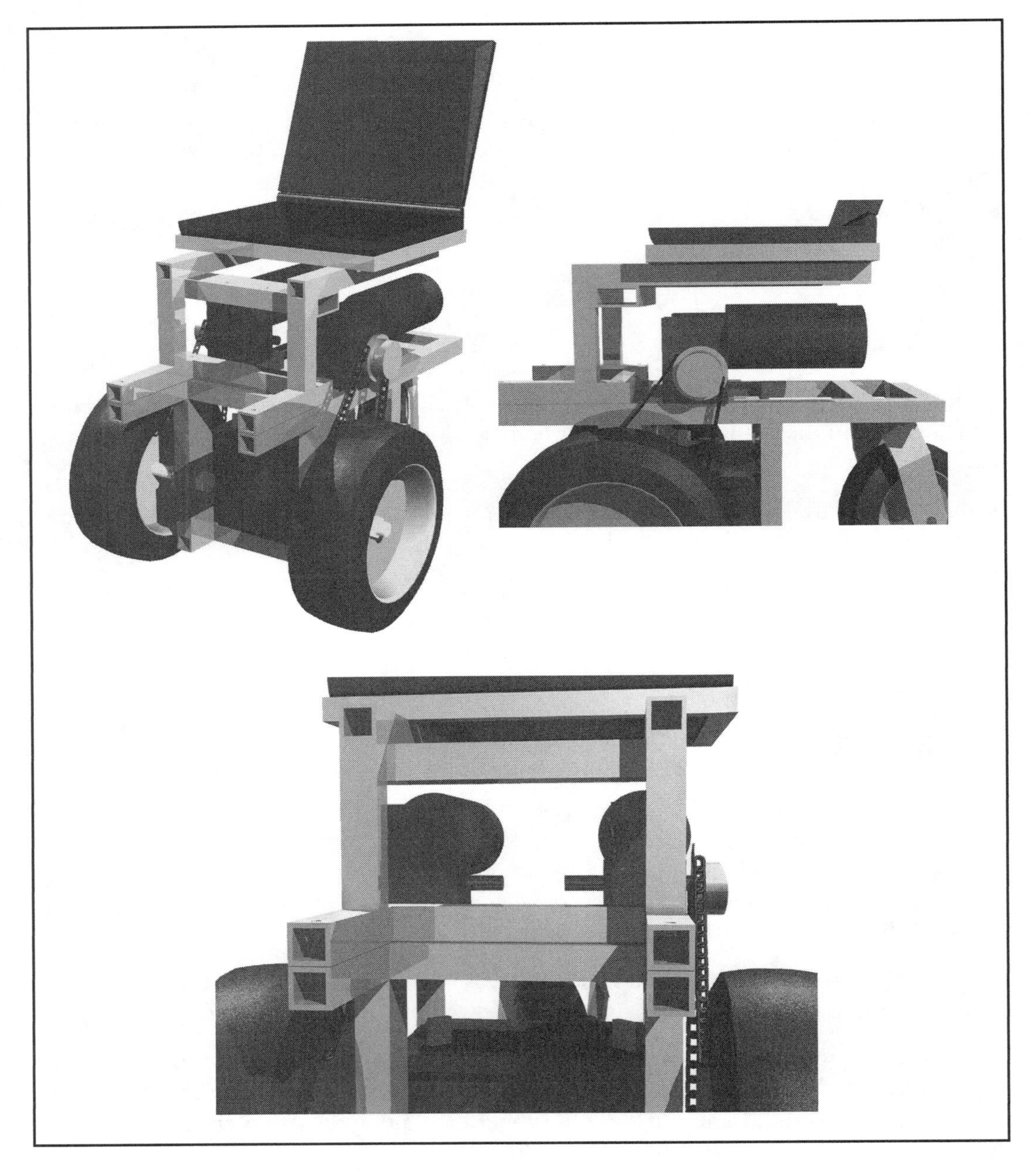

Figure 5-1 Several views of the upper frame

This project is broken into four parts, shown in **Figure 5-2**, which are assembled into the final upper frame. The Base is the part that bolts onto the lower frame. The Uprights connect the Base with the Head. The Head holds the camera, computer brain, and arm (so it's really a head/shoulder arrangement). The removable tray holds the computer, and can be changed for a new tray if you decide to upgrade the brains later.

BASE

The most important aspect of the base dimensions is that they must match the front part of the lower frame. The base is constructed, like most of the robot, from 1" square tubing.

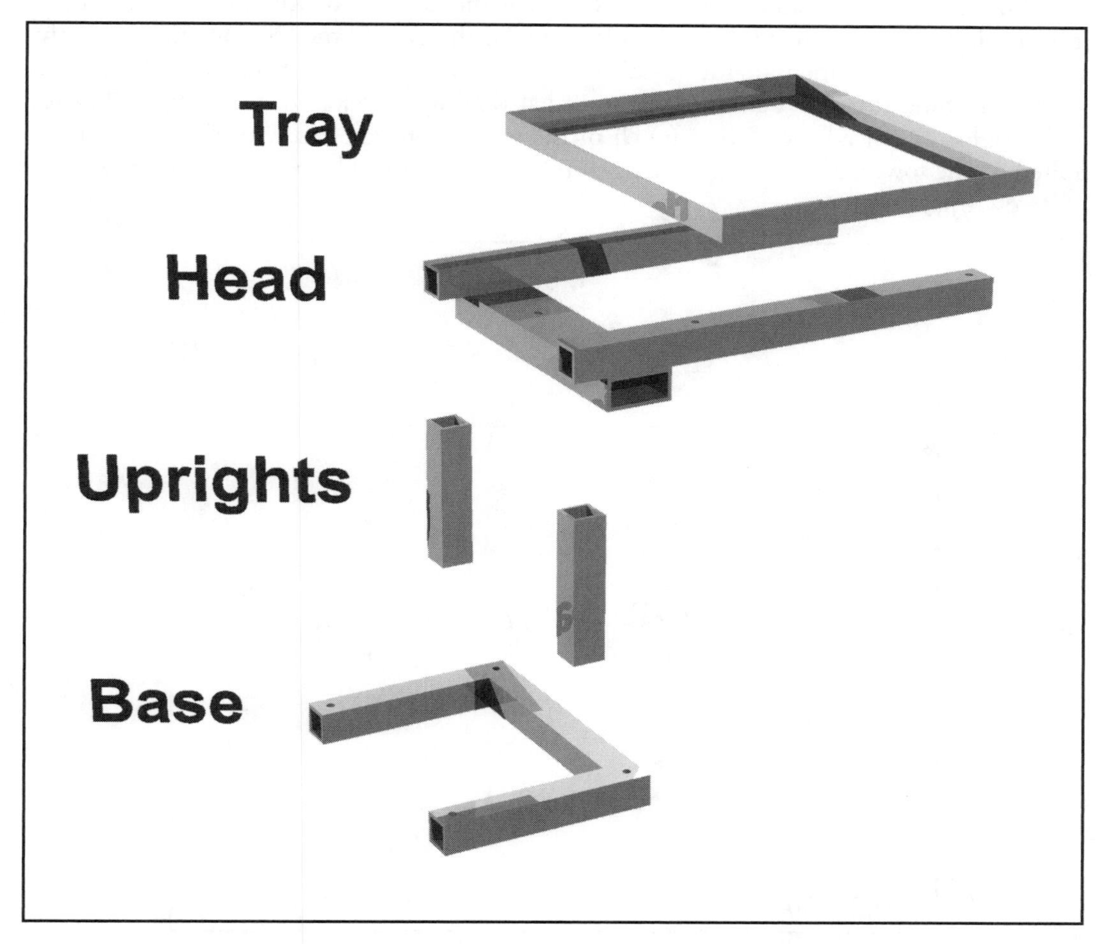

Figure 5-2 Upper frame, exploded view

The upright's position on the base is shown in dashed lines on the base diagram, **Figure 5-3**. You should scribe these lines onto the base, to help in assembly later.

The four holes are simply drilled to 1/4" diameter. Bolts fit through these and into the tapped holes in the lower frame. It may be wise to overdrill to 5/16", to allow for some slop in assembly.

You must be very careful to match the base to the lower frame's dimensions and hole placement, so it will fit. The best plan is to actually bolt the two base tubes onto the lower frame and fit the cross piece between them. Then firmly tack-weld it together, remove it from the lower frame, and finish welding.

Metal distorts in predictable ways when welded as the heating, melting, and cooling bend it around. This makes it doubly difficult to make the two frames fit together correctly, since you need to compensate for this distortion, or try to prevent it from occurring. Tack-weld everything together before doing any final welding, not just in this section, but for the uprights and head too. Do as much of the welding as you can with the upper frame bolted to the lower, and the tray bolted to the upper frame. This way, you use the robot as its own jig.

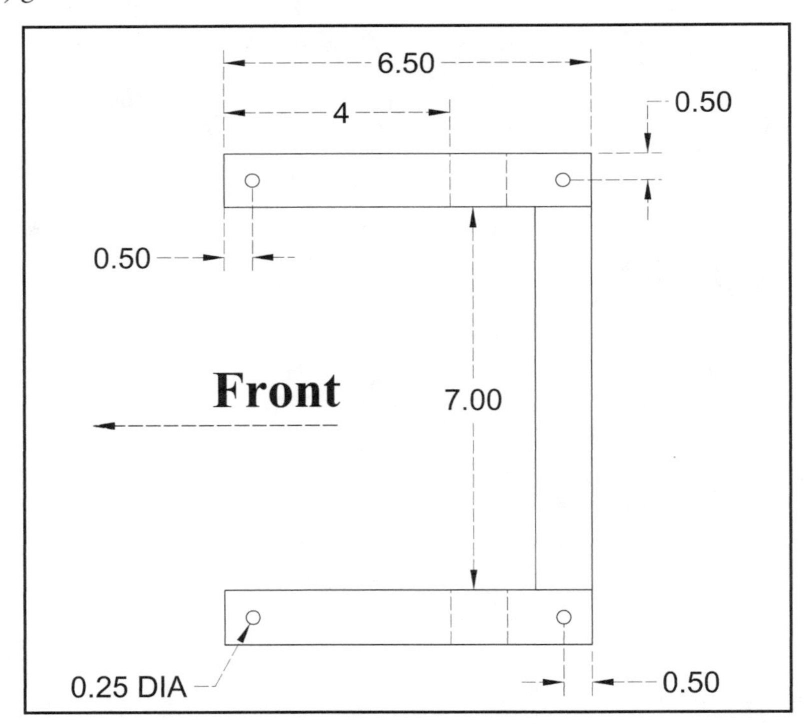

Figure 5-3 Base dimensions

UPRIGHTS

The two upright sections are almost embarrassingly simple (see **Figure 5-4**). A pair of 1" square steel tubes, five inches long. Even the 7" space between them is determined by the spacing of the lower frame's base.

Tack-weld the uprights over the scribed lines on the base.

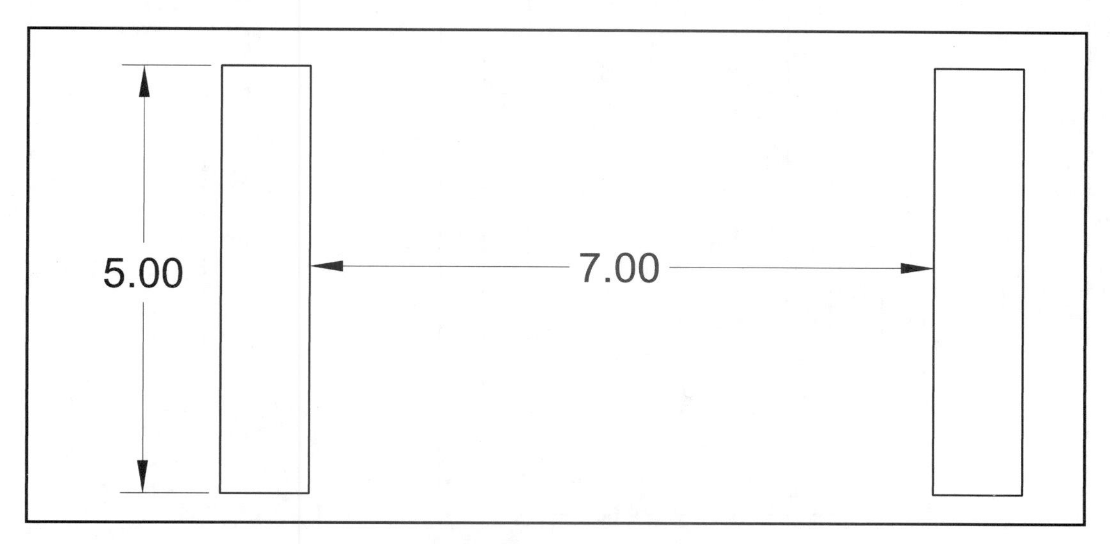

Figure 5-4 Upright dimensions

HEAD

Though the head (or brain-pan) consists of only three pieces, it is more complicated than the base, as you can see in **Figure 5-5**.

The first piece of note is the crossbar, a 9" length of 2"x1" steel tubing. The top of this bar is the mounting point for the robot's eye. This mount is two holes 2.5" apart, centered between the ends of the bar. They are drilled only 0.5" back from the front edge, rather than centered. These are drilled and tapped with 1/4"-28 fine-pitch threads.

The bottom of the bar has four holes, also drilled and tapped for 1/4"-28 threads. These holes make a centered box 3" long and 1" wide. This is where the arm will mount in the next chapter.

Note that the head and arm holes only go through one layer of the crossbar, though most of the holes in this book are drilled all the way through.

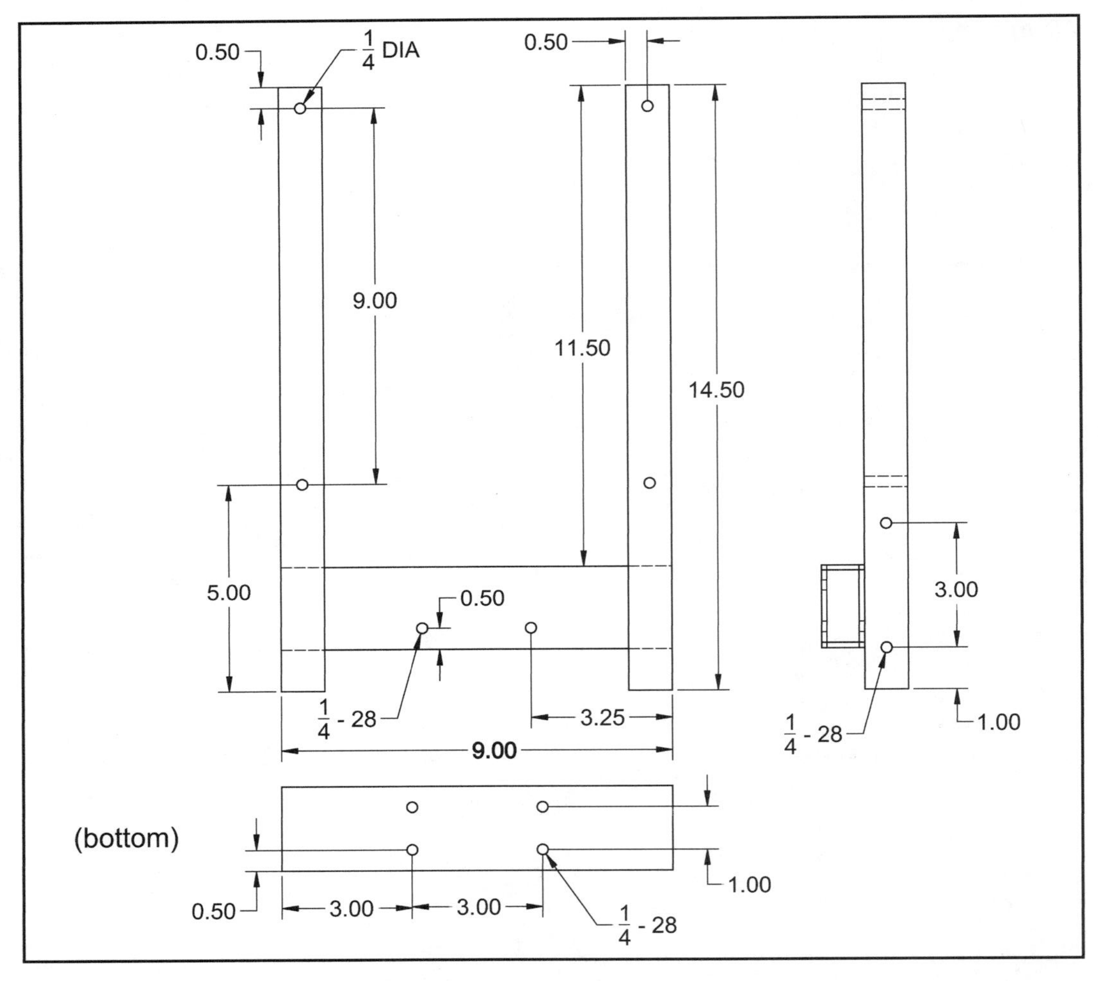

Figure 5-5 Head dimensions

Welded to the top face of the crossbar are the two brain-pan supports. These are 14.5" long 1" steel tubing. The two holes drilled 9" apart in the tops of each are for the tray. These are 1/4" holes, unthreaded. The two holes in the side of each bar are 3" apart, threaded to 1/4-28, are for mounting boxes for the electronics.

Be sure the brain tray crosspieces are bolted to the supports before welding, to ensure that they fit correctly *after* welding. The head's crossbar is welded 1" forward from the ends of the supports. The uprights fit into this 1" space and weld to both the crossbar and the supports, for maximum strength. The crossbar itself acts as a brace between the uprights and the support arms.

TRAY

The tray uses a new material: 3/4" steel angle. I wanted the low walls provided by the angle to make it easier to accommodate the odd curves of my laptop, and to make it easier to access the various ports and orifices in the computer. The bottom of the tray is also lined with a rubber strip, which raises the computer a bit more and provides some nice padding for it. Alternatively, you could set a piece of masonite or plastic into the tray as a shelf to set the computer on.

The tray itself is a simple 10" by 12.25" box—roughly the size needed to fit my computer (see **Figure 5-6**). Your tray may be different.

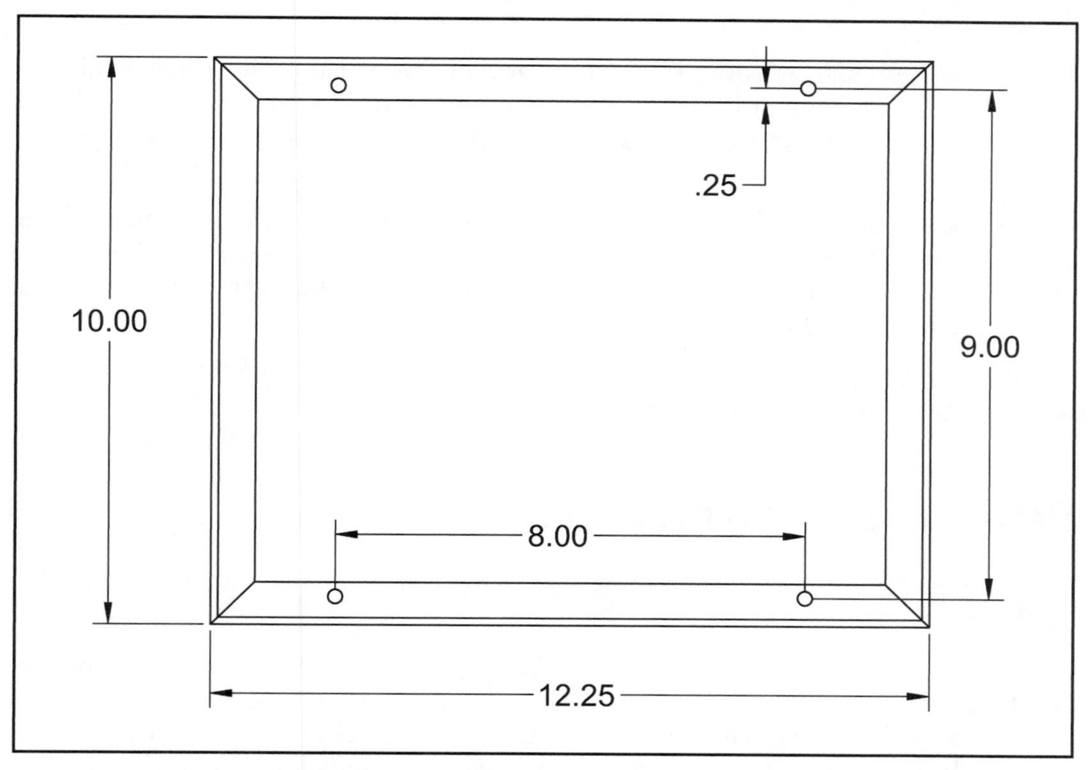

Figure 5-6 Computer tray dimensions

The spacing between the holes, however, must match the holes in the head supports. In this case, 8" wide and 9" long. Note that the holes are not centered on the channel, but are 1/4" in from the edge.

The holes are drilled and tapped with a 1/4”-28 thread. The bolts that hold the tray on to the head are threaded in from beneath. Cut and grind your bolts to sit just flush with the inside of the tray so they don’t make bumps under your computer.

ASSEMBLY

Bolt the two tubes (you know which ones) of the base of the upper frame to the lower frame. Fit the base crosspiece between these tubes and tighten the bolts. Tack-weld the crosspiece in place.

Next, place and lightly tack the two uprights onto the base. Assemble and tack the corners of the tray. Bolt the tray onto the head’s support tubes. Then, scribe lines at the edge of the uprights onto the support tubes.

Turning the head and tray over, lay the head/arm mounting bar on the support tubes at the scribe mark. Lightly tack-weld it into place, and then retry the fit to the uprights. Lightly tack the head onto the uprights. Visually inspect everything to make sure it is square and proper.

Once you are happy with the fit and trim of it all, heavily tack everything together. If you haven’t painted the base, you may want to do most of your heavy welding with it bolted together. Otherwise, once it’s firmly tacked, unbolt the upper frame from the lower and finish welding it together.

Grind, clean, and paint the pieces. Then reassemble it all. Apply any rubber strips, trays, bumps, or pads as desired. Set a laptop into the frame.

PAN-AND-TILT MECHANISM

A pan-tilt mechanism with a single camera “eye” is a *bare minimum* head, but still a valid starting point.

Looking at the pan-tilt, you would never imagine how much time it took to design (see **Figure** 5-7). However, this elegant pair of bent supports makes a two-axis platform for a small camera. The camera I am using is an Intel webcam that I picked up for $50. It connects to the computer via a USB cable, and can be accessed like any Windows video device. You could adjust it to fit a Lego camera, or any lightweight device of your choice.

The nifty thing about this camera, and something you should look for in the camera you purchase, is a threaded mounting lug in the bottom of it. My camera is conveniently designed to take a 1/4”-20 bolt.

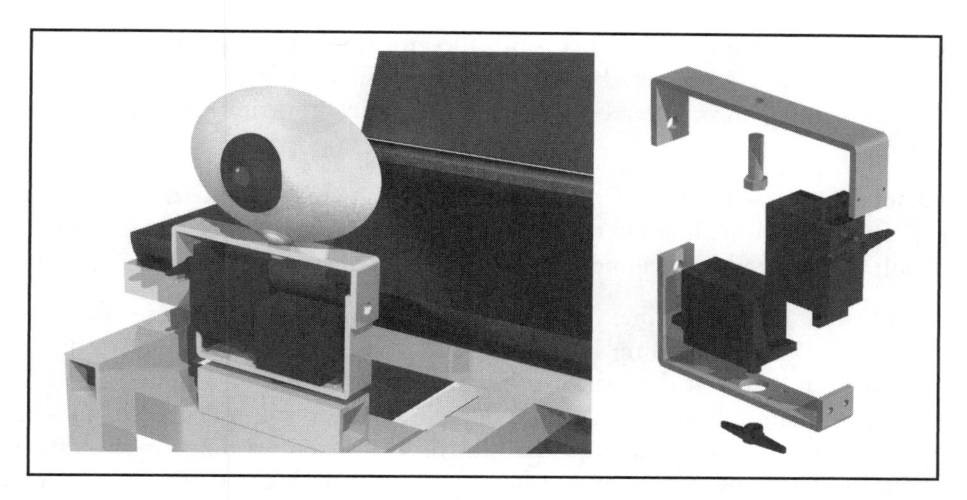

Figure 5-7 Pan-tilt mechanism

The pan-tilt bracket is designed around a pair of standard R/C servos. These little guys are designed in metric so measurements are going to be in metric in places. I will also mix in the occasional English measurements as needed, for clarity.

The two brackets are made from 1/16" thick aluminum strip that is 0.75" wide (about 19mm). The top bracket (see **Figure 5-8**) is 171mm long. This fits the servo and provides some space between the frames. A thin (1/16") washer can be used in the bolted joint to take up any slack later, if needed.

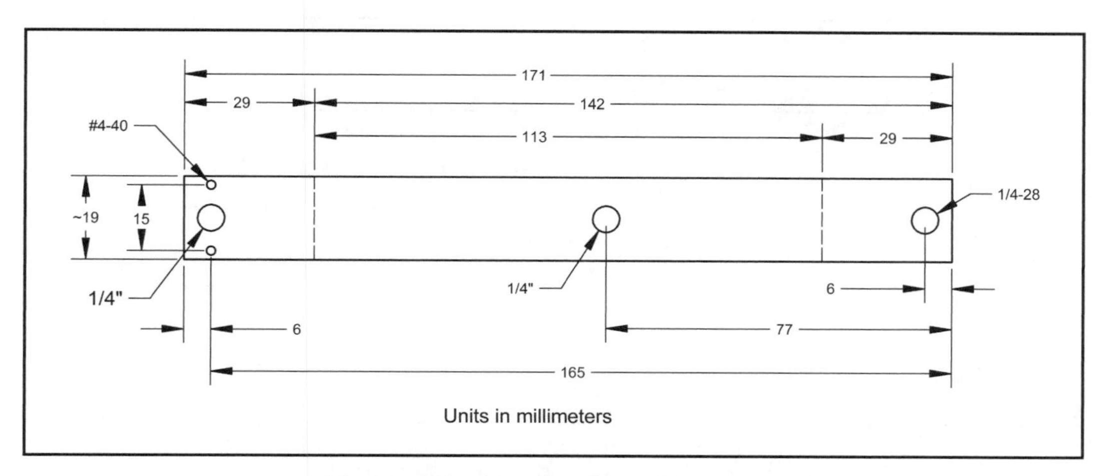

Figure 5-8 Top bracket dimensions

The right-most hole in the top bracket is threaded for the 1/4"-28 pivot bolt. The central hole is drilled for the 1/4" camera bolt. You can drill these to fit the bolts of your choice. Notice, however, that the camera bolt is not centered in the strip—the left edge extends out further.

The two small holes at the left of the strip are 6mm in from the edge. They are spaced 15mm apart, to match the holes in the metric servo horn. These holes are tapped to accept a #4-40 bolt. The servo horn will need to have its holes enlarged to pass this bolt through.

The strip is bent at the dotted lines. The dimensions given are for the inside of the bracket. When you bend the metal, clamp it in a vise with the bend mark just above the surface of the jaw. Bend it firmly at the mark, so it is square and you have a sharp bend (but not so sharp that the aluminum breaks). The outside of the bend will stretch, making the outside measurements somewhat larger than the inside. When bending, press on the metal with something hard and flat, so it will bend evenly. Whack the corner with a hammer to tighten the bend.

The bottom bracket is made in the same way as the top, and from the same material (see **Figure 5-9**).

The right-most hole is the matching hole for the pivot bolt.

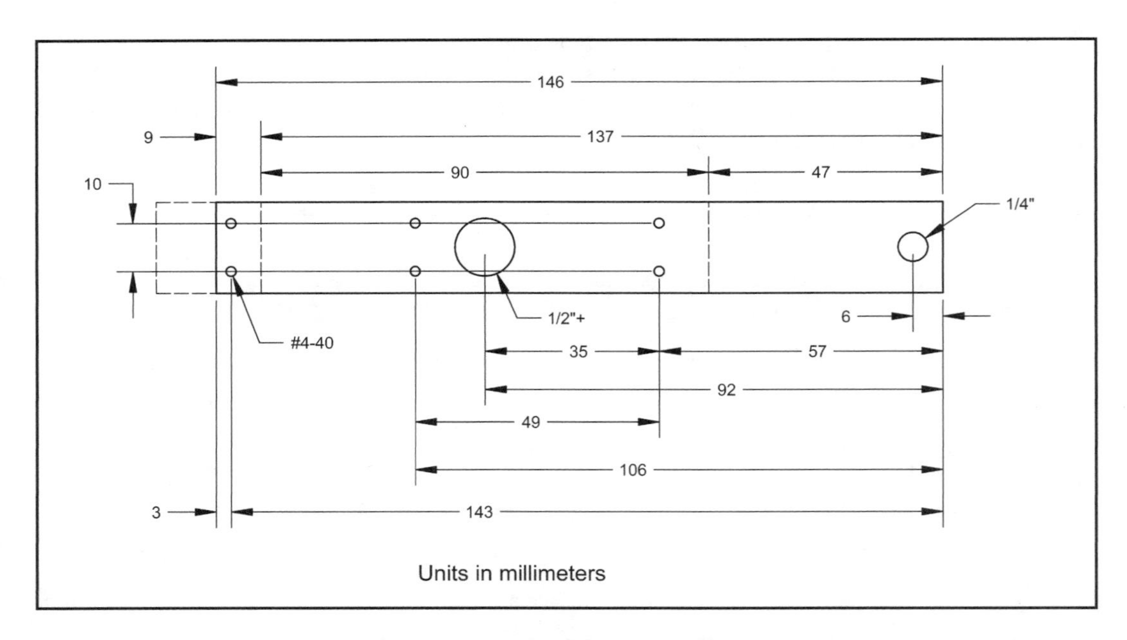

Figure 5-9 Top bracket dimensions

The five holes in the middle section are there to mount the servo. The large hole in the center is actually in the center. It is drilled to fit the bump on the servo. If you can drill it large enough so the servo face fits flat on the bracket, that is best. Otherwise, you will need to mount the servo to the bracket with spacers, to keep it firm and level.

The other four holes in the middle section are sized and spaced to accommodate the screws or bolts you use to fasten the pan servo to the bottom bracket. The best way to attach the servo is into tapped holes, with bolts that are just long enough to fit. This avoids clearance problems under the bracket.

The left pair of holes are for the tilt servo's mount. Only one edge of the tilt servo is fastened to the bracket. It is kept stable by its attachment (by way of the servo horn) to the top bracket.

The ghosted box off to the left of the bottom bracket is extra metal to make it easier to bend the left tip. Once bent, you can cut this metal off. It is not strictly necessary, but it can be quite difficult to cleanly bend the 9mm servo mount.

The pan-tilt brackets and servos go together as shown in **Figure 5-10**. There are a few tricks in store for you, however.

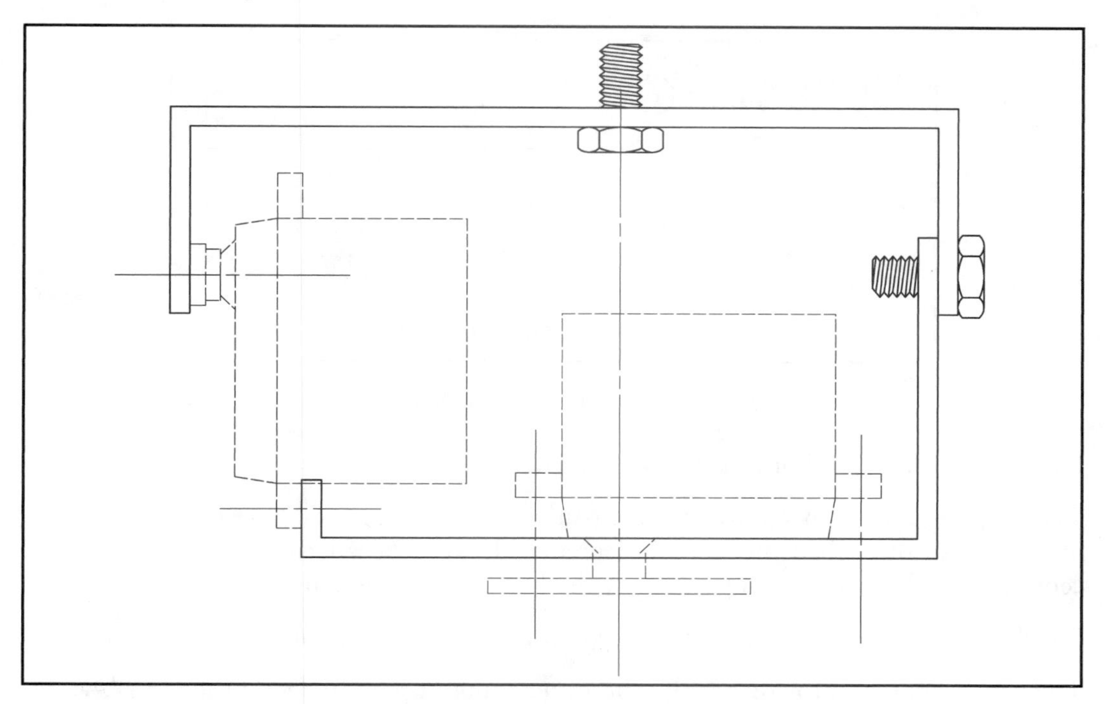

Figure 5-10 Bracket assembly

The first part is easy—bolt the servos to the bottom bracket. Bolt the camera firmly to the top bracket, then fit the top bracket into place and screw it into the tilt servo's horn. The pivot bolt is also fitted into place; it screws tightly into the threads in the top bracket and simply protrudes into the hole in the bottom bracket. Brass or nylon washers (or nylon shoulder washers) can be used to take up slack and provide a smooth pivot surface.

Fastening the pan-tilt mechanism to the robot is easier with an interface, shown in **Figure 5-11**. The outer pair of holes are used to bolt the interface onto the upper frame. The large center hole is used to get access to the bolt that fastens the horn onto the pan servo. The little holes are spaced and sized to match the outside holes on the servo horn or disk. These small holes should be drilled and tapped, to make the servo horn/circle easier to fasten down.

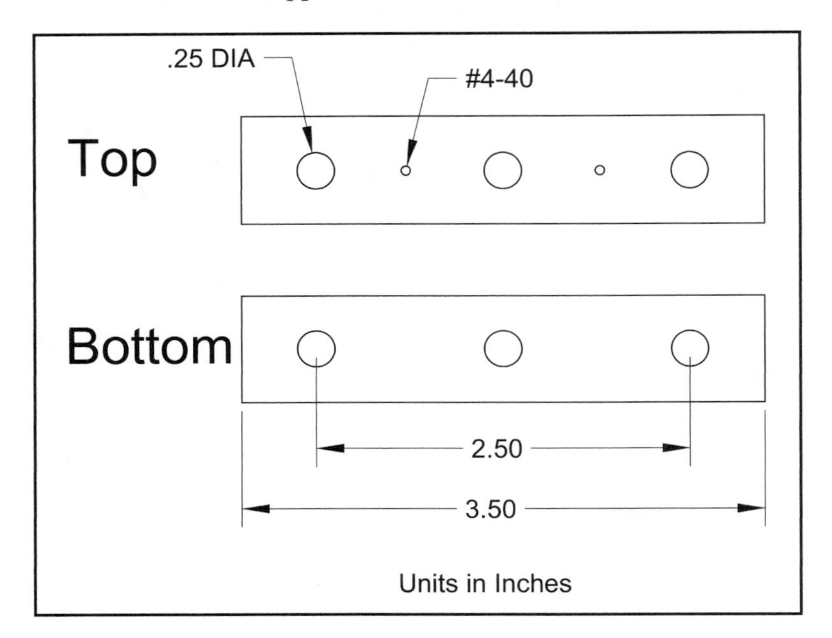

Figure 5-11 Top bracket dimensions

First, attach the servo horn or disk to the interface.

Then slip the hub of this horn onto the pan servo. Make sure you have the correct range of motion so that when the servo is centered, the camera is facing forward. Using the central access hole in the interface, bolt the horn onto the servo.

Finally, turn the pan-tilt out of the way and bolt the interface onto the upper frame.

The pan-tilt doesn't provide a rock-solid camera mount. It is, in fact, prone to vibration. It remains to be seen how this will affect image processing.

CHAPTER 6

ARM

The story of this arm is a long one. I started thinking about robot arms a long time ago—in fact, my first robot Boris (www.simreal.com/i_boris.html) was a six-legged machine, and arms and legs are similar in nature.

Looking around the various magazines and Web sites I could only find toy arms or servo-based arms, neither of which were capable of doing any real work. Of course, there are industrial trainers and other large arms, but these are not suited for a mobile robot and tend to be very expensive.

What I wanted was an arm that, at least in principle, would be able to open a refrigerator and fetch me a nice cold soda. Sure, it may be a lifetime (I hope not!) before the software is capable of this, but without the right foundation it will never happen.

As an early estimate, I figured a lightweight arm might weigh five pounds or so. The ability to lift two pounds of weight at the end of (say) a two-foot arm requires 9 to 14 pound-feet of torque at the shoulder (depending on the weight distribution). At the worst case, that is 2,688 ounce-inches of torque. Of course, this is just a starting estimate—the final weight and capability of the arm may ultimately be quite different.

This chapter does not achieve these goals—it is only a first step in that direction. An interesting, complicated, and difficult step. I spent a long time stuck in this project, trying to achieve perfection before I decided to get on with life and present to you here the not-quite-ready-for-prime-time robot arm. Please visit the robotics community at www.simreal.com for a current view of this research or to offer your own designs and suggestions.

Other People's Arms

An arm breaks down into three basic areas of motion: the shoulder, elbow, and wrist.

The human arm may seem to be a good place to start, but it describes a very complex motion in three-dimensional space. It is easier to use a simplified model.

The human shoulder consists of a ball joint (and the scapula, and the clavicle, and so forth) plus a bunch of muscles that conspire to create a wide range of motions. The human elbow, of course, has a simpler hinge motion.

The wrist is very complex and involves rotations of the radio-ulna joint that is the forearm, as well as a variety of motions around the wrist proper. For simplicity, it might be considered a ball joint like the shoulder.

Some interesting and innovative solutions to the arm-design problem can be found on the Internet, starting with the links at www.simreal.com.

Arm Actuators

There are a wide variety of devices available to create mechanical motion. These actuators can be roughly divided into linear and rotary mechanisms.

Linear actuators include such things as the classic electrical linear actuator which is a (rotating) motor driving a long threaded rod through a threaded nut. True linear actuators include pneumatic cylinders, McKibben air "muscles", Nitinol memory wire (a nickel-titanium alloy which can be set to shrink as it gets hot and relax upon cooling), and the new breeds of biomimetic polymer muscles. Some researchers are even growing biological muscles on artificial substrates for use in their robots.

Rotary actuators include your basic electric motor, as well as pneumatic and hydraulic motors. In terms of expense, availability, flexibility, and all-around capability, it is hard to beat the electric motor. Since most electric motors spin at a high natural speed, it is important to have gearing on the motor, much like what was seen in the electric wheelchair motors that drive the rolling platform.

Gearmotors

Looking at catalogs and Web sites I was able to find a number of gearmotors in the 500 oz/in range. With speed reduction I could easily get 2,000 oz/in out of them—more if I exceeded their ratings (the rating in some cases was for the gearbox; the motor could do more). The catch was, each gearmotor cost $150 or more, which was more than I wanted to pay for this project.

Of course, when custom parts fail to deliver there is often a mass-market alternative that can fill in. In this case it is the Black & Decker Pivot Driver cordless screwdriver, which was briefly discussed in Chapter 1.

Black & Decker rates the Pivot Driver at 40 in/lbs of torque—that's 640 in/oz or 3.3 ft/lbs. With a 4:1 reduction that becomes 13 ft/lbs or 2,560 in/oz of torque. Of course, that torque is probably the stall torque; the actual useable torque may be half that. These ratings are for 3 volts; higher voltages will increase the current flow (and hence, torque rating), giving the motor enough power to be useable.

The bendy Pivot Driver costs more than straight screwdrivers, however that pivot is a great addition for the hardware hacker. The motor and gearbox are efficiently encapsulated at the tip end of the product, with the batteries and switch (e.g. stuff we will throw away) packaged at the handle end.

One nice feature of cordless screwdrivers is the anti-backdrive mechanism. Once the motor is stopped, the shaft locks in place. This will be a great energy-saving feature in the arm.

I had hoped to use the motor module as it came, without taking it apart. Unfortunately the output shaft is too wide; I need a 1/4" shaft to mount the pulley or sprockets on, and the screwdriver comes with something closer to 7/16". The shaft, then, must be removed and replaced before we can use the motor modules.

Customizing the Motor

This is a simple customization—we are replacing the output shaft. The first step is to decide on what length shaft you need. In this example we put on a 1" shaft. For the actual shaft lengths, look ahead to the arm construction and the different places the motor is used.

The new shaft must be 1.125" longer than you need, to allow for the section of the shaft that is inside the motor. For this example we cut a piece of 1/4" diameter steel rod 2.125" long.

Removing the Motor Module

Starting with the electric screwdriver, we want to end up with a motor module plus junk, as shown in **Figure 6-1**.

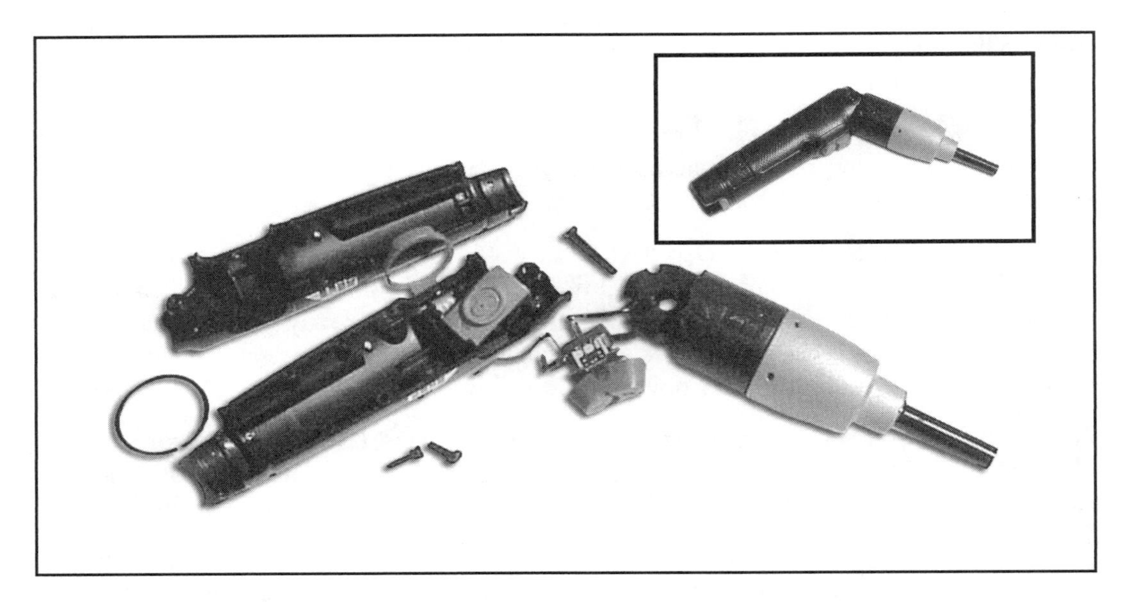

Figure 6-1 Electric screwdriver, with the motor module separated

First remove the long screw at the pivot point of the screwdriver. Then remove the two shorter screws at the center of the body. Finally, carefully using a strong sharp knife, cut the retaining ring at the base and slit the decal that is at the junction of the base's two halves. This done, you can separate the halves and it should all fall into pieces in your hands.

The motor is fastened to the switch with two quick-disconnect plugs, so slide them off now. Set everything but the motor aside.

DISASSEMBLING THE MOTOR MODULE

The motor module shell consists of three pieces; the two plastic halves that contain the motor plus the cast metal gearbox. Two friction pins hold the plastic motor case and the gearbox together.

With an appropriate tool (or even an inappropriate tool; I used an allen wrench from my jeweler's screwdriver kit) and a light touch with a hammer, tap out the two pins. Carefully work the plastic motor section off of the gearbox. Keep the halves of the motor case together. Setting the motor aside for a minute, carefully remove all of the pieces in the gearbox (see **Figure 6-2**).

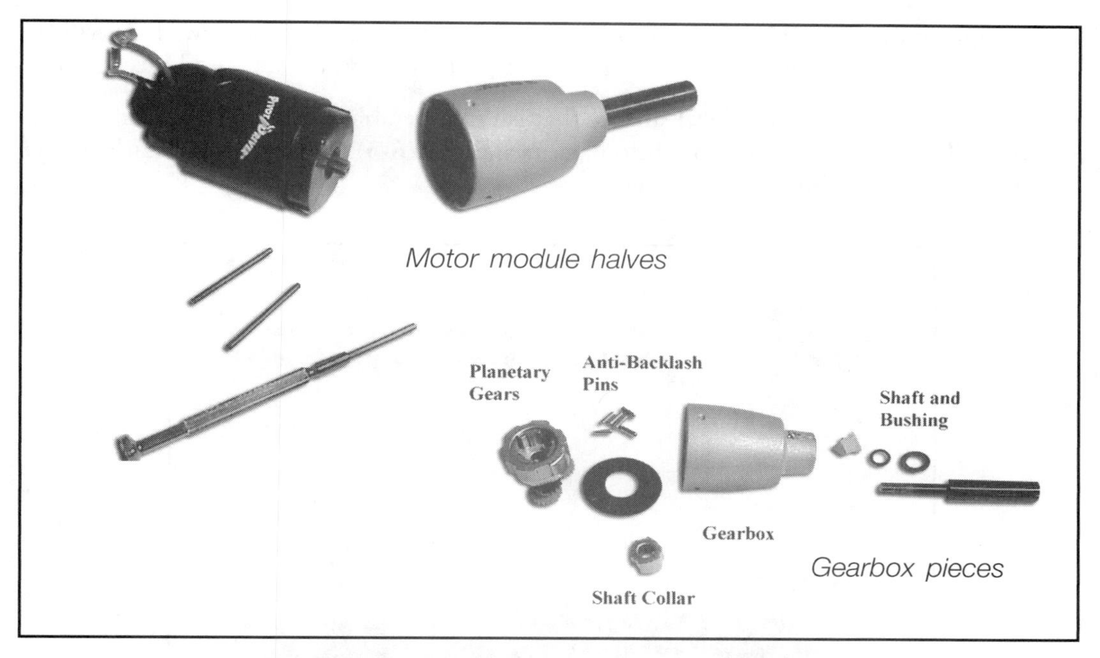

Figure 6-2 Taking the motor module apart

The good thing about taking it apart *carefully* is that you can see how it is supposed to fit together again. A hint here; pay careful attention to where the anti-backlash pins at the very bottom go.

Replacing the Output Shaft

The shaft is locked into place by the eccentrically shaped shaft collar. Take the gearbox over to a steady surface (such as the partly open jaws of a vise). Place the gearbox, opening up, so the shaft is clear of obstruction. You will be doing some heavy pounding soon, so the support under the gearbox should be firm.

Using your hammer and some round thing that is smaller than 1/4" in diameter (e.g. a Phillips-head screwdriver you no longer love, or a 3/16" diameter steel rod), pound the shaft out of the shaft collar. Be careful, however, not to damage the shaft collar.

Once the shaft gracefully slides out of the delicate grip of the shaft collar (if you're lucky!), a flat washer, a wavy spring, and a light plastic bushing will fall out of the gearbox. Keep these, especially the plastic bushing. Hopefully you didn't destroy this bushing while beating up on the shaft.

Cut an appropriately sized length of 1/4" diameter rod. 1.125" of the new shaft will be lost inside the gearbox, so for our desired 1" shaft protrusion we cut a 2.125" piece of rod.

Set the shaft collar on a solid, level surface that isn't vulnerable to damage. Center the rod over the hole in the collar and tap it into place (see **Figure 6-3**). It should go in fairly easily, yet still be gripped firmly by the collar.

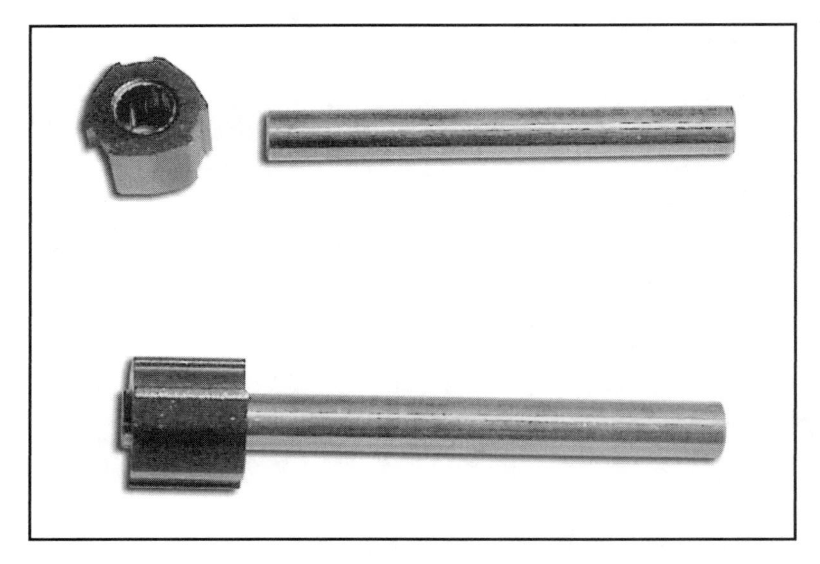

Figure 6-3 New shaft

REASSEMBLING THE MOTOR MODULE

Slide the new shaft with collar into the gearbox. Then slide the steel ring around the shaft collar.

If you remember your recent study of the anti-backlash pins, you will remember where they go (e.g. between the shaft collar and the outer ring, in the flat spaces of the collar, between the notches). Using needle-nose pliers, put the six pins into the gearbox. Once the pins are in place, the black gear's platform locks onto the shaft collar. Populate the platform with the three black gears. Before going further, and at each later step in assembly, make sure that gearbox still moves smoothly.

Carefully place the red gear platform, toothed nubbin first, into the black gears. Populate it with the three red gears.

Once you are sure the gearbox still works, place the thin plastic retaining washer into place.

Finally, carefully work the motor module back into the gearbox. Make sure the holes for the friction pins are lined up between the motor and gear halves! You will probably have to back-drive the shaft with a pair of pliers, as much as the pins will let you, to get the motor's drive gear lined up with the red gears. Take it slowly and don't force things too much.

With the motor back into place give it a quick power-up test to make sure it still works. If it does, drive the pins (carefully, yet with authority) back into the motor, locking the gearbox into place. If it doesn't, take it apart again until you can locate the source of the problem.

TRANSFER OF POWER

With a source of power in hand it becomes important to find a way to transfer this power to the arm.

LINEAR ACTUATORS

A linear actuator can transfer power like the human muscle, pulling (or pushing) directly on the bones to both sides of the joint. This strategy is seen in the giant construction machines you see tearing up (er, repairing) our highways, and in the legs of Project Boris (**Figure 6-4**). Each joint is a triangle composed of the two fixed-length structural pieces, plus the variable-length actuator. Each corner of the triangle is a pivot so that when the actuator changes length all of the angles change to compensate. The forces and angles can be calculated with simple geometry.

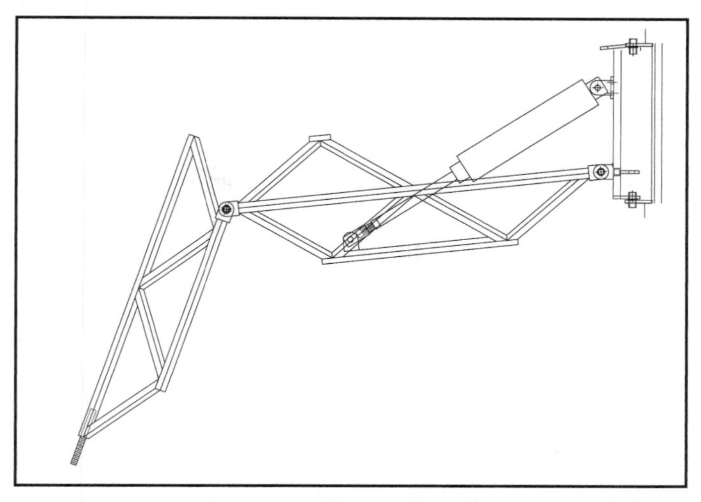

Figure 6-4 Boris leg

A somewhat more compact method is used by the SoftArm project using a pair of McKibben pneumatic muscles (**Figure 6-5**). The McKibben muscles can only pull, so they are placed in opposing pairs. The ends of the muscles attach to a chain that wraps around a sprocket at the joint. As the pressure is balanced between the two muscles, the chain shifts and the joint rotates. Presumably, stiff springs inline with the muscles can compensate for when they are both slack and to provide additional compliance at the joint.

Cable wrapped around a drum may also be used instead of chain; slipping can be eliminated by fastening the cable to the drum and not simply wrapping it.

Note that this cable attachment method also works well with R/C servos—the cable attaches to the two ends of the servo horn instead of to two separate actuators. This technique will be explored when it comes time to move the fingers in next chapter's hand.

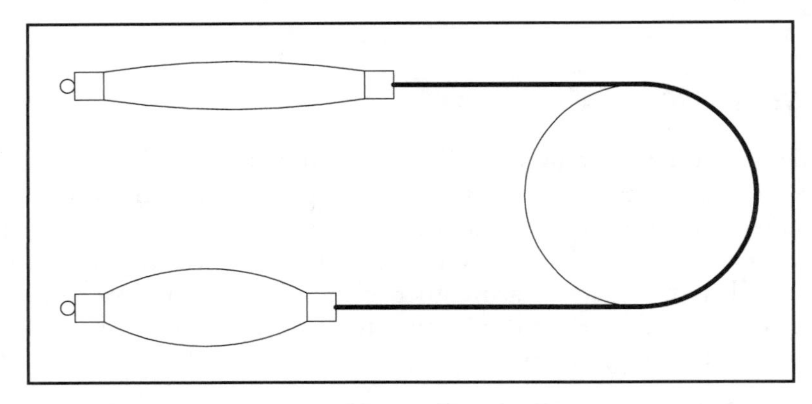

Figure 6-5 SoftArm joint

ROTARY ACTUATORS

Transferring rotary motion from a motor to the (rotating) joint can be done many different ways. Traditional methods include gears, chain and sprocket, timing belts, pulleys, and so forth. These are so common in everyday machines that they don't really warrant illustration.

Industrial robots tend to use direct gearing, including the expensive but really interesting harmonic gearing.

The Cog project at MIT has an interesting solution, shown in **Figure 6-6**. The small circle to the right is the motor's output drum, and the large circle to the left is the joint being turned. A cable is wrapped around the motor and anchored at the far side of the joint.

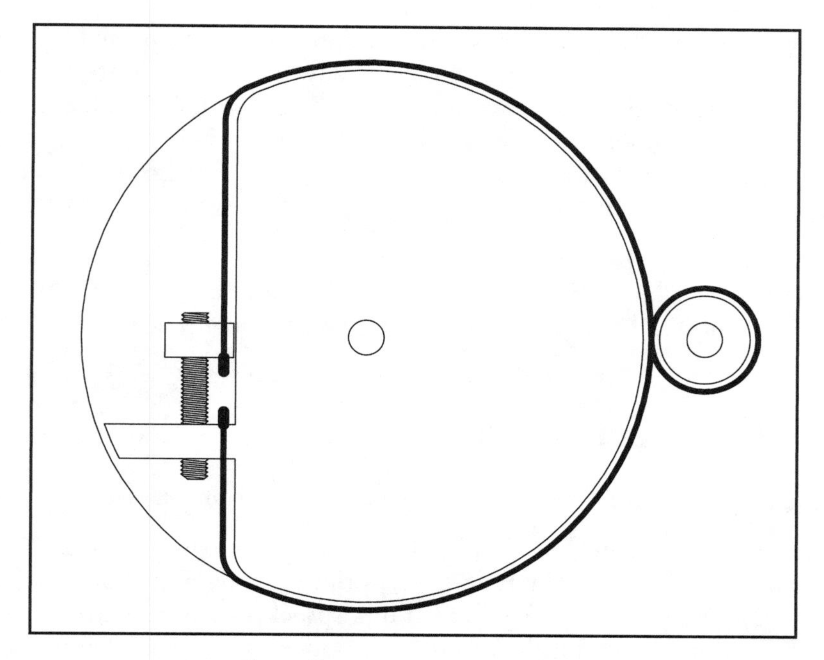

Figure 6-6 Cog joint cables

A variation of this joint puts the motor at 90 degrees to the joint, as shown in **Figure 6-7**. Both of these mechanisms put the motor shaft under a lot of strain, so it is important to support the tip end of the shaft in a bearing or bushing.

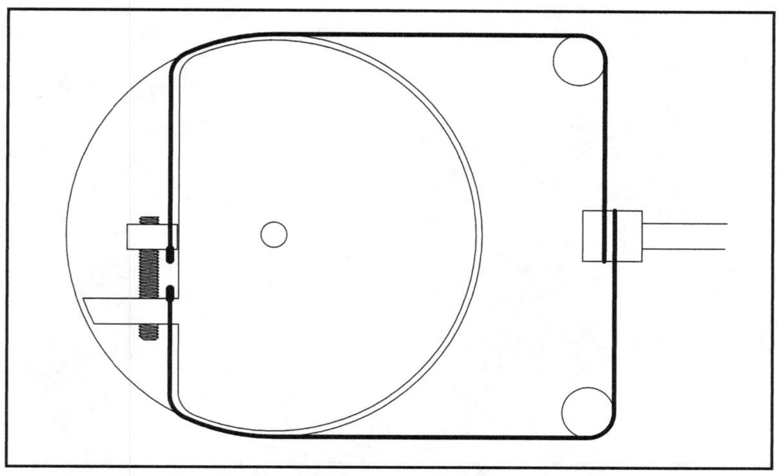

Figure 6-7 Cog joint with 90-degree motor

The arm project in this chapter chooses the simplest methods possible. The wrist is directly attached to the relevant motor, and the elbow and shoulder are attached to the motor with a chain and sprocket arrangement. The original design called for a toothed timing belt and matching pulleys; however, the strain the arm subjected them to made the teeth "skip." The sprocket allows a much stronger pull. Timing belts, pulleys, chains, and sprockets are all available from Small Parts.

The motors receive a small 9-tooth sprocket with a 1/4" bore (Small Parts #U-RCS-9), and the joints receive a large 48-tooth sprocket with a 3/8" bore (Small Parts #U-RCS-48). These are connected by a loop of #25 roller chain (Small Parts #U-RRCS-25).

PROJECT 6-1A: SHOULDER

Now it's time to get down to business and build an arm, starting with the shoulder.

The shoulder box will ultimately bolt under the crossbar of the upper frame. The shoulder is shown in context to the left in **Figure 6-8**, just prior to being placed on the frame. Note that the arm rendering shows a pulley instead of a sprocket. The renderings were created during the initial arm design and the change from pulley to sprocket happened after the first arm was built and tested.

I recommend the use of a good 3D design and rendering system to help in project design. It helps highlight issues before going to metal. I am using the Moray design package that renders using POV-Ray. The dimensioned CAD figures are done using DeltaCAD. All of these packages can be found on the Internet.

The image to the right in Figure 6-8 shows the shoulder components in more detail.

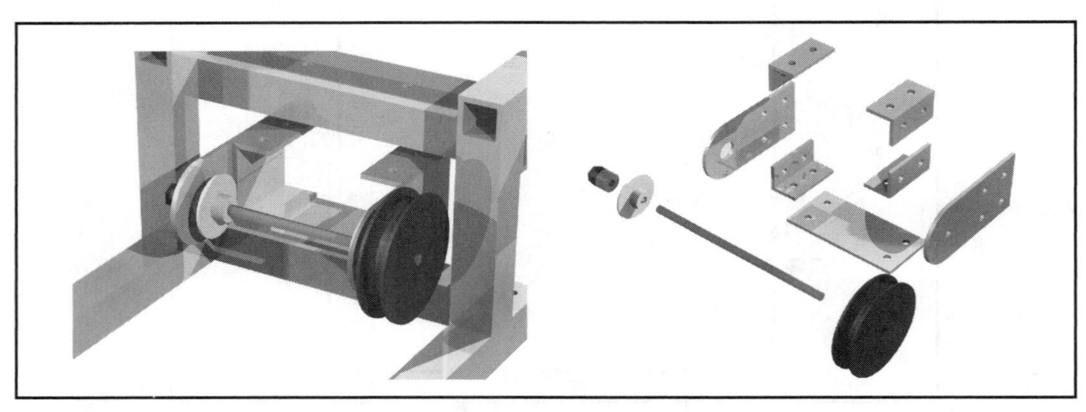

Figure 6-8 Shoulder

SHOULDER PIECES

This section talks about the construction of the aluminum shoulder box. Additional components, such as the Trantorque keyless bushing, the sensor gear, and the timing pulley are discussed in the next section on assembly.

In basic outline, the shoulder consists of three sides of a box held together by four pieces of angle. Of course, two of these sides protrude and have pivot holes to the support the rest of the arm, and two of the angles protrude the other direction to mount the shoulder to the framework. **Figure 6-9** shows the basic dimensions of the shoulder box. Though this diagram (and, unfortunately, the rest in this chapter) looks very confusing at first glance, remember that the arm consists of a series of boxes. Fancy boxes with lots of holes, but just boxes.

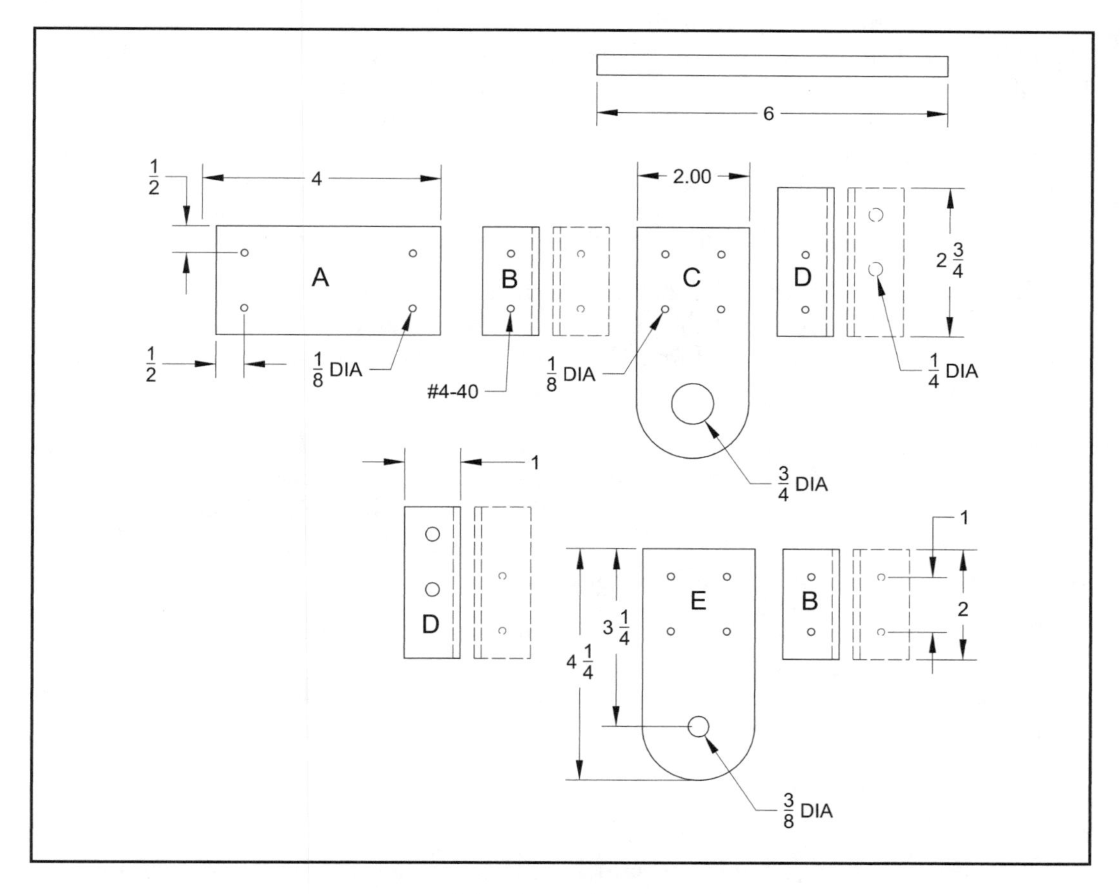

Figure 6-9 Shoulder dimensions

The four smaller pieces of the shoulder (B and D) are made from 1" angle aluminum with 1/8" thick walls. You can use steel here, too, since there is no weight issue with the shoulder. Aluminum is easier to tap.

The two B angle sections are 2" long, while the two D sections are 2.75" long. The B pieces are drilled and tapped with #4-40 threads on both faces. The two D pieces have #4-40 threads on one face, and 1/4" holes on the other; note that the large holes are on the other end of the bracket from the small, tapped holes. The D pair is used to mount the shoulder onto the upper frame.

The bottom panel A is 4" long and is cut from 2" wide aluminum that is 1/8" thick. The four holes in it allow the #4 bolts through to the B angle brackets.

An aside, here, on the subject of drills and taps. When you buy a tap at the hardware store, they often come shrink-wrapped with a suitable drill bit. But for the record, a #4-40 tap should be predrilled with a 3/32" drill. The clearance hole to pass a #4 bolt takes a number 32 drill, which measures 0.1160" diameter. A 1/8" drill is easier to find and, while it works, will make for a looser fit. This may actually be a good thing, depending on how accurately you are able to place your tapped holes! Also, it's not critical that you use #4-40 bolts to hold things together; anything that is convenient should work.

Back to the bottom A panel. The four holes are all placed a half-inch in from the edges.

The side panels C and E hold the shaft where the arm mounts to the shoulder. Each side panel is cut to 4.25" long out of the same 2" wide aluminum as the bottom A panel. The four mounting holes are arranged in a 1" square at one end of the panel, in 1/2" in from the end and the sides.

At the other end of each side panel make a mark centered in the panel 1" in from the end (best measured 3.25" from the other end). From this mark scribe the 1" radius semicircle that makes the pivot end of the panel. Cut this arc out with a bandsaw or other tool of your choice (I had the good fortune to be able to use a small mill and rotary table—gotta love nice tools). File it smooth if it is not already nice.

The **E** side panel has a 3/8" hole and the **C** panel has a 3/4" hole drilled at the center mark. The 3/8" hole provides direct support to the 6" long shaft and the 3/4" hole allows the shaft to be locked into place with a Trantorque keyless bushing. No bushing is needed in the **E** panel since the shaft will not rotate there.

TRANTORQUE

For the arm, one half of each joint must lock securely onto a shaft and the other must pivot freely around this shaft. Providing free motion is easy; there is a large array of bronze, brass, and plastic T-bushings, sleeves, and shoulder washers for this purpose. But I was stuck on how to lock the shaft into place! There seemed to be no elegant solution, short of welding or threading a bar on the end of the shaft, giving it a "T" shape, and then bolting this onto the fixed side of the arm joint.

After spending more time than I care to remember, I stumbled across something called Trantorque (see **Figure 6-10**) in my Small Parts catalog (get yours at www.smallparts.com). I rejected this device several times, due to its $20 cost, but I finally broke down and bought a set. Though it wasn't designed to lock into something as thin as the arm's 1/8" pivot plate, it should work sufficiently well for our needs.

You can find out more about Trantorque from its manufacturer at Fenner Drives (www.fennerdrives.com) or Fenner Industrial (www.fennerindustrial.com). Though expensive, it's a great little device.

Figure 6-10 Trantorque

ASSEMBLY

Loosely bolt the top brackets to the upper framework. Loosely bolt the lower brackets to the bottom plate and the two side plates. Then loosely bolt this assembly to the top brackets. Tighten everything.

The pivot for the arm consists of several important pieces. There is the 6" length of 3/8" steel bar that is the pivot itself. One side of this extends 3/4" through the side plate, providing a mounting point for the Trantorque. The other side extends 1" out as a mounting point for the large shoulder sprocket.

The Trantorque attaches at the far end of the shaft, locking it into the side plate. Don't lock it in place until the next arm piece is ready—and when you *do* lock it down, be sure it protrudes through the side plate by 1/16" or so, so it gets a good grip on the thin plate.

Once assembled, the upper arm nests inside the shoulder (and the sensor gear sits inside the arm).

The sensor gear locks onto the shaft and meshes with a matching gear that is, in turn, attached to a rotary sensor. The gear I am using is a Delrin spur gear I found in the Small Parts catalog. I am using the 48-pitch gear with 72 teeth and a 1.5" pitch diameter. These do not come with any method of locking them onto the shaft, so I drilled and tapped them to take a setscrew.

At the far end of the shaft from the Trantorque is the shoulder sprocket, again, from the Small Parts catalog. Though Small Parts, Inc. is far from inexpensive, it is not always easy to find equivalent parts elsewhere. The shoulder sprocket is a 1/4" pitch roller-chain sprocket with 48 teeth with a pitch diameter of 3.822". The sprocket itself is fiberglass-reinforced nylon with a steel setscrew.

Since the sprocket will be experiencing significant forces trying to twist it off the shaft, I recommend grinding a flat or drilling a spot on the shaft to provide a better grip for the setscrew.

When it comes time to assemble the shaft, fit the shoulder and upper arm together and slide the shaft through them. Be sure to place the sensor gear inside the upper arm! Then, lock the shaft to the shoulder with the Trantorque. In the process of locking it down, the shaft will shift a bit more than 1/16". Then, adjust the rotary sensor and slide the gear down the shaft so it meshes with the gear on the sensor. Lock the gear into place on the shaft. Finally, fasten the shoulder sprocket to the far end of the shaft.

PROJECT 6-1B: UPPER ARM AND ELBOW

The arm consists of two main sections. The midarm (or "humerus," though I find nothing funny about it) is the mounting point for the shoulder and elbow motors and is the interface from the shoulder to the lower arm. The lower arm, counterpart to our radius and ulna, holds the wrist motor and is the mounting point for the hand.

There are a number of issues to address with the arm. Even though this is a simple machine (not in the same class, say, as NASA's new Robonaut system), it is still a large and complex project.

Mounting the motors is complex since they are not designed to be used in this kind of project. Instead of being able to bolt them directly to the arm we need to clamp them into place between two sides of the arm. One side will support the round trunnion at the shaft end of the motor pod and the other side will lock around the rectangular projection at the base end to keep the motor from rotating.

The motors themselves will need to be fitted with 1/4" diameter shafts that are 2.125" long, 1.125" inside the motor, 1" outside for the gear.

The coupling from the motor to the arm's pivot is done through a #25 roller-chain and two sprockets. A small 9-tooth sprocket fastens onto the motor, and a large 48-tooth sprocket attaches to the joint. Where a timing belt would allow a slight amount of give in the arm and also allow for inaccuracies in mounting the motor, it is unable to handle the strain applied to it. The use of a chain, or a timing belt, for that matter, requires a complication of the arm design—we need a way to tension the chain once it is in place on the sprockets. This is done by making part of the arm slide, giving a variable distance between the two sprockets.

At each motor-driven joint we also need to attach a rotational sensor. This is coupled to the joint with a gear. The sensor could be a cheap potentiometer or an expensive rotary encoder, depending on your needs and budget.

Upper Arm

Let's build the upper arm and motor mounts now. Remember, even though this part of the arm consists of twelve unique pieces of metal, not counting the motors, shafts, gears, and so forth, it is still, essentially, a box with holes in it.

Figure 6-11 shows a conceptual rendering of the arm attached to the shoulder, including the motor and sensor gear placement. The exploded view should give you a better idea of the pieces and their relative placement. Refer to this image when you get lost in the following diagrams.

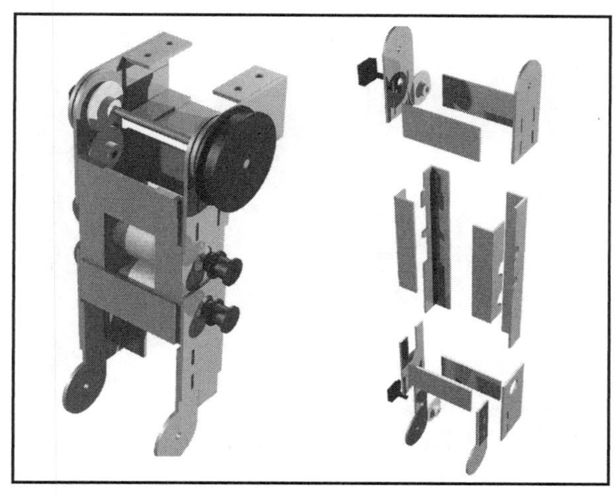

Figure 6-11 Upper arm

Figure 6-12 shows all of the upper-arm pieces, laid out in their correct relationship to each other. There are twelve pieces in this picture, but only four types of piece. First, there are the angled supports D, E, F, and G. These provide the framework for the box. Second are the fixed panels C, I, and J that separate the supports and hold them together. Third are the sliding shaft-supporting panels A, B, H and K that provide the motion needed to tension the chains. Finally, there is one spacer piece L that can be used to help support the motors. With any luck, these diagrams and later dimensions are all still correct; be sure to double-check your work in case an error has worked into the design.

The upper-arm pieces A and B in **Figure 6-13** are the interface between the upper arm and the shoulder. These pieces are made of 2" wide aluminum that is 1/8" thick.

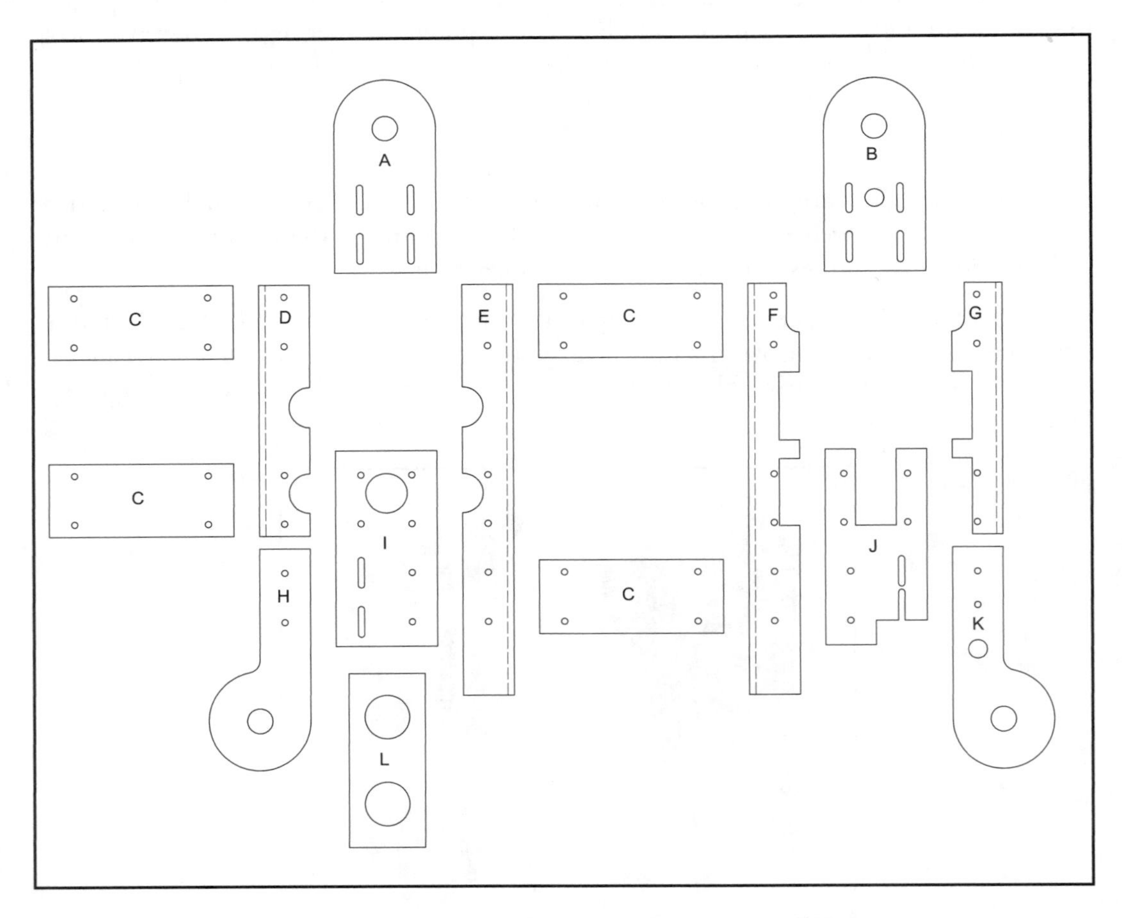

Figure 6-12 Upper-arm pieces

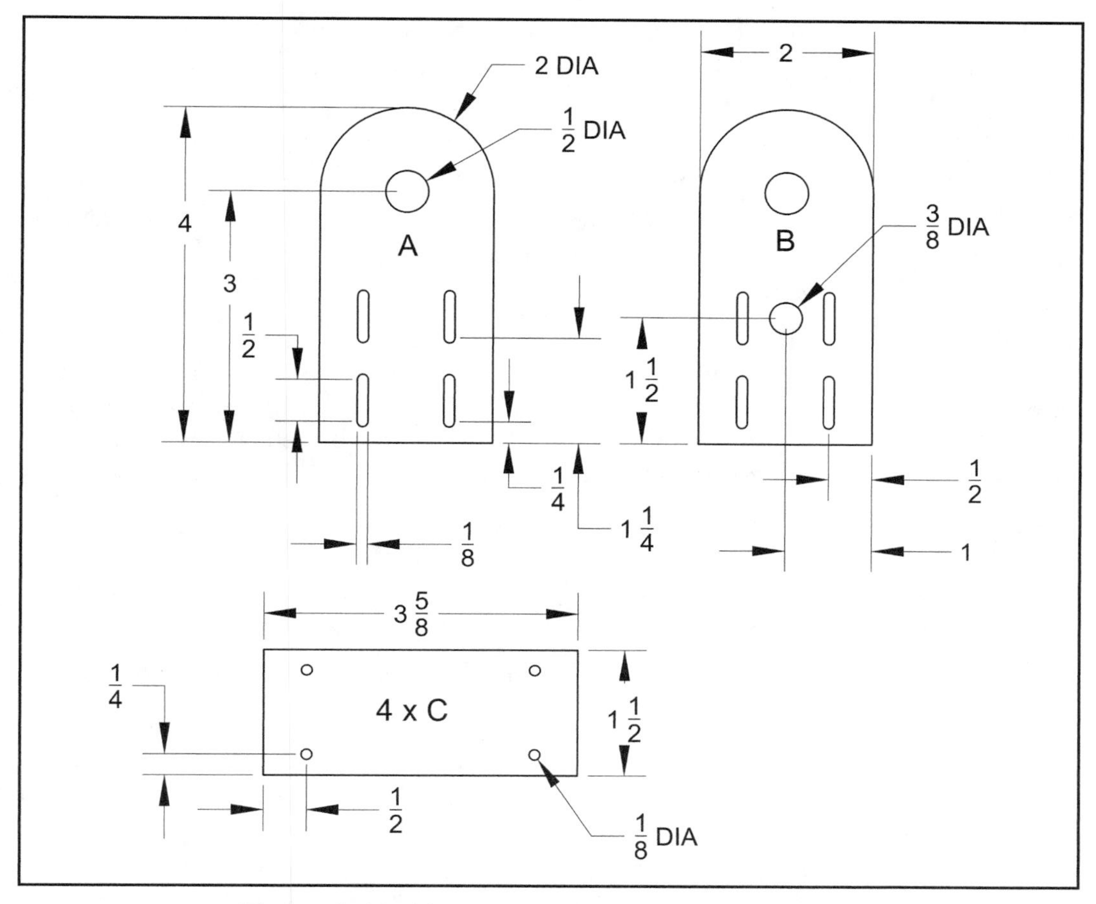

Figure 6-13 Upper-arm dimensions A, B, and C

The A and B pieces have 1/8" side slots cut into one end to allow them to slide along their angle mounts (D, E, F, and G). The 1/2" hole at the rounded end allows for a flange bushing with a 3/8" inner diameter, which in turn supports the shaft through the shoulder and provides a smooth surface for the arm's rotation.

The 3/8" diameter hole in the middle of the B piece allows the insertion of a potentiometer or other rotation sensor. This sensor should have no larger than a 1/2" body. If the sensor doesn't have a 3/8" mounting hub, adjust the hole accordingly. The center of this sensor must be *precisely* 1.5" from the center of the pivot shaft. If it is not, the gears will not mesh correctly.

The large end radius is cut in the same manner as on the shoulder.

The four C panels provide the width of the arm. These are 3.625" wide, which you may notice leaves a 1/8" total gap (or two 1/16" gaps) between the outside of the A and B pieces and the inside of the shoulder. This gap gives the extra space for the Trantorque to grip the shoulder wall, and also provides space for the face of the shoulder bushings. If you want to adjust this spacing, change the width of the C panels.

The angle mounts D, E, F, and G are made from aluminum angle with 1" wide sides that are 1/8" thick. These are complicated bits of metal, as **Figure 6-14** shows. They provide the mounting points for most of the pieces in the upper arm.

All of the small holes in the angle mounts are drilled and tapped for #4-40 threads. These holes are all 1/2" in from the edges (e.g. in the middle of the angles) and start 1/4" in from the ends. All of these mounting holes come in pairs and are 1" apart. The mounting holes are in the same positions on both sides of the angles. Though there are many measurements to make, this conformity should make it easier.

The large semi-holes in D and E are sized to provide a friction-fit around the angled hub at the shaft end of the motors. Though they are marked at 13/16" diameter, they should be sized to grip the motor hub tightly with about 1/8" of the hub allowed to protrude through. It is better to start too small and then file to fit.

The A piece will ultimately attach across the small holes in D and E at the top end, and the I piece across the bottom.

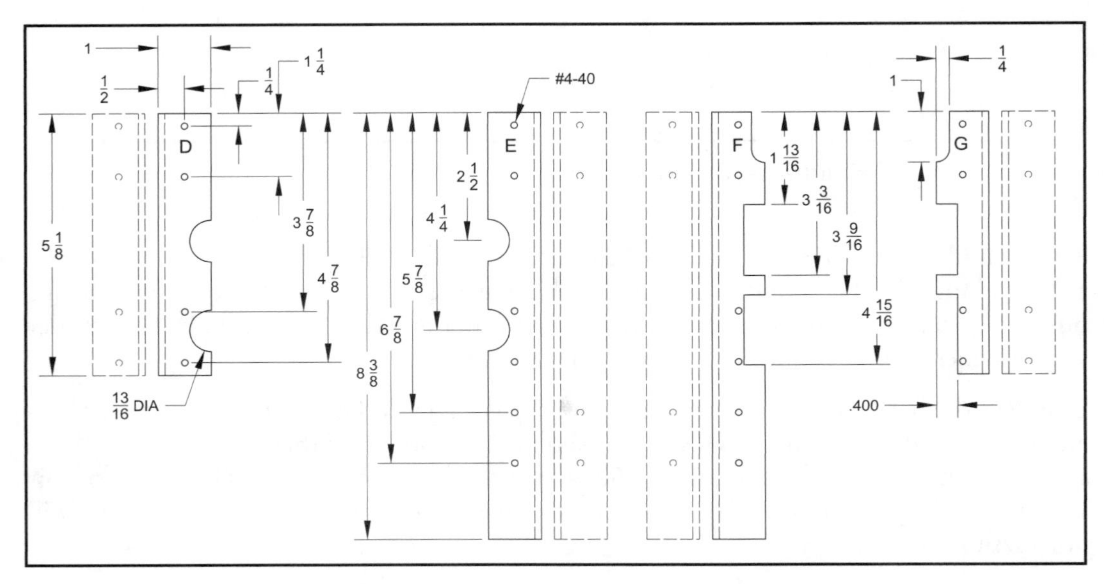

Figure 6-14 Upper-arm dimensions D, E, F, and G

The F and G pieces have rectangular sections cut out of them to fit the square end of the motors. These are 0.4" wide on each half for a total width of 0.8", and 1.375" long. Be sure to measure your motors to make sure they will fit. The rectangular holes come perilously close to some of the #4-40 threaded mounting holes, but should not interfere with their operation if you are careful.

There is an additional 1" long slot cut into the top of F and G. This is to provide clearance for the shaft and mounting nut on the rotary sensor, as B slides up and down F and G. Though this slot is shown in the diagram with a rounded end, it could also be cut square.

The final five pieces are particularly complicated, as you can see in **Figure 6-15**. Though they contain fewer holes than the angle supports, the holes are not laid out as uniformly. All of the little holes and slots in these pieces are 1/8" diameter or are drilled and tapped with #4-40 threads.

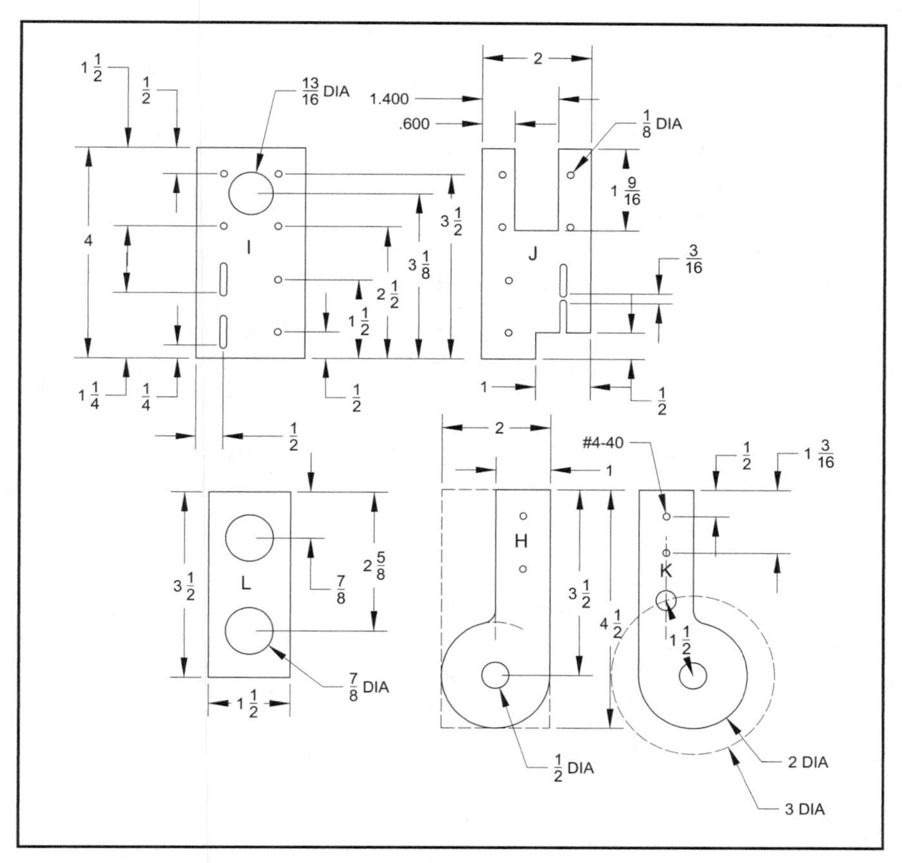

Figure 6-15 Upper-arm dimensions H, I, J, K, and L

The H piece is the prototypical elbow joint. It is constructed from 2" wide aluminum plate that is 1/8" thick. Starting with a 4.5" long rectangle, mark a center point that is 1" in from both sides and 3.5" down from the end. As part of this process scribe a line down the center of the entire piece. Then, scribe the 1" radius circle that will make up the rounded end of the joint. Note where this circle intersects the center line, and cut out the elbow shape.

Drill the 1/2" hole in the rounded end, where you will later mount a bushing to support the 3/8" shaft, and the drill and tap the two #4-40 mounting holes in the skinny end of the piece. These mounting holes are 1" apart, and start 1/2" in from the end.

The K piece is similar to H, but different in two important ways. The body of the piece is marked and cut in the same way as before, but the mounting holes in the skinny part are 5/16" close together. This is because the body of the rotary sensor requires a large slot in J which, in turn, interferes with the placement of the mounting bolts.

The 1/2" hole in the round end is the same, but be sure to drill it *last*.

There is one additional hole in K. This is the mounting hole for the rotary sensor. We have already created one sensor mount, in B. The spacing in B was easy to achieve, because the two holes were in a direct line with each other. In K, the holes are not in line. Starting with the mark in the round end of K (assuming you haven't drilled the hole yet), scribe a 1.5" radius (5" diameter) arc across the skinny end of the piece. Where this mark crosses the center line of the skinny piece, 1/2" in from the edge is where you drill the sensor hole. Don't try to calculate the spacing.

The plates I and J mount to the angle supports like A and B do, but they do not slide like A and B. Instead, they allow H and K to slide. I has a large hole to pass the motor hub through. It doesn't need to be a tight fit. J has a slot for the motor's back end.

Note that the 1/8" slots in J that fasten to K are 5/16" close together than the slots in I that fasten to H. This is to make room for the honking huge chunk taken out of J to make room for the rotary sensor's body and still provide a firm mounting spot for the bolts.

The last piece, L, is actually optional. If the motors are already a snug fit, it is not necessary. Otherwise it can be used as a spacer to help hold the motors in place. The two holes are designed to fit over the motor hubs.

ASSEMBLY

Start by attaching the rotary sensors to B and K. I found a pair of small-bodied potentiometers for this job, but you can use anything that will fit. The trick is to find a sensor with only a 1/2" body.

You can also fasten the small sprockets to the shafts of the motors.

There are a bunch of metal pieces to put together here. First, attach the C pieces to their respective angle supports D, E, F, and G.

If you are using the spacer, slip it over the two motors. Lay the motors in place and sandwich them between the DG and EF assemblies.

Loosely fasten A and B onto the angle supports. Be sure to use a washer between the bolt and the sliding pieces to provide more friction when it comes time to tighten everything into place.

Firmly fasten I and J to the other end of the angles. Set H and K under I and J and loosely fasten them in place, again, using washers.

With the bulk of the upper arm assembled, you can fasten it to the shoulder at any time, using the 6" long 3/8" diameter steel shaft and a Trantorque, as described earlier. Once it is in place on the shoulder, you can put a large sprocket on the shaft and loop the chain around the sprockets for the shoulder.

Note that with the large shoulder sprocket in place, it becomes very hard to reach the bolts that hold the sliding piece B in place! I had to cut down an Allen wrench to fit in that space. Slide the upper arm pieces A and B so that the chain is tight and the arm is even, and tighten them into place.

Warning! Do not test the shoulder motor yet! In fact, you may want to leave the sprockets loose, for easy manipulation until everything is in place.

PROJECT 6-1C: LOWER ARM AND WRIST

This is the last segment of the arm, and the box-with-holes form is very similar to the upper arm but without the cutouts for the motors. Another difference lies in the wrist. A motor is mounted inside the lower arm, and its shaft provides a mounting point for the hand.

The lower arm also provides mounting space for the R/C servos that move the fingers. In this design we only use three servos, but there is room for six. We'll talk more about the hand and fingers in the next chapter.

The left side of **Figure 6-16** shows the rendering of the lower arm, attached at the elbow. You can also see how the motor is boxed into the arm making a rotating wrist. The right side shows the exploded view of the lower arm.

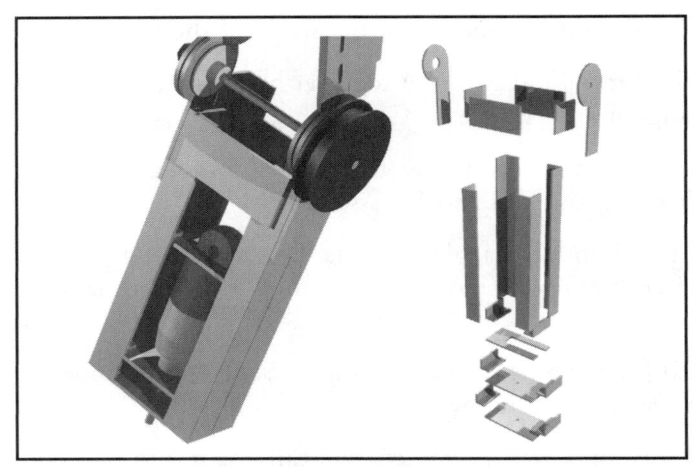

Figure 6-16 Lower arm

Note that the wrist motor will need a modified shaft, like the shoulder and elbow motors. This shaft should be about 3.375" long, 1.125" inside the motor and the remaining 2.25" to make up the wrist.

Figure 6-17 illustrates the various pieces used in the lower arm. Though there are more actual pieces here than in the upper arm as shown in Figure 6-12, there are fewer *varieties* of pieces and they are simpler in form.

The piece types C, E, F, and G are the long angle supports that the rest of the pieces bolt on to. Pieces A and D, are the flat panels that hold the top of the box together, while B and H make up the lower elbow. The angled I pieces both hold the arm together and provide mount points for the motor box J, K, and L.

Pieces B and H are made from 2" wide, 1/8" thick aluminum. They interface to the elbow shaft—B locks onto the Trantorque and H supports the other end of the shaft. These pieces bolt securely onto pieces C and G through the bolt holes. These holes are 1" apart, 1/2" up from the end of the piece (see **Figure 6-18**).

The two D pieces are 1.75" lengths of 1.5" wide aluminum plate, again 1/8" thick. These connector plates fit *inside* the angled arm pieces so their corner holes are threaded for #4-40 bolts. Because these plates are narrower and are interior to the angle supports, the corner holes have the odd spacing of 1/4" in from the end, and 3/8" in from the edges.

The one A piece is a 3.75" length of 0.75" wide aluminum (yup, it's 1/8" thick. But then, it *all* is). The two bolt holes are 1/8" diameter, and are 3/8" in from the edges. You might expect that there would be a second A piece for the other side of the arm, but it would interfere with the sensor gear there.

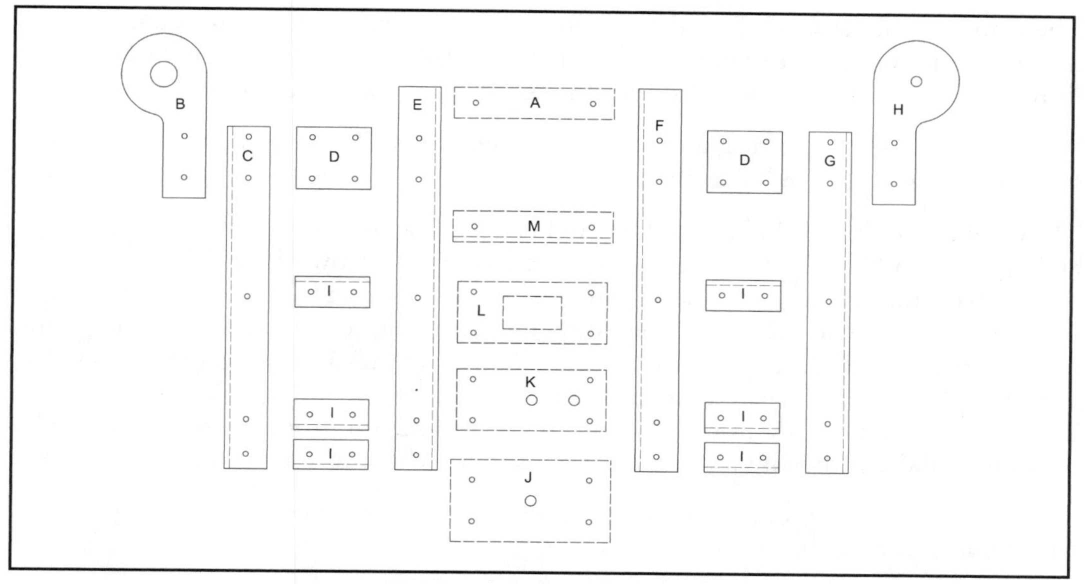

Figure 6-17 Lower-arm pieces

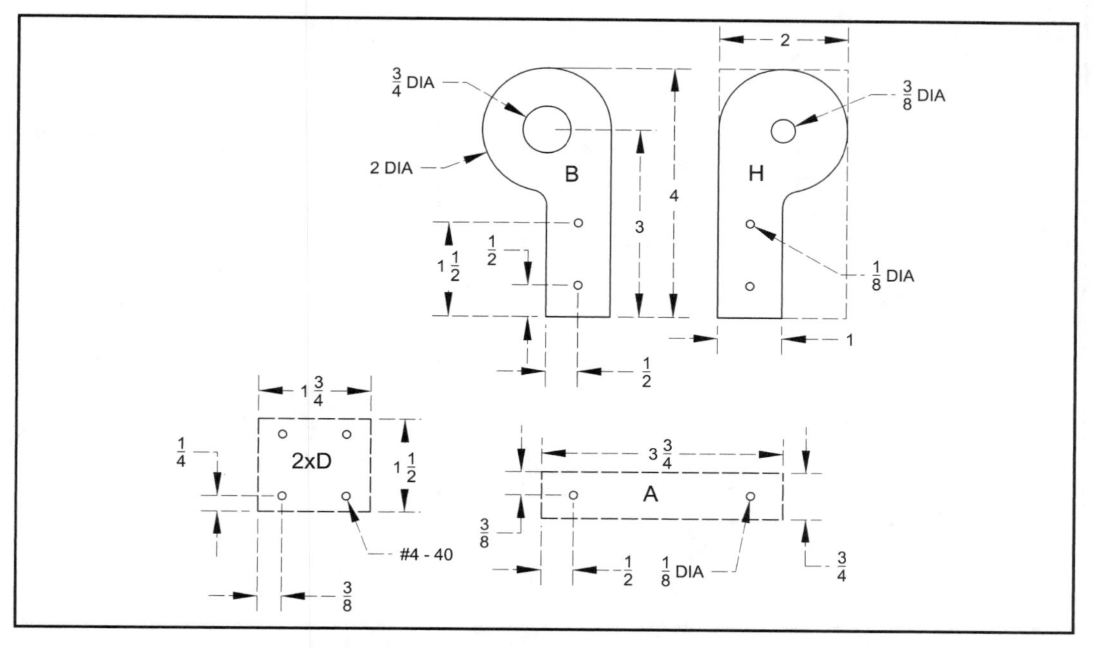

Figure 6-18 Lower-arm dimensions A, B, D, and H

The main arm pieces C, E, F, and G are made from 1" angle aluminum, the same as in the upper arm. Unlike the upper arm, few of these holes are tapped. Most of the bolts pass through these brackets to threaded metal inside the arm (see **Figure 6-19**).

Pieces E and F are mirror images of each other. They are 9.375" long and the holes in one face are all drilled to 1/8" diameter.

The other face of E (and F) has two types of holes. One set is placed very close to the edge and are used to hold the R/C servos into place. Holes are shown for a full complement of six servos, though only three servos are used (on pieces E and F, ignoring the holes in C and G). These servo holes are drilled with a #52 drill bit, for an exact pass-through for a #4-40 bolt, or 1/8" drill bit, which has the drawback of making the material at the edge very thin. The holes are spaced 10mm apart to match the mounting holes on the servos. The servo hole centers are placed 2.5mm (or about 3/32") in from the edge. You could try to drill and tap these holes for #4-40 bolts, but I chose to use locking nuts instead.

The other two holes in the center of the servo side of the E and F pieces are tapped for #4-40 bolts, to hold on pieces A and M.

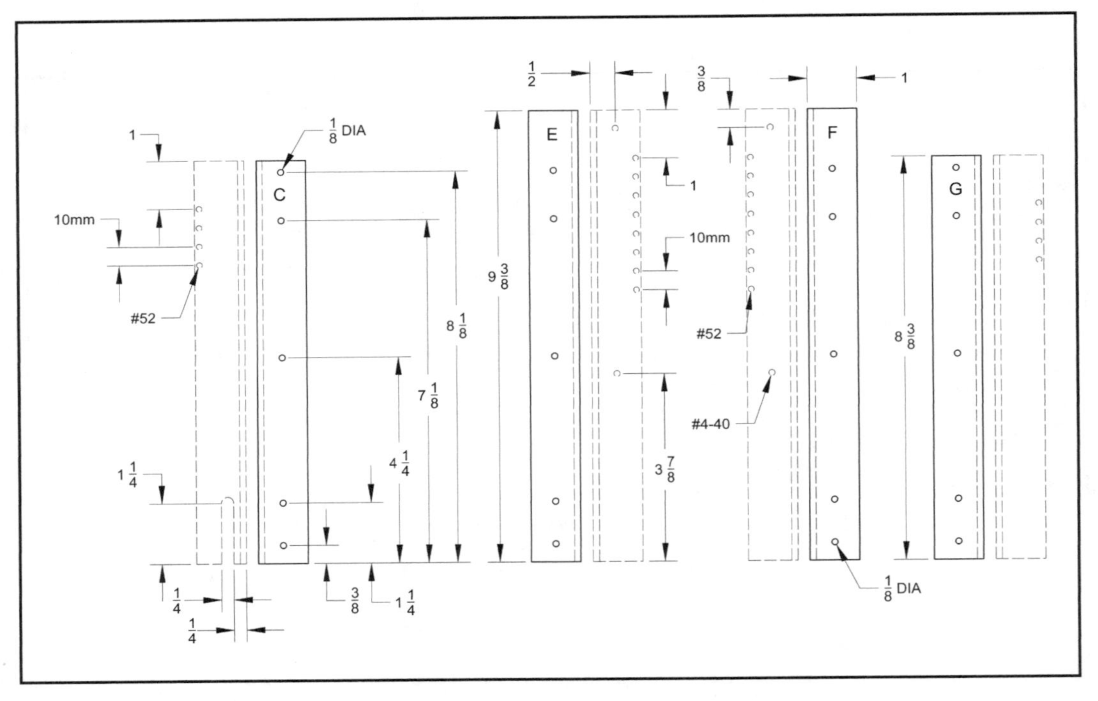

Figure 6-19 Lower-arm dimensions C, E, F, and G

Pieces C and G are essentially shorter versions of E and F. There is a 1/4" slot cut 1.5" into C to make room for the sensor gear from the upper arm. The gear on the sensor should fit right into this slot so the arm can close completely. Any servos mounted in C/G will fit right into the upper arm cavity next to the sensor.

The six I pieces perform dual duty—they hold together the arm pieces C, E, F, and G and they provide mounting shelves for motor mount panels J, K, and L (see **Figure 6-20**). These six pieces are 1.75" lengths of 3/4" angle aluminum. They have two holes centered in each side and set 3/8" in from the ends. These holes are all tapped for #4-40 bolts.

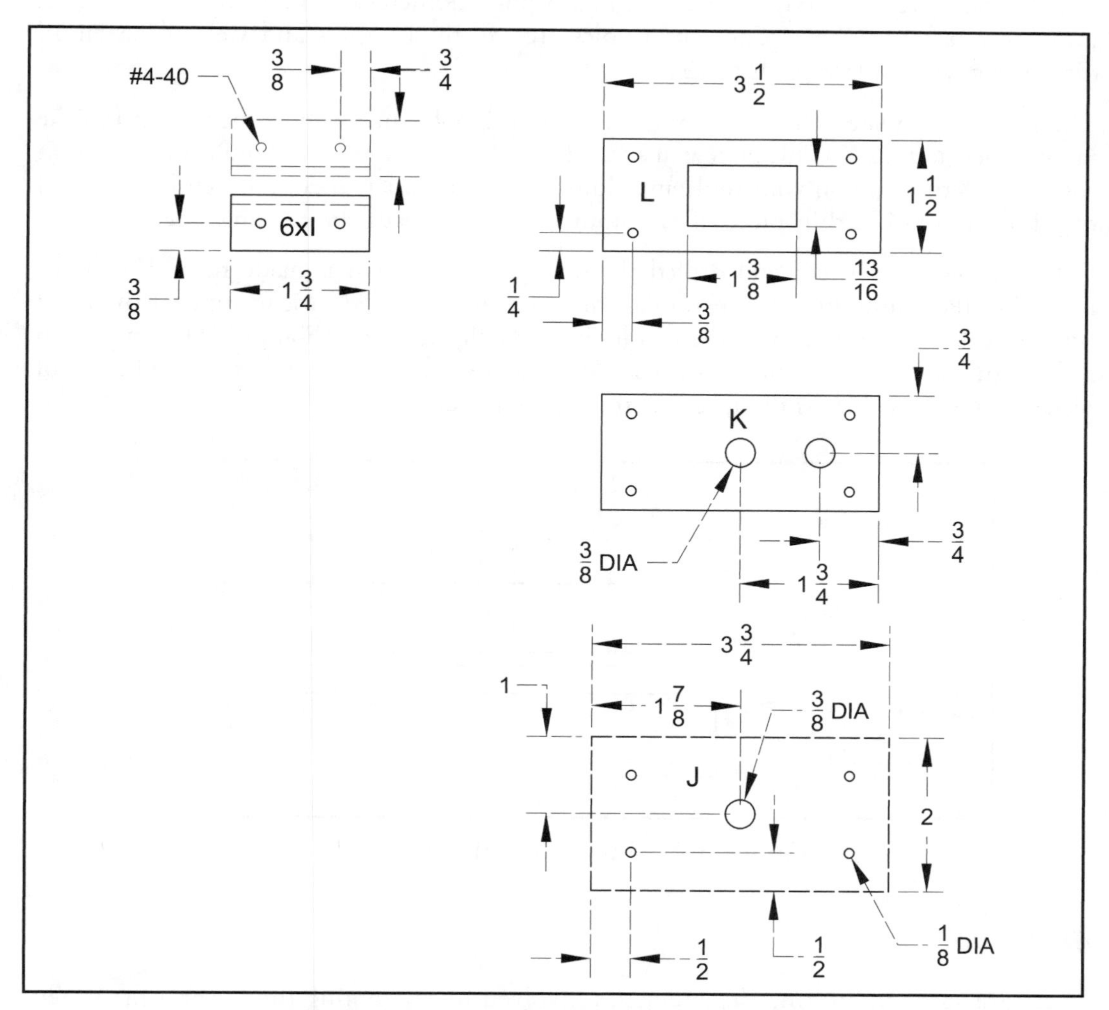

Figure 6-20 Lower-arm dimensions I, J, K, and L

The wrist piece J is a simple 3.75" length of 2" aluminum with the familiar 1/8" diameter corner holes plus a 3/8" diameter hole in the center of the plate. A flange bushing will fit here, reducing it to 1/4" diameter and the motor's shaft will then pass right through. Any number of attachments can be clamped onto this wrist shaft.

The sensor piece K provides the resting place for the business end of the motor. This piece is cut from a 3.5" length of 1.5" wide aluminum plate and it has the usual corner holes. To preserve the 1" spacing between the holes, they are 1/4" in from the sides; the holes are also 3/8" in from the ends. The hole in the center of K is 3/8" to allow for a bushing and the motor rests directly on this bushing. A potentiometer or rotary encode will be set into the offset 3/8" hole. These two holes are exactly 1" apart and we will fasten 1" diameter gears in place to drive the sensor.

The motor end plate L has the familiar rectangular hole cut into it, otherwise it has the same dimensions as K. This plate will sit at the back end of the motor and prevent it from rotating. Depending on your tools and skills, it may be easier to cut a notch in L instead of a hole and bolt a thin piece of aluminum to it to replace the missing side.

Cable bulkhead M is cut and drilled the same as A, except it is made from 3/4" angle instead of flat aluminum. The extra face of M has holes drilled in it to support the hand actuator cables. The spacing of these holes is essentially arbitrary (**Figure 6-21** shows them 3/4" apart), and their diameter is chosen to be a close fit for the cable tubes. The use of piece M and its cables is discussed in the next chapter.

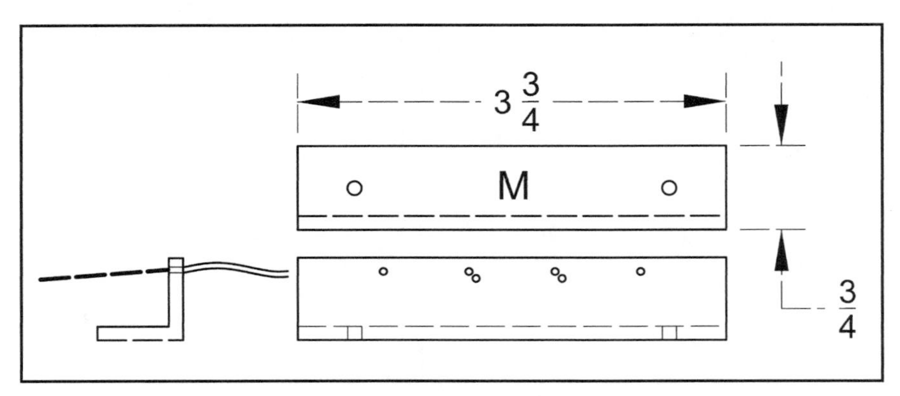

Figure 6-21 Lower-arm dimensions M

ASSEMBLY

There are some interesting dependencies involved in assembling the lower arm. When putting the parts together remember to fasten the bolts loosely at first and, when all of the pieces are in place, tighten everything down. There are exceptions to this standard process that are noted below. Note that these instructions may not be definitive; feel free to experiment.

Start by loosely attaching two I pieces to the back motor support L. Now loosely attach the long supports C, E, F, and G to these I pieces. Note that L is near the elbow, and the open parts of the I angles point down the arm toward the wrist.

Fasten the cross piece A to E and F. Fasten the two D pieces to E and F. Fasten B and H with bolts that pass through E and F and lock into D.

Now the elbow end of the lower arm is assembled loosely. Tighten the bolts to firmly lock the motor support L into place. Though these bolts are hard to reach now, they will be harder to reach later!

Fasten a pair of I supports to the set of holes one-in from the wrist, where the K plate will fasten. These angles are placed so they open up towards the elbow (the K plate will be on the wrist side of the angle). Thread the wires in the motor through the L plate and secure the rectangular end of the motor in that plate.

Securely fasten a sensor to the K plate and a 1" gear on the sensor. Slip the K assembly over the motor shaft. The sensor's body will be on the same side of the plate as the motor. Securely bolt K onto the two I brackets.

Now securely fasten a gear on the motor shaft so it engages the gear on the sensor.

Finally, the wrist plate J and its two I brackets can be bolted into place.

Fit your three R/C servos over their holes in E and F and bolt them into place. Start with the servo nearest the wrist and work your way back towards the elbow. Fasten the servos firmly in place, but not so tightly that you damage the rubber gasket or the plastic tabs on the servos.

Fasten the M bulkhead into place in front of the servos, with the upright part closer to the wrist than the elbow, so that the angle opens back to the servos.

Once the servos are in place, tighten all of the bolts in the lower arm.

The elbow joint is essentially the same as the shoulder joint. The lower arm is narrower than the upper arm, so the elbow shaft can be 5.75" long. Assemble and tighten the elbow the same way you did the shoulder.

When the mechanical bits are together you can run all of the wires for the motors, servos, and sensors up through the hollow core of the arm to the control system(s) of your choice.

Figure 6-22 shows the finished arm with the hand attached.

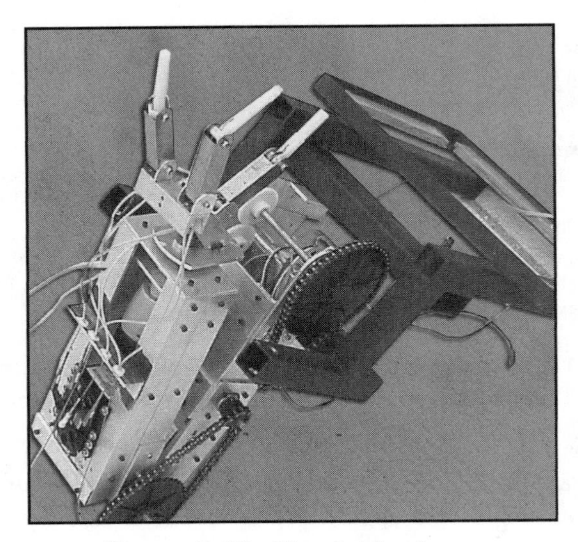

Figure 6-22 The finished arm

ARM POWER AND OPERATION

Most of the commercial R/C electronic speed controllers run off of 6 volts or 12 volts. Since I don't want to put in two 6-volt batteries to get both 6- and 12- volt high-current power, I'm pretty much limited to using the 12-volt ESCs. At 12 volts, the arm motor is running at a factor of 4 above its design limit. This can cause excessive arcing inside the motor and lead to an early failure...not to mention the massive overheating possible when the motor is under load.

On the bright side, the arm is quite strong at 12 volts—quite strong enough to break itself, as well as too fast to safely control by hand. I have a cracked gear to attest to this point.

The arm will have to be managed by a microcontroller, not just because of its speed and the damaging strength of the arm, but to be sure the motors are driven intermittently so they do not overheat.

The process of programming the arm drivers is not addressed in this book, so you are on your own there.

This is technically a three-axis arm, but due to its mounting position between the robot's drive wheels, spinning the robot around its center also happens to spin the arm near its center. The robot body acts as a fourth axis of motion. It's an imperfect system, to be sure.

Next up, a gripper to fit on the end of the arm.

CHAPTER 7

HAND

The need for a hand, or *some* type of gripping assembly, comes as a natural extension of the arm-building project. The purpose of a robotic arm is to manipulate the environment—and the part that does the manipulating is usually the bits attached to the end of the arm.

Most "hand" projects seem to be simple two-finger grippers, either parallel grippers or angled grippers. Sure, these are simple, having a single axis of motion, and can be made quite strong, but I was hoping to go a little further up the path of technology.

The human hand, you may notice, has many possible motions and degrees of freedom. Each of the four fingers and thumb can spin in little circles in space, as well as close and open with some limited independence across the joints. Most of the muscles that power the hand are placed in your arm, with power carried to the digits by tendons.

Although most industrial robots have fairly specialized end-effectors for their arms, one exception is Barrett Technology's BarrettHand BH8 series graspers (www.barretttechnology.com), designed for "programmably flexible part handling and assembly." The mechanisms inside this hand are too complex a project for a book of this scope, though I encourage the interested reader to learn more about it.

The project in this chapter is actually a simplified version of a 1990 NASA patent (inventors Carl F. Ruoff and J. Kenneth Salisbury Jr.), illustrated in **Figure 7-1**. I urge you to examine this patent, as well as later patents that reference this one, via the U.S. Patent and Trademark Office (www.uspto.gov) or the Delphion Intellectual Property Network (www.delphion.com). For example, some interesting variations are found in US Patents #5,062,673 and #6,247,738. Especially when you are working with mechanical projects, make a point to cruise the available patents for inspiration.

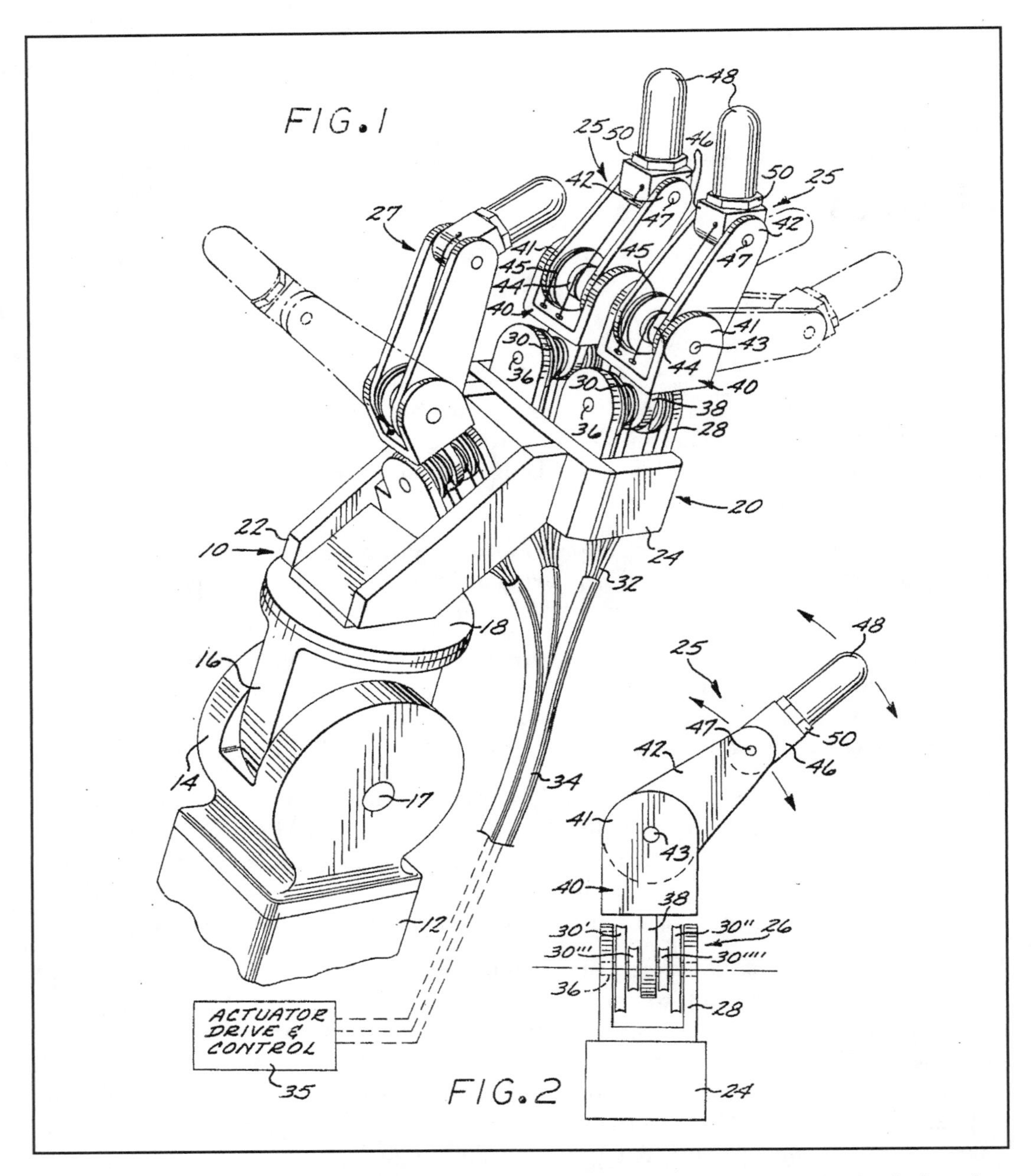

Figure 7-1 U.S. Patent #4,921,293 Figures 1 and 2, multifingered robotic hand

PROJECT 7-1: HAND

Figure 7-2 shows a couple of renderings of the hand, minus a few details such as the actuating cables, as well as an exploded view of the major components of an early incarnation of this hand.

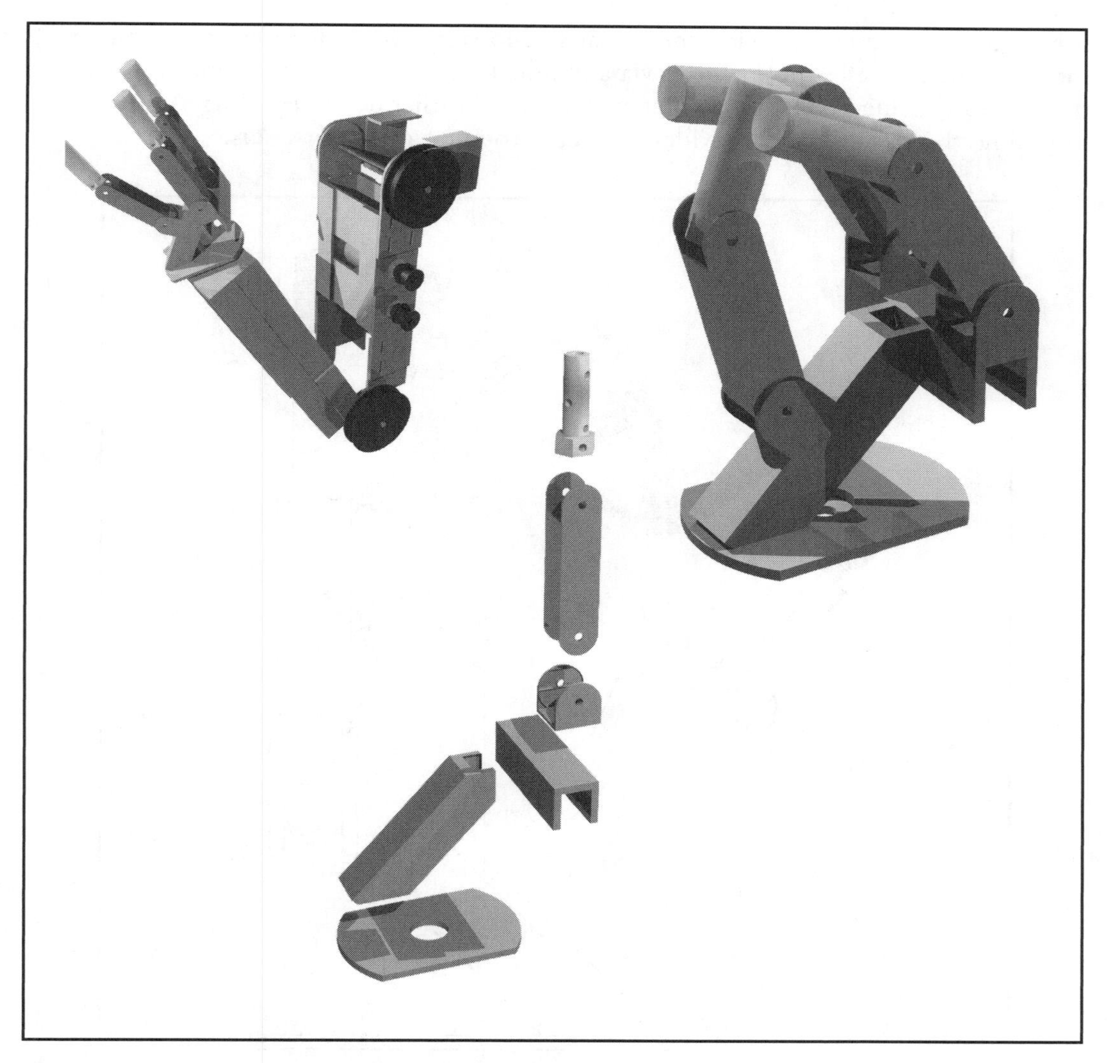

Figure 7-2 The hand

The hand consists of three fingers—two on one side of the "palm" and the third in an opposing "thumb" position. As you can see in the renderings, each finger has two moving segments that move through a simple pivot joint. The motion of the fingers is illustrated in **Figure** 7-3. Each finger is like a miniature arm, with the motors moved off-site due to the fingers' small size. Fingers have the same design problems as arms but with the added complication of being really small.

Advanced hand designs provide additional side-to-side motion for the fingers, either by rotating around the central palm or via additional cable controls and a universal joint at the base of the finger. The NASA patent referenced earlier describes a finger of the latter sort while the BarrettHand provides for finger rotation of the first sort.

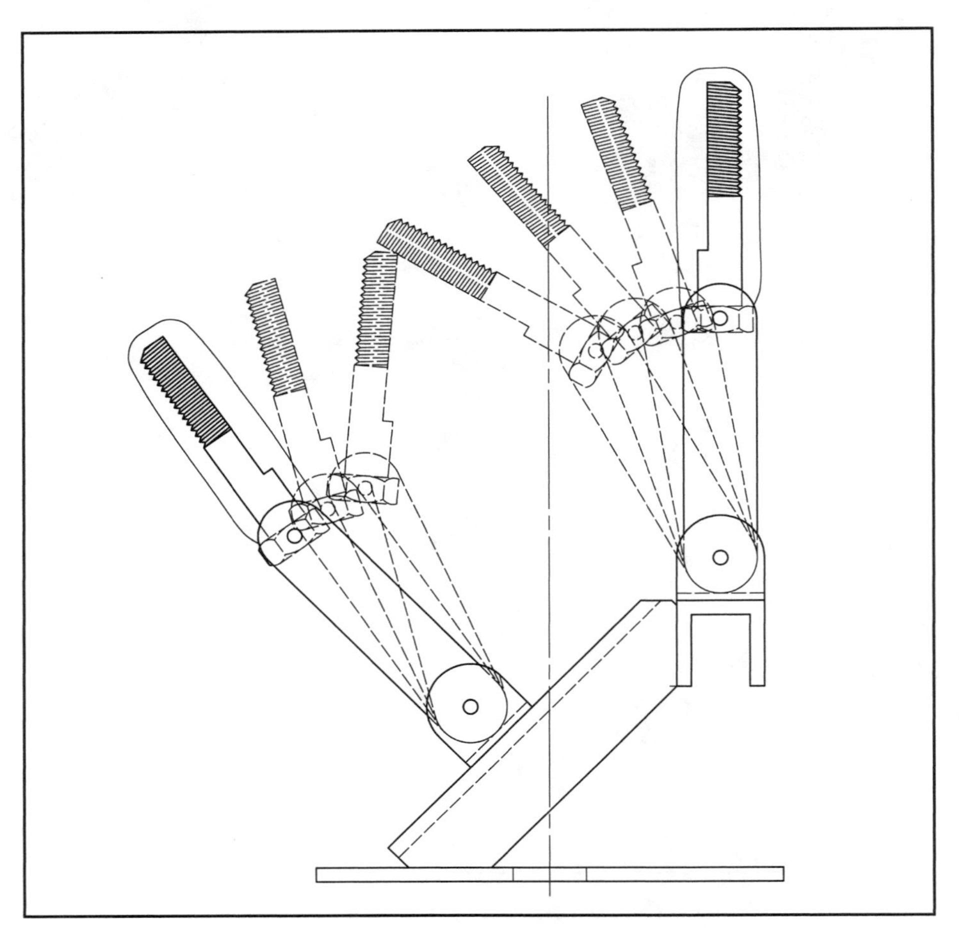

Figure 7-3 Hand motion

Each finger in the design in this chapter is powered by a single R/C servo that controls the motion of both joints. A simple extension of this design provides a separate servo for the second finger joint, but this extra servo provides minimal additional capability. The BarrettHand only uses four motors—one per finger, plus one for finger rotation—and a complex system of clutches to give somewhat independent motion to each finger joint.

The finger strength is limited by the power available in the R/C servos used to drive them, mitigated by the leverage involved. Unfortunately, the inexpensive servos are not very strong and the finger mechanism itself applies a fairly large mechanical reduction to that strength. Stronger servos are available, for a price. Some finger designs make use of small linear actuators to increase the available power. The extreme gear ratios used to drive linear actuators can increase the modest power of a small motor to useable levels.

The hand, as described here, is not capable of grasping anything with more than a few ounces of force. Fairly straightforward (though expensive) improvements in the muscles can increase this grasping strength with minimal mechanical redesign.

The various parts of the hand are illustrated in **Figure 7-4**. The hand will be described (and constructed) from the bottom up, starting with the wrist and culminating at the fingertips.

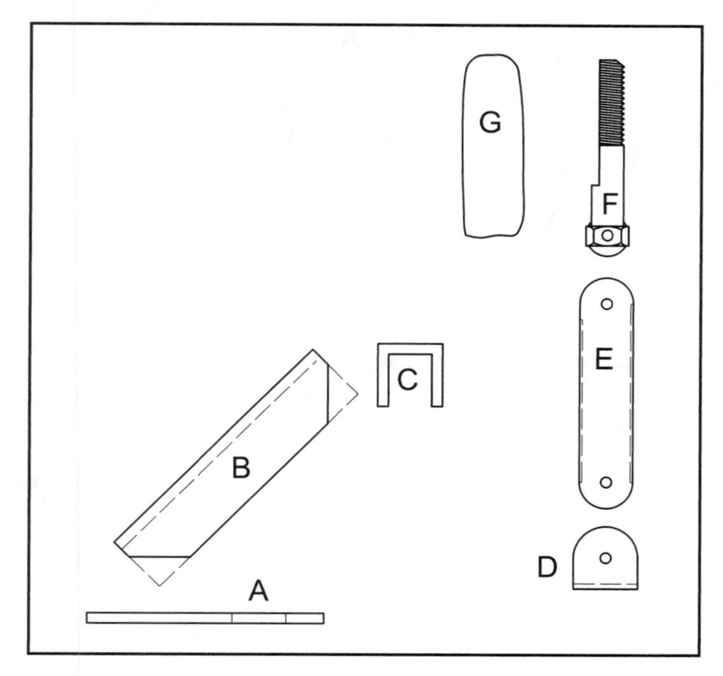

Figure 7-4 Hand parts

WRIST

When we left the story of the wrist in the previous chapter, it was nothing more than a short section of 1/4" steel rod protruding from the end of the arm. The hand contributes to the wrist with a simple flat plate A with a hole drilled in it (**Figure** 7-5). This is held onto the wrist motor with the ubiquitous Trantorque fitting.

The plate itself is cut from a 2.75" length of 2" aluminum plate. The ends are rounded for aesthetic reasons; the end curves shown in **Figure** 7-5 have a radius of 1.5". The center of these arcs is 1.5" in from the ends of the plate, on the centerline.

The large hole is 5/8" diameter to fit the Trantorque that will interface this plate to the 1/4" wrist shaft. The small holes are 1/8" diameter to pass the #4-40 bolts that will be holding palm support B onto wrist plate A.

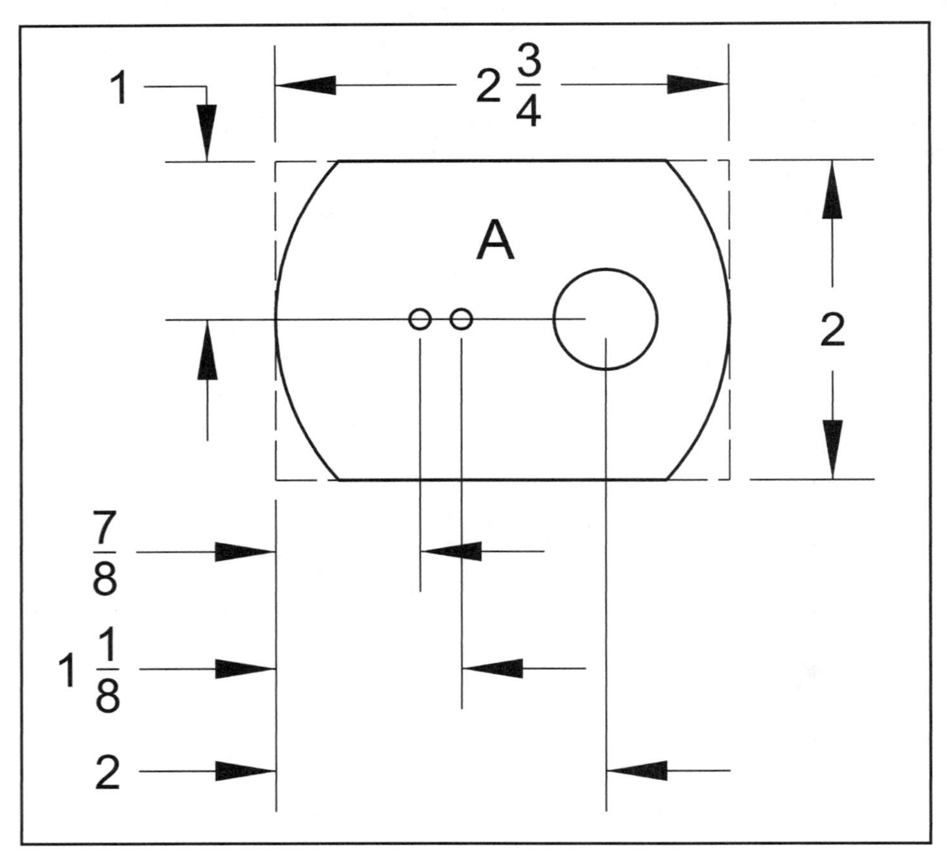

Figure 7-5 Wrist plate A dimensions

PALM

The palm consists of two major pieces B and C, plus a few painfully tiny mounting blocks that should prove a good challenge to your machining skills. The first palm piece B attaches to the wrist by way of the mounting blocks. The second palm piece C mounts to the first by another set of mounting blocks. The finger base pivots D then mount directly onto B and C using conventional means.

Figure 7-6 shows the dimensions for piece B. It is a 3.25" length of 3/4" U-channel aluminum with 1/8" thick walls. The thick walls are necessary for the two threaded finger-mounting holes in the top face of the piece. The ends of the channel are cut at a 45° angle about 1/2" in from the corner, giving a 3/4" wide flat spot. These are the surfaces that mate to palm pieces A and C later.

Both ends of piece B have a pair of 1/8" mounting holes that pass through both walls of the channel. These holes are 1/4" apart and centered in the angled ends of the piece, 1/8" in from the edge. Bolts pass through these holes and thread into mounting blocks.

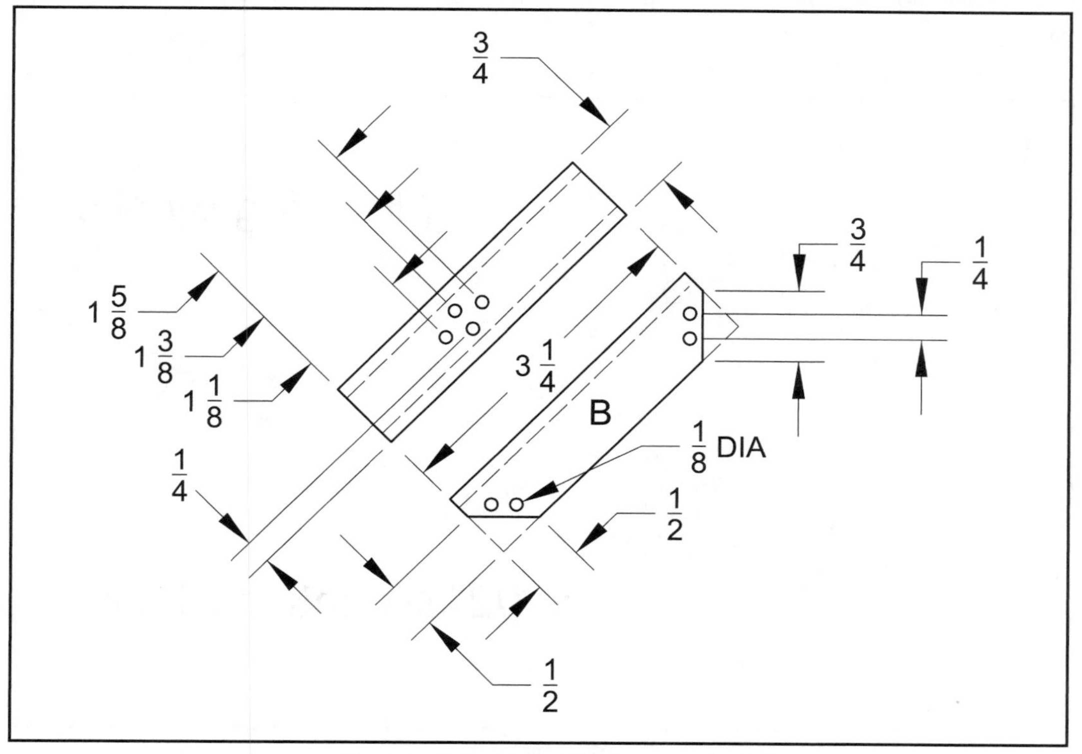

Figure 7-6 Palm B dimensions

The top face of B has four holes in a diamond pattern. The two holes at the tips of the diamond (1.125" and 1.625" from the end) are drilled to hold the actuator cable tubes, described later. The other two holes, across the piece at 1.375" from the end, are tapped to accept #4-40 bolts from the finger base D. The spacing of the holes in the diamond matches the holes in piece D.

The mounting blocks that fit into palm piece B are illustrated in **Figure** 7-7. There are two ways you can make these blocks. The first method, shown at the top of Figure 7-7, is to cut 1/2" lengths of 1/4" square steel rod. These lengths must be cut and carefully filed so that they fit snuggly between the sides of the 3/4" U-channel. Then, drill through the center of the blocks once lengthwise and again crosswise through the center, making a T-shaped hole in the block. Carefully tap these holes to hold #4-40 bolts. First tap the long hole, and then each half of the cross hole. Make two of these blocks for each end of piece B, for a total of four blocks.

The second method, shown at the bottom of Figure 7-7, is to cut 1/2" lengths of 1/4" by 1/2" rectangular steel plate. These are fitted, drilled, and tapped in a similar manner to the smaller blocks. You would need two of these larger blocks, one for each end of the palm piece B.

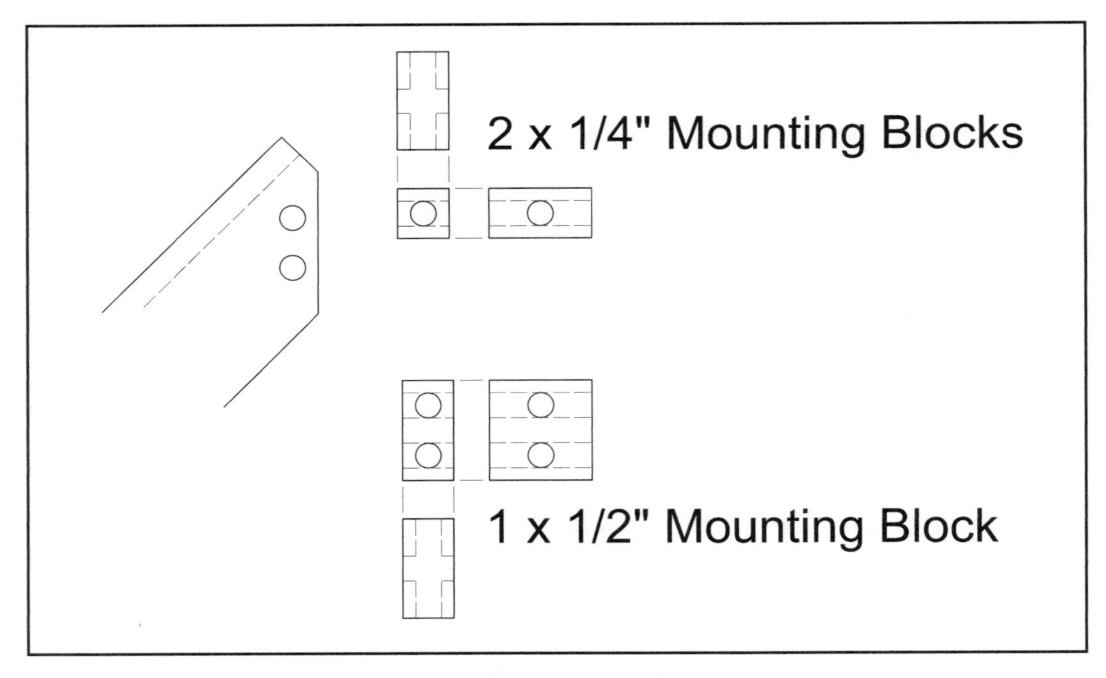

Figure 7-7 Mounting blocks

Palm crosspiece C is a 2.5" length of the same thick-walled U-channel aluminum as B (see **Figure 7-8**). On the top face, at each end of the piece, are sets of diamond-patterned mounting holes. The ends of the pattern are moved inward by 1/16" so the cable holes do not interfere with the walls. As before, two of the holes are tapped for #4-40 bolts, and the other two are drilled to accept the cable tubes.

The sides of this piece each have two additional holes, 1/4" apart and centered. One set of holes is drilled to 1/8" and are used to mount piece C to the mounting blocks of piece B. The other set of holes, directly opposite the first set, are drilled to an arbitrary size to allow access by a screwdriver or Allen wrench; a necessary consideration so that you can tighten the mounting bolts.

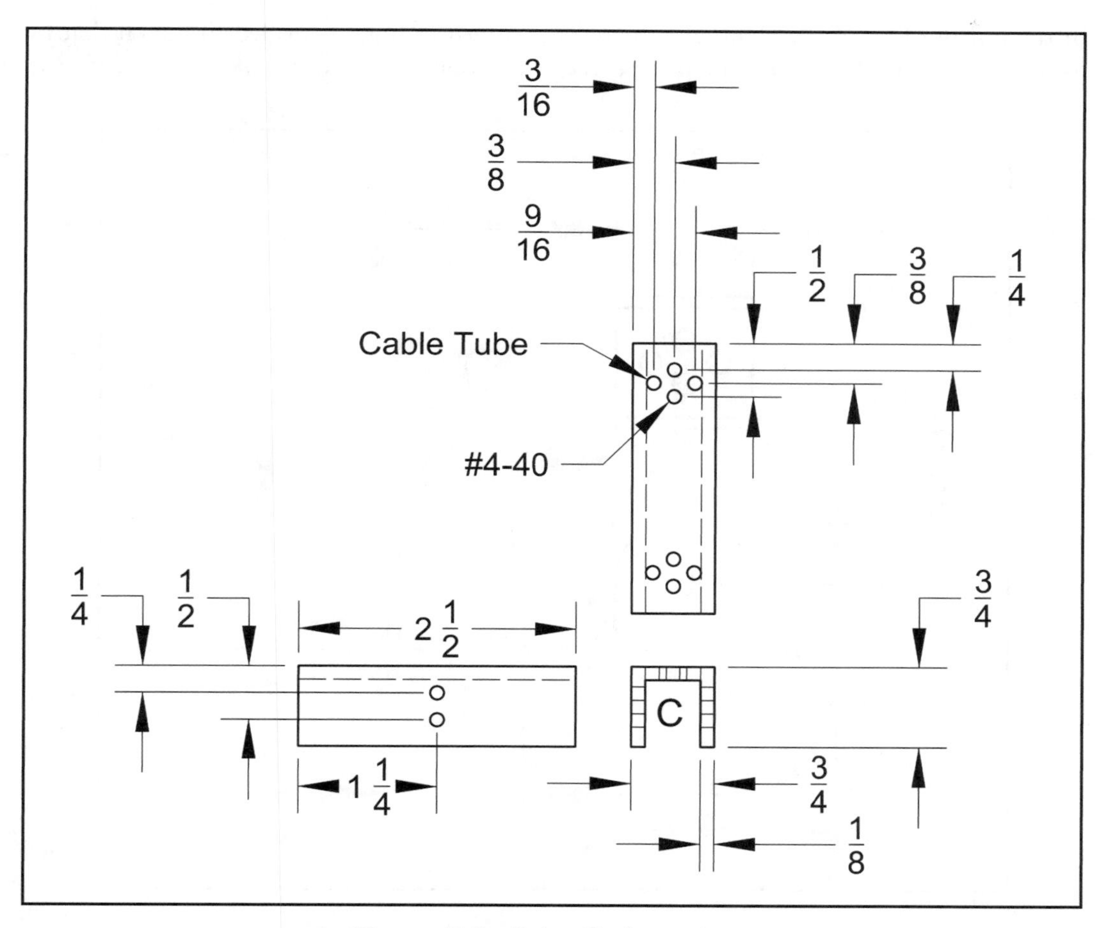

Figure 7-8 Palm C dimensions

FINGERS

The fingers are made from a careful selection of materials. The finger base D is made from U-channel aluminum with 1/16" thick walls. The finger tube E is made from very thin-walled square brass tubing (with optional pulley or spacer tube, described later). The fingertip F is made from a 3/8" nylon bolt and a 1/2" pulley. The fingers are all held together with #4-40 bolts with friction-lock nuts and driven by metal cable appropriated from the radio-control airplane hobbyists.

The finger base D is a simple 3/4" cube of U-channel aluminum with 1/16" thick walls (see **Figure** 7-9). The open end of the channel is rounded into a 3/4" diameter curve, and a 1/8" hole is drilled in the center of the side faces to hold a #4-40 pivot bolt.

In the bottom face of the channel, a diamond pattern of holes is drilled to match that in palm pieces **B** and **C**. Where the palm pieces were tapped for #4-40 bolts, the finger base

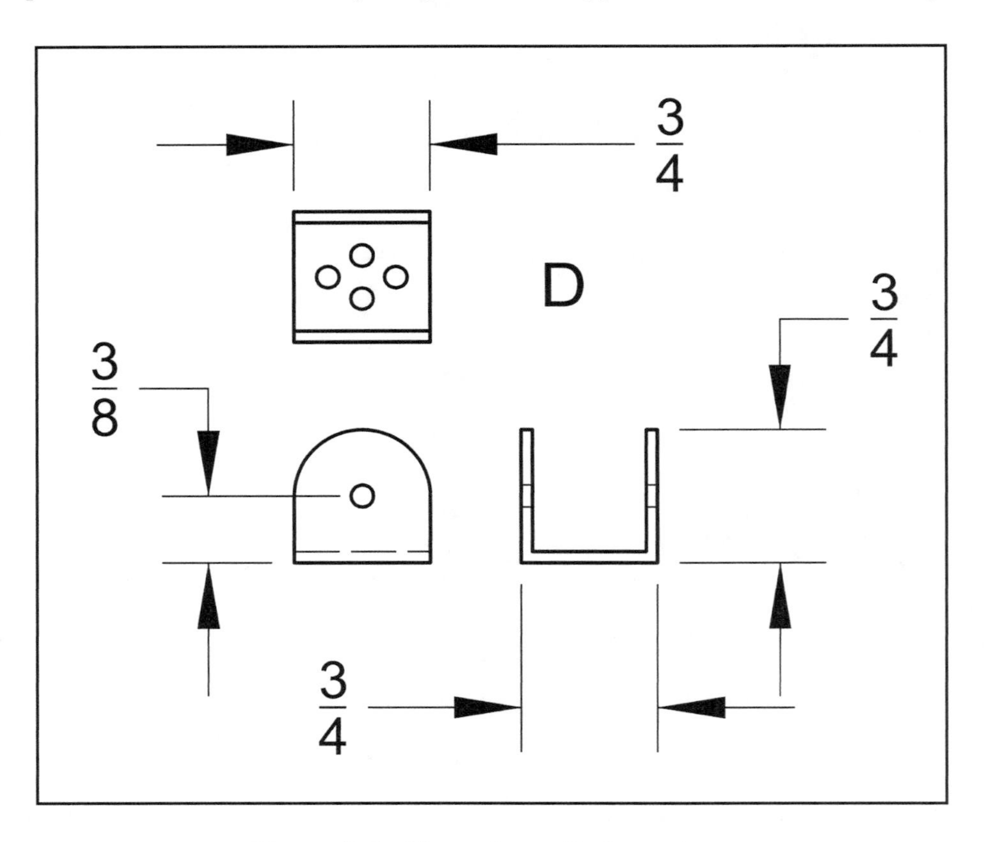

Figure 7-9 Finger base D dimensions

is drilled to 1/8" to pass the bolt. Where the palm pieces were drilled to hold the cable tubes, the finger base is drilled slightly larger to allow for misalignment during mounting.

Note that palm pieces B and C are shown with slightly different hole patterns. You can either modify the design so that piece B matches C, or be careful to create the finger base pieces D to match the appropriate set of holes.

The body of finger piece E is made out of square brass tubing found in the Small Parts catalog (see **Figure 7-10**). It is the so-called "Thick Wall" tubing, with a wall dimension of 0.028". The outside dimension is 5/8" and the piece is cut to 2.75" length.

Mark a center point 5/16" in from the ends and edges at both ends of the tube. Round both ends to 5/16" radius, though only half of one end is rounded, as shown in **Figure 7-10**, and then drill 1/8" holes at the center marks.

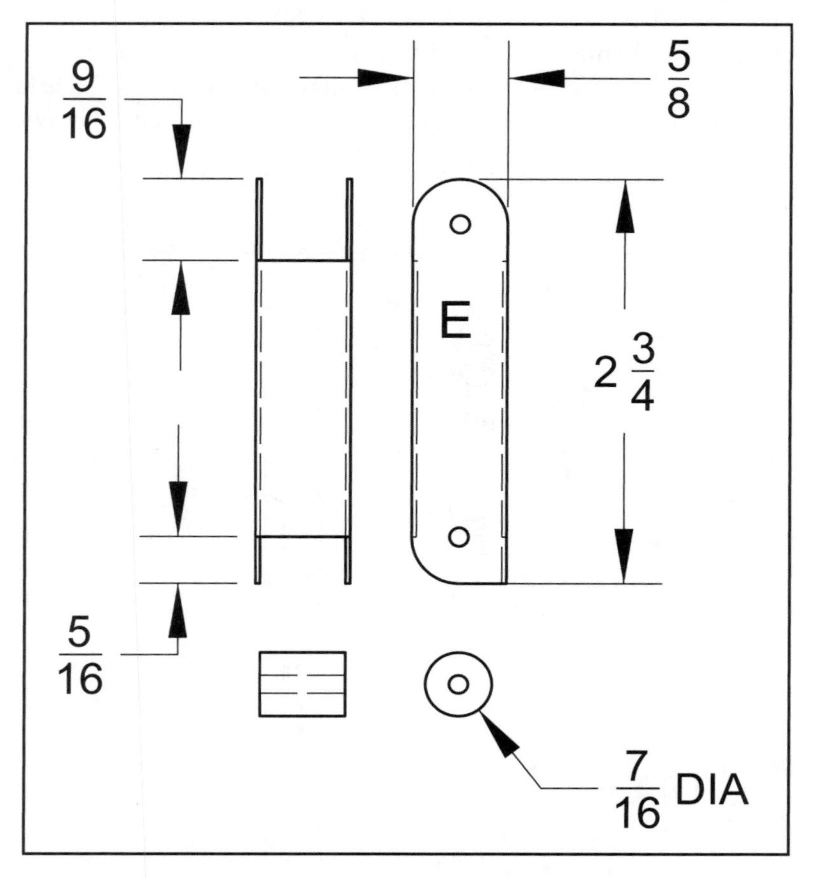

Figure 7-10 Finger piece E dimensions

The act of rounding the tube ends will cut back the walls to about 5/16" in from the ends. You should take care to clean and square these edges. One edge at the end of the tube needs to be cut back to 9/16" from the end on both sides. This provides room for the pulley in the fingertip **F** as well as leaving room for the fingertip to bend into. Note that the lower-rear corner of the finger is *not* rounded. This corner creates a stop to keep the finger from bending backwards.

An optional piece of the finger is a 7/16" pulley or bushing placed at the bottom pivot piece **E**. This pulley or bushing helps maintain the spacing of the finger tendons. I made my bushing out of nested plastic tubes with a final layer of brass tube. I found these various tubes at my local hobby store—the plastic came in the form of nesting polystyrene tubes and I browsed the brass tubing until I found one that fit my largest plastic tube. Though not strictly necessary this cable guide improves the strength and control of the finger.

The fingertip **F** is made from a modified 2" long by 3/8" diameter nylon bolt that I found at the hardware store (see **Figure** 7-11). The flats of the head of this bolt fit perfectly inside the 5/8" brass tube used for finger piece E. You may have to rummage around a bit to find a bolt that fits, or you can take a slightly oversize bolt and cut it down as needed.

A 1.5" section of the threads of the bolt is cut flat to support a small pressure sensor at the fingertip.

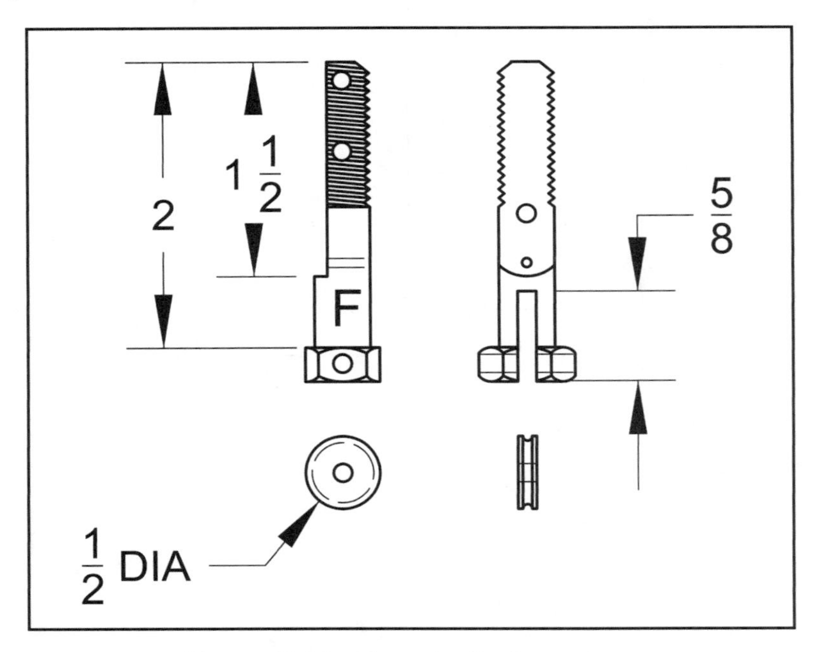

Figure 7-11 Fingertip F dimensions

Cut a 1/8" wide slot 5/8" into the head of the bolt to accept a small pulley. This slot bisects the head of the bolt from tip to tip of the hex facets. A variety of other 1/8" holes are drilled crosswise through the tip of the bolt to hold the finger pad G into place later.

Finally, a hole just large enough to pass the control cable is drilled right above the slot in the flat section of the bolt.

The pulley is a 1/2" diameter Delrin pulley from the Small Parts catalog. It is glued inside the fingertip's slot with epoxy, and kept centered with a well-oiled piece of metal through the pivot holes. This pulley is very important to the overall strength and control of the fingertip; it acts as a guide for the cable through the full rotation of the fingertip. I originally tried to make the fingers without this pulley but I was not pleased with the result.

Once the epoxy has dried, you can do a test fit of the fingertip F in the finger piece E. It should be able to bend all the way forward, to at least a 90° angle and straighten back up again. If there are problems, trim or grind things until it all fits.

The fingertip is where all of the action in the hand occurs. The tendon for the fingertip feeds through the small hole above the pulley and is kinked sharply at both sides of the bolt (see **Figure 7-12**). The cable is then crossed through the slot and run down the inside of the finger by way of the pulley. When it is finally connected, the tendon wire will remain tight and stay in the pulley's groove.

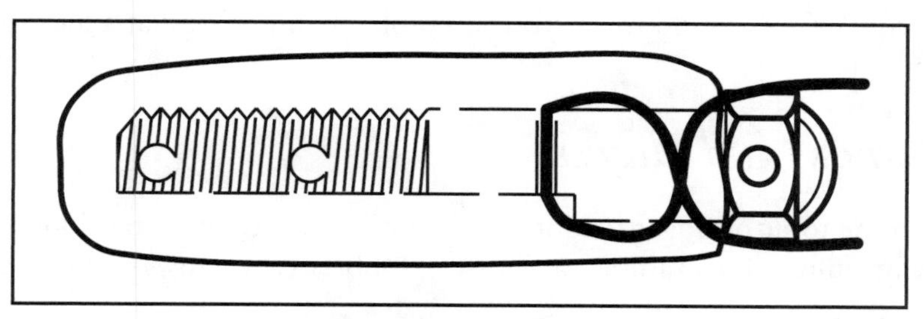

Figure 7-12 Fingertip assembly

If you want to give the finger a sense of touch you need to glue a pressure sensor or two on the flat tip. Pressure sensors, however, can be hard to find. Interlink Electronics (www.interlinkelec.com) sells flat force-sensing resistors in a variety of shapes and sizes. Though the company requires a minimum order, its Force Sensing Resistor Design Kit fulfills that requirement. This kit is not easy to find on its Web site—it is not really linked by the Web pages—so talk to a representative at Interlink to get the order form. This kit is worth every penny.

Anyway, force-sensing resistor FSR #400 is a 0.3" diameter circle that is perfect for the fingertips.

With the tendon cable and pressure sensor in place, you now have the option to cast a soft fingertip around the bolt. The cross holes allow the cast material to pass through the bolt, providing a mechanical lock on the otherwise slippery nylon. Question: What can you use to get a fingertip that has friction, is compliant enough to pass pressure through to the sensor, and is also easy to use? Answer: soft fishing lures.

You sometimes find the strangest solutions to robot problems, and the solution to the fingertip happens to be soft fishing lures. These are the slimy silicone worms and bugs sold at the sporting goods store. If you chop them into tiny pieces, put them in a can, and then heat them to a couple hundred degrees (Fahrenheit), they melt. They will also smoke and make your house smell terrible—I recommend doing this in a well-ventilated area.

First you need to make the female, or negative, mold. This could be a cast of your own finger (or thumb), or just a tinfoil-covered cardboard tube. Set the fingertip into this, heat the fish bait, and pour. Yeah, I'm leaving most of the details to your imagination here...but you can find entire books on casting and mold-making that could do a much better job of guiding you. One of the better books is *The Prop Builder's Molding & Casting Handbook* by Thurston James (Betterway Books, 1989).

If you want to get serious about this silicone rubber, you can buy the casting material and coloring agents in liquid form from a variety of sources (do a Web search on "Soft plastic lure making" for a variety of relevant Web links, such as Zeiner's Bass Shop at www.zeiners.com). Call around town, you can probably find a lead at a local bait shop or sporting goods store.

FINGER TENDONS AND MUSCLES

Going back to the topic of finger tendons, you can find a variety of solutions at your local hobby store, or online, for example, at Tower Hobbies (www.towerhobbies.com).

What you are looking for is a control rod set. These are used for transmitting push and pull forces from an R/C servo in the hull of an airplane to the control linkages out in the wings or tail.

Control rod sets consist of an outer tube that acts as a guide and an inner piece that can range in stiffness from a solid metal rod to a thin metal cable. You can see this same control arrangement on bicycle brake and shift cables.

For this project I used a Sullivan Gold-N-Cable #506, "very flexible."

The outside tubing runs between the servo bulkhead illustrated in Figure 6-21, to the holes in the palm shown in Figure 7-6 and Figure 7-7. When you run the tubes you need to leave enough slack for the wrist to turn. Once they are measured to a happy length, cut the tubes. Vigorously clean the metal at the palm and arm where the tube fastens, and roughen the tube itself with sandpaper. Then, epoxy the tube into place. Be sure to keep the holes clear so the wire can pass through the tube and its metal mounting points.

Attach the fingers together with #4-40 bolts with friction-lock nuts. Keep them loose enough so that the pressure from the nuts does not bind the finger motion. The cables from the fingertips thread straight through the middle of the finger, past the (optional) bushing at the finger base, through the holes in the palm, and through the control tubes. From there, the wires protrude out of the bulkhead on the arm and over to the servos. Pull them firm and cut the cables to size.

Finally, solder the wires into the clips that come with then and clip them onto the servo horns on the servo. The servos, of course, provide the motion for the fingers, by way of the cables.

Though these cables, in airplanes, both push and pull, in this configuration we are only using the pull force to move the finger.

The pulling force on the inner cable will make the fingertip bend inward, but it will also make the middle finger piece bend. The reverse is also true, straightening the finger.

We're getting closer to a working arm (see **Figure 7-13**).

Figure 7-13 Hand and arm, assembly in progress

CHAPTER 8

REFLEXES

We are at the midpoint of the book now. All of the major mechanical systems have been built—the rolling platform, the pan-tilt head, the arm, a hand. From here on the chapters will be less cohesive—less of a multicourse meal that fits together as one organized piece and more of a buffet where you can take the pieces you want and build your own custom meal, er, robot. At this point we also leave the arm and hand behind—the complexities of programming them for useful work are too great for this book.

In Chapter 9, we explore a wide range of sensory systems. To make use of the sensors to come, we need a way to control the hardware that we have. This calls for a set of robot reflexes.

The reflex system is not really AI (or, as I prefer to call it, simulated intelligence; it's not artificial, but it's not "real" either). The reflexes form a first layer in the hierarchy of simulated intelligence. We will talk more about this later. Suffice it to say that the robot needs low-level control over its basic hardware functions. This control needs to be fast, automatic, and designed to keep the robot out of immediate trouble.

The reflex processing can be done in microcontrollers (MCUs) that communicate to the main computer as needed. These MCUs can do most of the bookkeeping involved with managing actual hardware, leaving the PC for more abstract tasks.

There are many excellent control systems on the market, including embedded versions of the PC you have on your desk. When it comes to embedded control solutions, I urge you to purchase rather than build when possible. It will almost always be easier and cheaper to buy a product and extend it than it is to make your own. This chapter explores some of the available control options and then ignores its own advice and constructs a robot reflex system from scratch.

EMBEDDED CONTROL OPTIONS

This section discusses some of the things other people have done to microcontrollers to make them easier for you to use. Some products focus on mechanical form, while others simply preload the chips with easy-to-use languages. This discussion is *far* from complete—I only talk about some of the better-known representatives in each class. There are many other products available.

STAND-ALONE BOARDS

The classic robot control system is the Handy Board, (www.handyboard.com) which contains Motorola's 68HC11 MCU surrounded by useful interfaces. The Handy Board was designed for robotics control and is used at MIT for educational and research projects.

Marvin Green's BotBoard series of controllers are full-featured 68HC11 controllers, similar in principle to the Handy Board. You can find Marvin's work through the links page of the Portland Area Robotics Society (www.portlandrobotics.org).

Even the Lego Mindstorms (www.legomindstorms.com) computer can be used for many simpler projects. Its main limitation is the three inputs and three outputs—not enough I/O for a larger robot.

MicroMint (www.micromint.com) also has a wide range of embedded solutions. I am just touching the tip of the iceberg here. The hardest part about finding an MCU system to use is not in *finding* one, but in *deciding* which one best fits your needs and budget.

The MCU-on-a-board product is common and provides any experimenter a good head start.

BOARD AND BACKPLANE

A similar product concept is the MCU board that is designed to plug into a backplane. A backplane is simply another circuit board that is used to connect different boards together using a common interface...much like the peripheral slots and boards in all personal computers.

One of the largest proponents of this architecture is DonTronics (www.dontronics.com), an Australian company. Its SimmStick product line provides a compact and consistent form for MicroChip PIC and Atmel AVR microcontrollers. These boards fit into 30-pin SIMM sockets like those used for many years in PC motherboards for memory sticks.

The only real drawback to the SimmStick boards is that, to fit the memory sockets, they are produced on thinner circuit boards than most prototyping houses support. This makes it hard to build your own extensions to the product line, though DonTronics does sell blank SimmStick prototyping boards. They also have fat-to-skinny PC board adapters.

I encourage you to browse through DonTronics' large and intimidating Web site.

SYSTEM ON A CHIP

Moving down the size scale we see that several products are in the form of a single DIP chip (or hybrid board in the shape of a DIP chip) preprogrammed for ease of use.

One of the earlier products, and the most widely available one (since it is sold through the Radio Shack catalog now), is the Parallax (www.parallaxinc.com) Basic Stamp series of products (and their imitators). The Basic Stamp is a small microcontroller programmed with the PBasic interpreter. They also include some support circuitry, making these small devices nearly plug-and-play for your projects.

Imitators of the Basic Stamp include the BasicX (www.basicx.com), whose claim to fame appears to be increased speed and storage space. Both companies continue to improve their product lines, so check them out and see what they can do for you.

On a slightly different tack, the OOPic (Object-Oriented Programmable Integrated Circuit, as found at www.oopic.com) takes a MicroChip PIC MCU and gives it code to make it act something like a hardware component. Programming the OOPic is like defining a circuit. Scott Savage of Savage Innovations, the creator of the OOPic, has provided programming interfaces that look like Basic, C, or Java so you can use familiar syntax.

As an aside, Wirz Electronics (www.wirz.com) even sells a Basic Stamp in SimmStick format!

PROJECT 8-1: REFLEX CONTROL SYSTEM

In *Applied Robotics* I outlined a microcontroller system based on the Atmel AT90S8515 MCU. This was a fairly generic control board, with external SRAM for the interpreted Fuzbol program, some I/O buffers, and a bunch of connectors to hook things up. The AVR architecture of the Atmel chip is nice to program in, and Atmel was one of the first companies to provide extensive in-circuit programmable Flash for its MCUs.

Of course, since then, MicroChip PICs have sprouted a growing line of Flash chips, as have other manufacturers. None of them had a device that would lure me off the AVR chips—many had improvements, to be sure, but I was happy on the AVR.

Then, to my grand annoyance, I was lured away by one of the oldest MCU architectures on the planet—an 8051 MCU embodied in a Cygnal (www.cygnal.com) C8051F015 system on a chip. Again, while not necessarily a spectacular chip in the constellation of MCUs (after all, many other chips share the features of this system), the Cygnal chip simply fit my needs just right.

Cygnal is a new company with a somewhat modest, but growing, product line. One of its nicer features is that it will sell you chips directly, making its parts easier to get than those of some other companies. It also is located in Austin, Texas, where I live, so I can always run down to the main office and personally harass the personnel. They have, in fact, been very responsive to my various technical questions.

In the end, I convinced myself to move to the C8051F-series chips for the projects in this book.

I'm not an MCU snob. Most of these various MCUs do essentially the same things—sure, some have more Flash, others have more RAM, better timers, or more registers. But in the end, given a few peripheral support chips, they can all be made to do the same things. So read around and find the chip or system that feels comfortable to you.

I present the MCU system in this chapter in all of its detail for its educational value.

Lifting a page from the SimmStick form factor, the control system is designed to use a passive backplane with multiple PCB slots in it. The working boards plug into this and communicate with each other to get their jobs done.

I chose the size and form of all the boards to fit into a stock Hammond box.

The programmer for this system is orders of magnitude simpler than the programmer used in *Applied Robotics*. And, using the debugger/programmer from Cygnal, you can monitor every aspect of the chip's operation, including putting the chip into single-step mode while watching it progress through the source code. It's a very nice system to develop on and not very expensive either.

BUS DEFINITION

The heart of backplane architecture is the bus definition. The backplane's bus is how all of the cards communicate, and the choices and tradeoffs you make here affect the entire future of the system.

My first two choices involved what type of connectors to use and how many pins they should have. The best connectors on the market are designed for SIMM and DIMM memory cards. These little numbers are small and have respectable pin counts (72 pins), and carry a reasonable price tag (about $4). Unfortunately, they also require that you use a different PC board thickness (0.050") than the prototyping houses usually provide.

Going back into the prehistory of today's computers, you can find really inexpensive PC-XT bus connectors. Though larger in dimension than the 72-pin SIMM socket, 62-pin PC-XT edgeboard sockets run about $1. The best part is that they can accept standard 0.062" thick boards.

So, both questions are answered. The backplane connector will be the dirt-cheap PC-XT edgeboard connector, which has 62 pins. Now the question is, how to configure those pins in the most useful layout?

There are two sides to the connector, A and B. **Table 8-1** and **Table 8-2** list the pin definitions for both sides of the connector.

Pin	Signal	Description
A1	U+1	Unregulated Plus (3x5 = 15A)
	U+2	
	U+3	
	U+4	
	U+5	
A6	VCC	Regulated 5V Logic Power
A7	DO 0	Digital Master Output 1 - 8
	DO 1	
	DO 2	
	DO 3	
	DO 4	
	DO 5	
	DO 6	
	DO 7	
A15	DI 0	Digital Master Input 1 - 8
	DI 1	
	DI 2	
	DI 3	
	DI 4	
	DI 5	
	DI 6	
	DI 7	
A23	AI 0	Analog Master Input 1 - 8
	AI 1	
	AI 2	
	AI 3	
	AI 4	
	AI 5	
	AI 6	
	AI 7	
A31	SCL	I2C Serial Clock

Table 8-1 Connector Side A

Pin	Name	Description
B1	U-1	Unregulated Ground (3x5 = 15A)
	U-2	
	U-3	
	U-4	
	U-5	
B6	GND	Regulated Logic Ground
B7	DO 8	Digital Master Output 9 - 16
	DO 9	
	DO 10	
	DO 11	
	DO 12	
	DO 13	
	DO 14	
	DO 15	
B15	DI 8	Digital Master Input 9 - 16
	DI 9	
	DI 10	
	DI 11	
	DI 12	
	DI 13	
	DI 14	
	DI 15	
B23	AI 8	Analog Master Input 9 - 16
	AI 9	
	AI 10	
	AI 11	
	AI 12	
	AI 13	
	AI 14	
	AI 15	
B31	SDA	I2C Serial Data

Table 8-2 Connector Side B

The pins are named from a "master" point of view, using a master/slave terminology. An MCU card would be the typical master device, and any supporting or I/O cards would be slaves. The backplane would normally only have a single master device, though it is possible to have multiple "masters" so long as they do not try to over-drive each other's output signals. Some cards could also act as both master and slave—receiving signals from one area and then sending additional signals out to another.

The first five pins of both the A and B sides are unregulated power from an external source, such as a battery. Since each pin in the connector is capable of withstanding 3 amps of power, the five pins in parallel can manage 15 amps—more than enough for most applications.

The next A and B pin is a more modest 5-volt regulated power source, to run the various active electronics.

There are a total of 16 digital outputs, split between the A and B sides. Then there are 16 digital inputs—giving 32 total digital lines. While most microcontrollers have configurable I/O pins that can act as either input or output, I chose to lock the backplane lines into a single direction. This makes it easier to set up buffers so that the MCU doesn't need to power these digital lines directly.

Finally, there are 16 analog lines.

The last two lines are for the Inter-IC (I2C) communications lines. I2C is a multimaster protocol developed by Philips Semiconductor. This two-wire serial bus allows an easy way to network controllers between cards, as well as providing communication between backplanes and devices external to the control system. For example, the OOPic system uses I2C networking, allowing it to network naturally with the reflex control system (though I have not had the opportunity to try this yet).

BACKPLANE

The backplane itself is amazingly uninteresting. The bulk of it is composed of five PC-XT connectors that have all of their pins bussed together. There is also a copy of the 5-amp switching power supply we explored previously. Add a nice PowerPole connector for the battery, and that's about it.

For readability, this schematic (and the others, later) is broken down into multiple sheets, each with a single aspect of the circuit on each sheet. Named arrow connectors indicate junctions between sheets. The first sheet of the backplane circuit, **Figure 8-1**, is simply the connectors and their traces. The second sheet is the power supply, **Figure 8-2**.

Figure 8-3 shows the dimensions of the backplane board itself. The part placement and trace layout is difficult to provide in a useful manner in book illustration form, so I avoid it entirely.

The backplane board is designed to fit into either Hammond Manufacturing (www.hammondmfg.com) box 1591U or 1591V. These boxes measure 4.7" by 4.7" square. The U box is 2.2" tall and the V box is 3.5" tall—the one you use depends on how hard you want to work on the part layout. The taller box makes for a more relaxed circuit arrangement.

The dimensions for the two plug-in cards, regular and short, are given in **Figure 8-4**. Note that the boards are slightly wider than the backplane. This extra width allows the boards to lock into the vertical slots inside the Hammond boxes. The height of the boards is designed to just fit inside the boxes when plugged into the backplane.

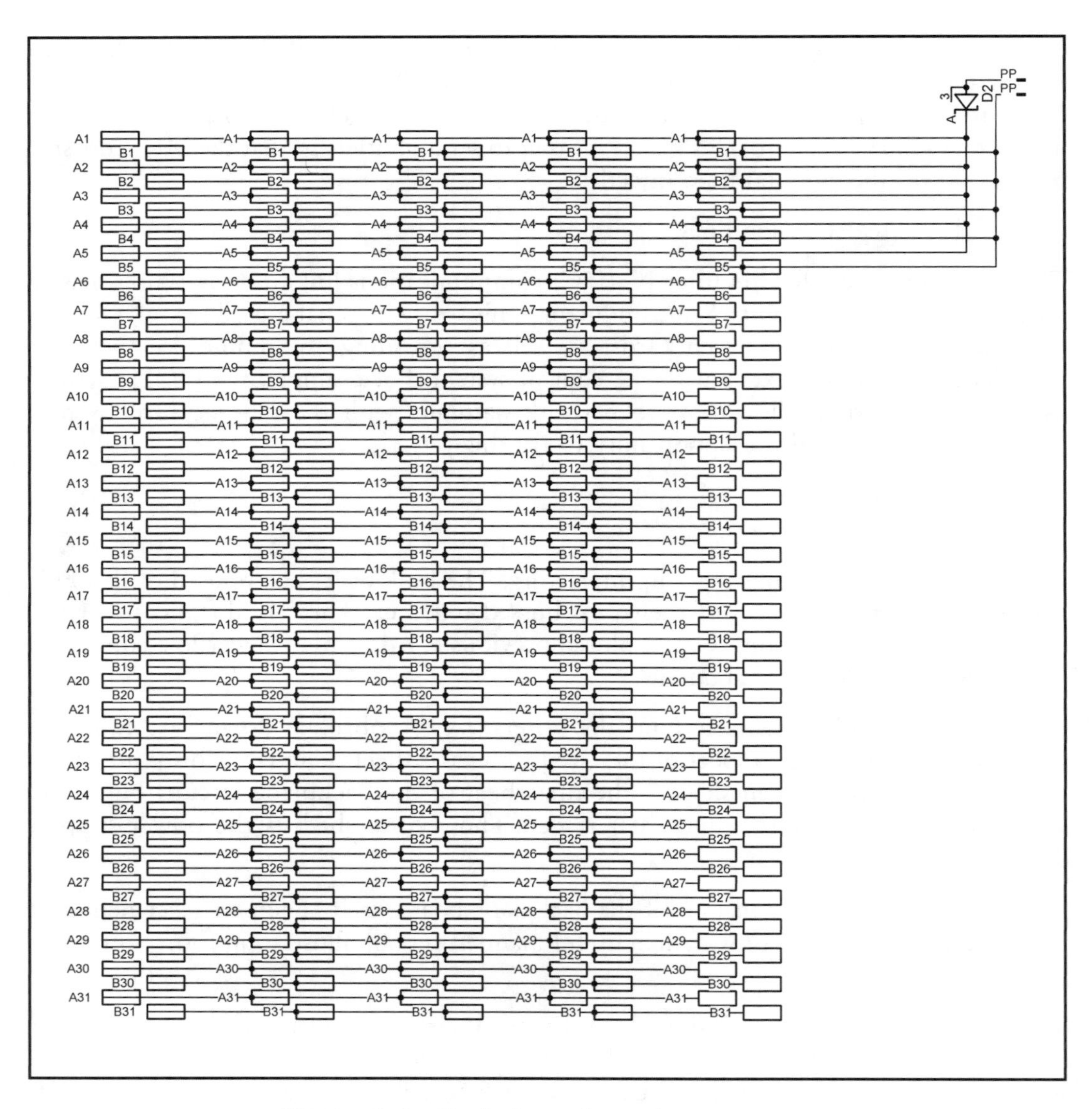

Figure 8-1 Backplane sheet 1, connectors

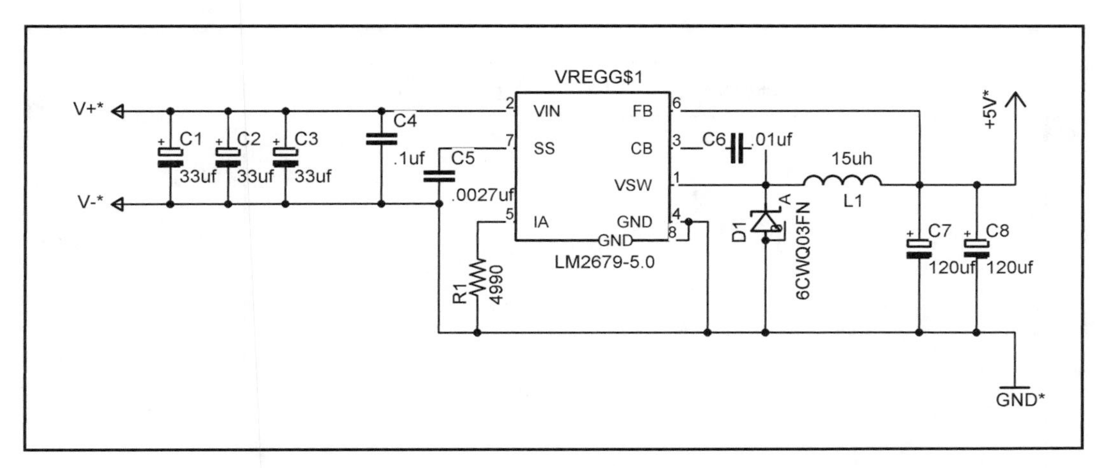

Figure 8-2 Backplane sheet 2, power supply

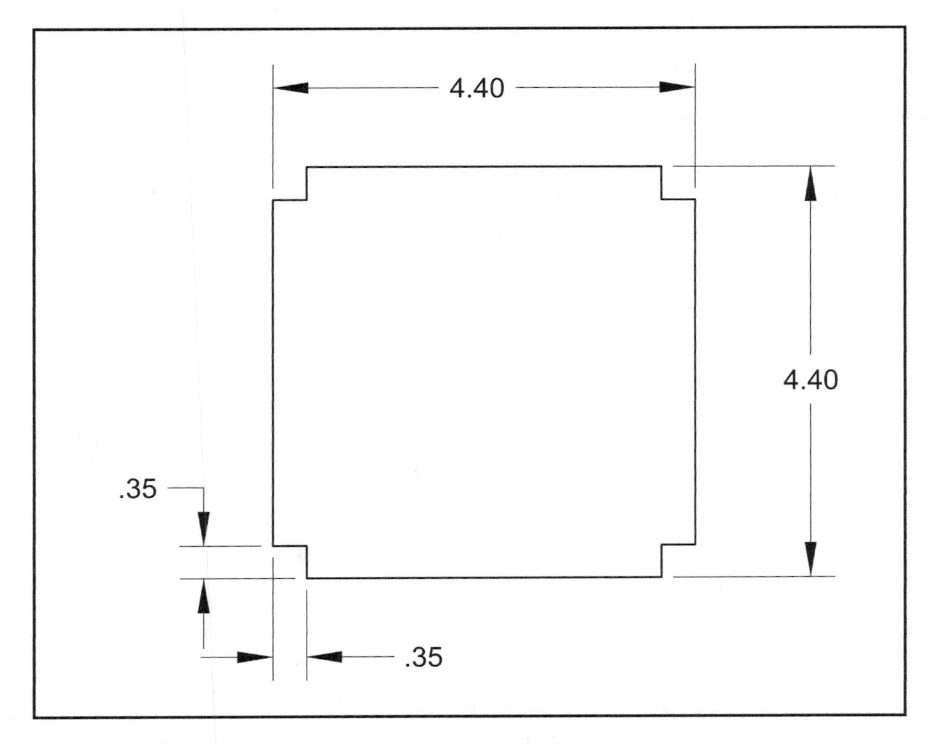

Figure 8-3 Backplane dimensions

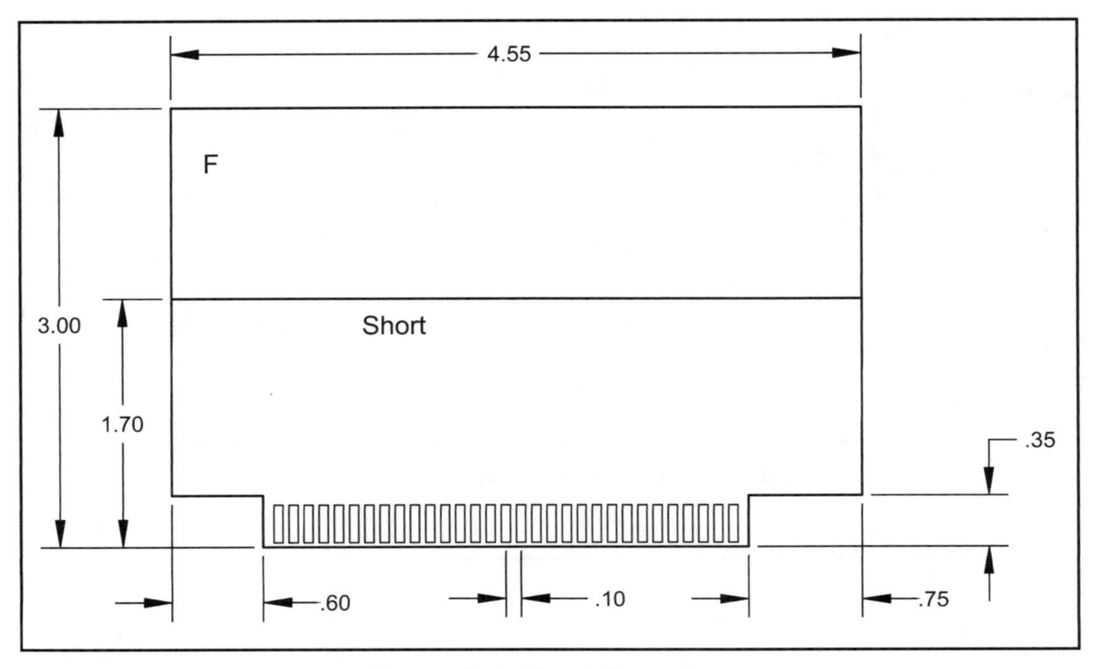

Figure 8-4 Card dimensions

These dimensions, while not terribly relevant to the technology contained in the boards, do give a mechanical frame of reference for what comes next.

MCU BOARD

Ahhh, the Cygnal system-on-a-chip. What do they mean by that? Let's look at a comparison. First, **Figure 8-5** has the schematic for a complete Fuzbol system as used in *Applied Robotics.* Sorry for the small size; I know it's hard to read.

This system features 32K of external SRAM for the interpreted program and program data storage and 8K internal Flash memory for the interpreter. For I/O it has 8 each buffered digital inputs and outputs, plus 8 analog inputs. There is a programming port, some decoding logic for the memory-mapped I/O, an RS232 converter and plug, an 8MHz crystal, and some other bits and pieces. In short, a system.

The new MCU board based on the Cygnal C87051F015 is illustrated in **Figure 8-6**, **Figure 8-7**, and **Figure 8-8**. This system has essentially the same capabilities as that in Figure 8-5 with a lower part count, though it was drawn using a more spread-out schematic style.

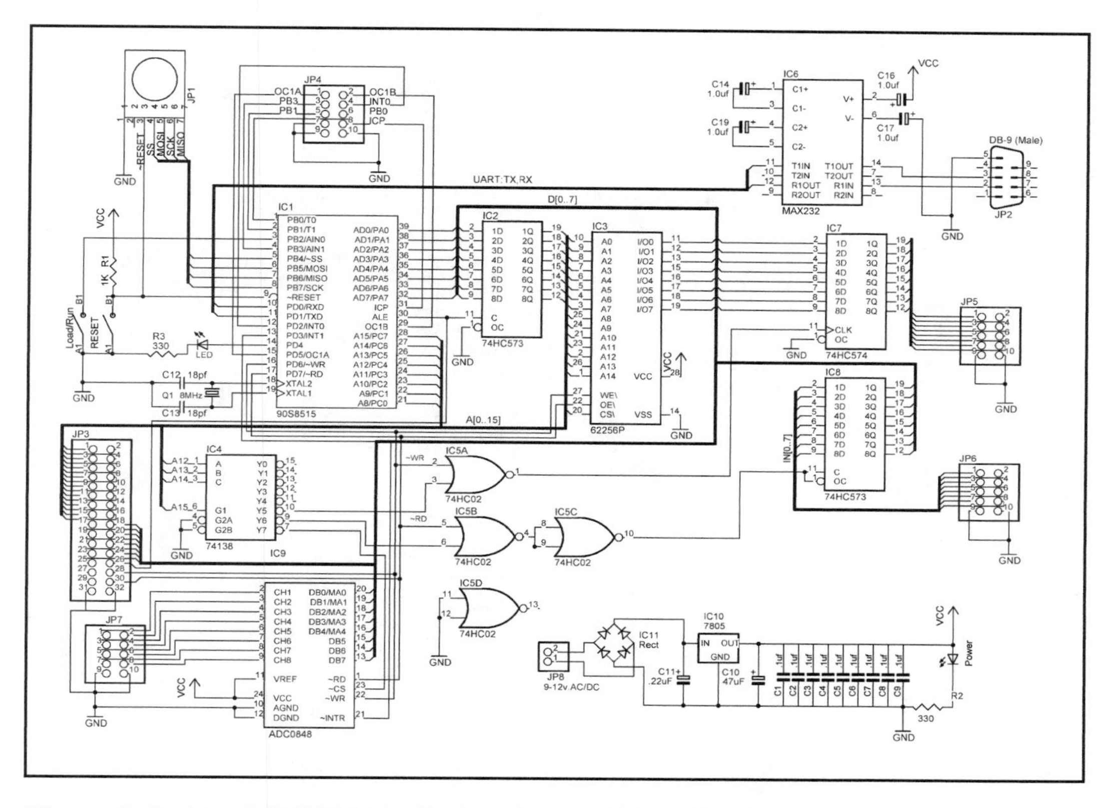

Figure 8-5 Atmel AT90S8515 Fuzbol System (*this graphic appeared as Figure 10-7 in Applied Robotics*)

Figure 8-6 is the MCU itself, and a few bits of supporting circuitry, such as the reset button and some output LEDs. Discussions of microcontrollers are best done with the official data sheets in hand (and, at 170 pages of documentation for the C8705015 MCU, I can not reproduce it here), so you may want to visit the Cygnal Web site and get a copy for yourself.

Internal to this chip is 32K of program Flash, 2K of internal data SRAM plus another 256 bytes of RAM used for the registers and stack space, and a 16MHz oscillator which is fast enough that I don't feel a need to apply an external crystal or oscillator.

The ports on the chip include eight analog inputs and thirty-two general-purpose I/O ports that have been assigned particular purposes for this board. A handful of ports are targeted at RS232 and I2C communication (though I decided to leave off the RS232 voltage-level conversion chip), four ports are dedicated pulse-width modulation (PWM)

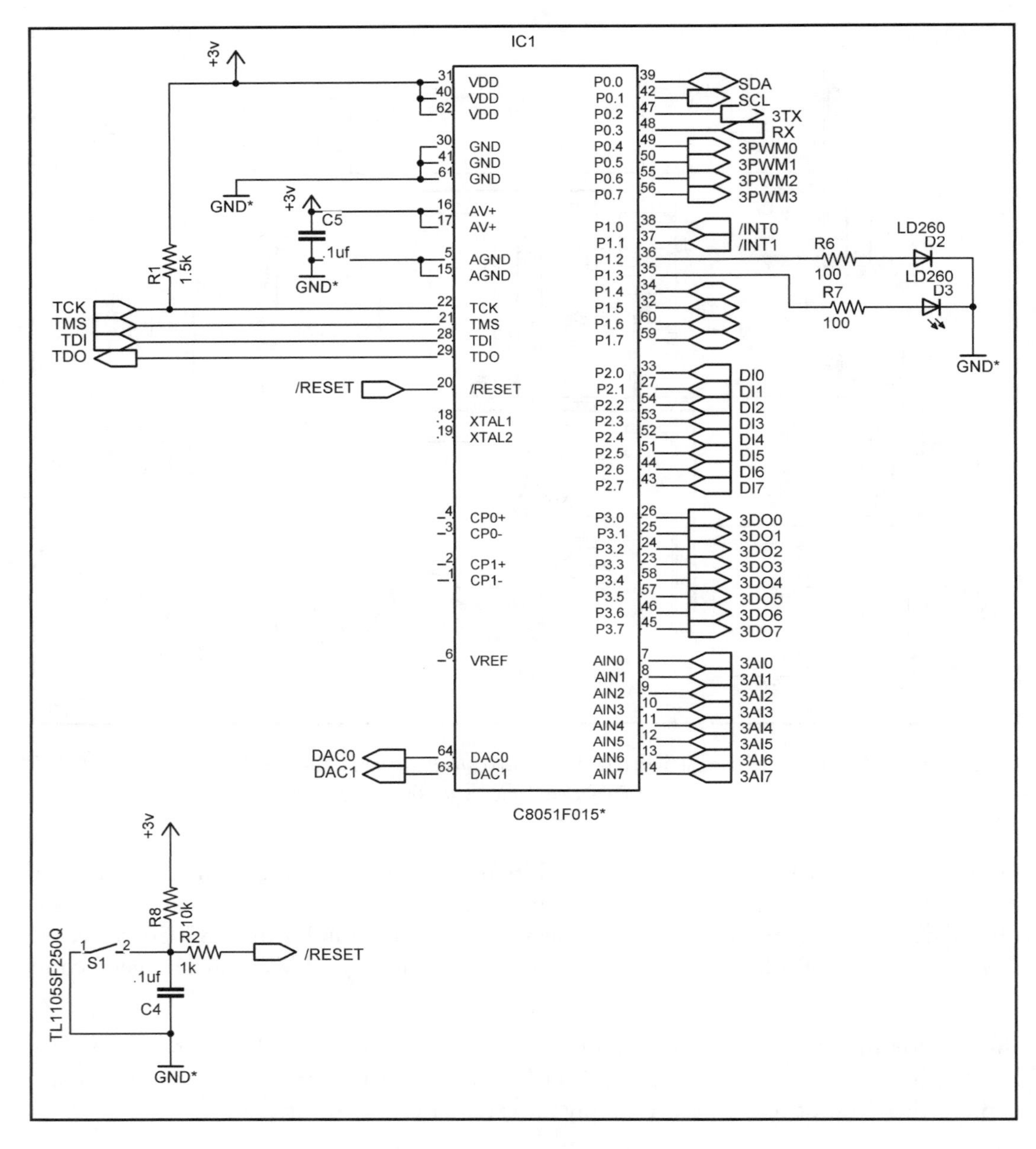

Figure 8-6 MCU (sheet 1)

outputs for motor control, and eight each for digital input and output. There are even ports left over with no particular use.

This chip costs about $15 in single quantities (though prices change over time; check for current rates), but eliminates several chips from the older design (e.g. address decoders, external SRAM, and external A/D converter). Note that I am not implying that the Cygnal chip is unique—only that it fits my needs. Several manufacturers make chips with similar capabilities, including Atmel.

Figure 8-7 illustrates some I/O buffers and the power regulator. Note that this MCU operates at 3 volts and not 5 volts, like more familiar systems. All of the I/O ports can accept full 5-volt inputs. The two 74HC541 chips convert the MCU's 3-volt output signals into more robust 5-volt levels. Though not strictly necessary for all circuits, these buffers give the board more power to talk to a wider variety of external systems.

In the middle of the schematic is an essentially unreadable blob that is a resistor bridge. This performs a passive analog conversion from external (0 to 5v) signals to the MCU's preferred 0 to 3v levels.

The final sheet (**Figure 8-8**) is simply the mechanical hookup diagram. Central to this diagram are the edgeboard fingers. Around it are arrayed a variety of jumpers that allow the user to customize the outputs of the MCU to a certain extent. Finally, there are the network, RS232, and programming connector, all lumped together to minimize the board footprint and component count.

Techniques for programming this controller are described later.

I/O BOARDS

Though all of the relevant signals from the MCU and its peripherals are led to the edge of the board and, hence, to the backplane, there are no connections to the outside world. One or more additional boards can be called upon to make these inputs and outputs available for external hardware.

I've found that in robotics, the most annoying aspect of the hardware is the connectors. Most MCU boards throw up their hands and just provide a bunch of pin-headers (those stiff wires we place jumpers on). I don't like these. Their two advantages are large—they are inexpensive, and they don't take up much board space. Their disadvantages mostly lie in the crimp-connectors you must build to hook up to them.

For many situations I prefer the ancient RCA connectors—those round metal plugs you find at the back of your TV, stereo, and DVD players for analog audio and video signals.

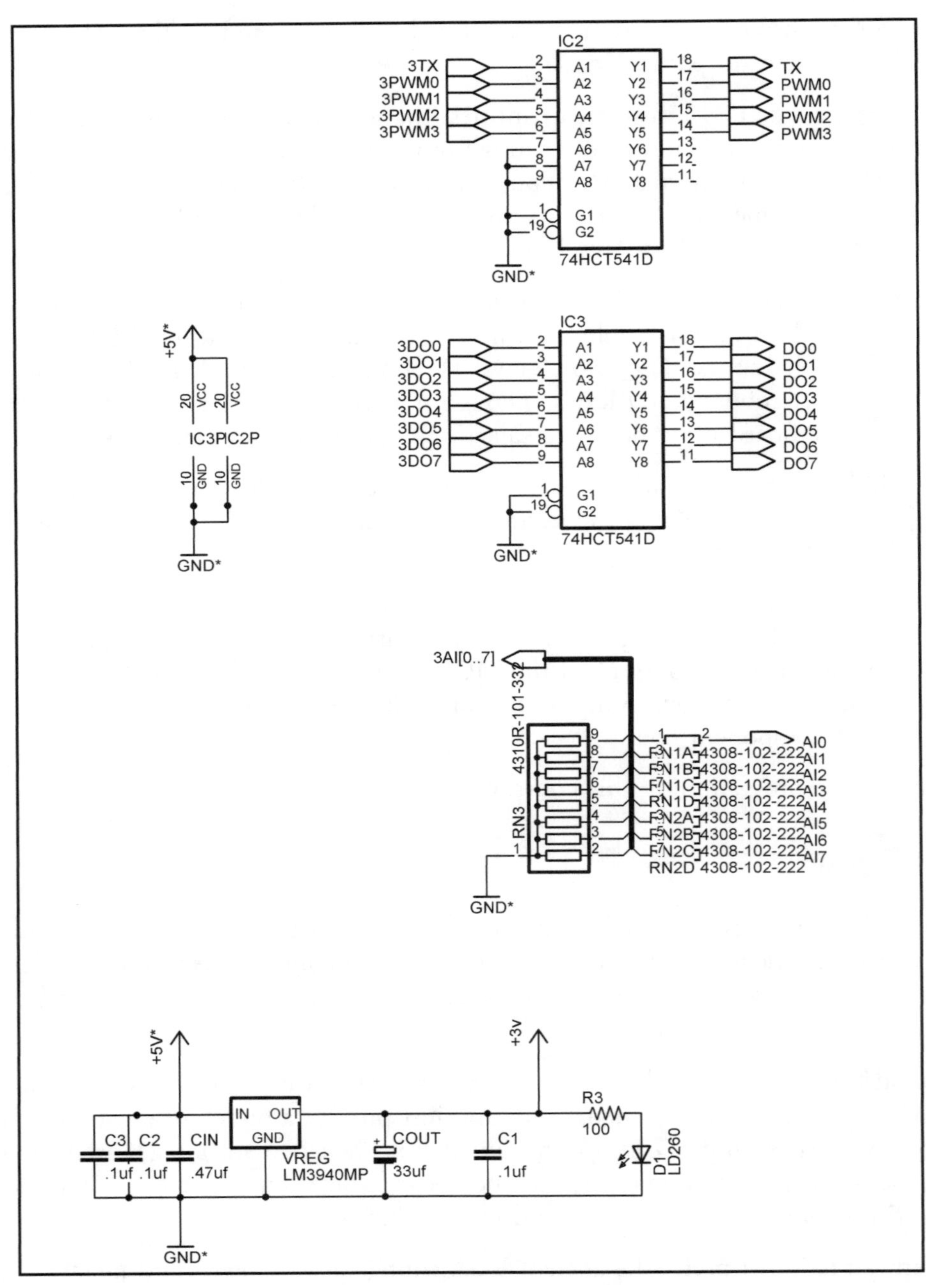

Figure 8-7 MCU (sheet 2)

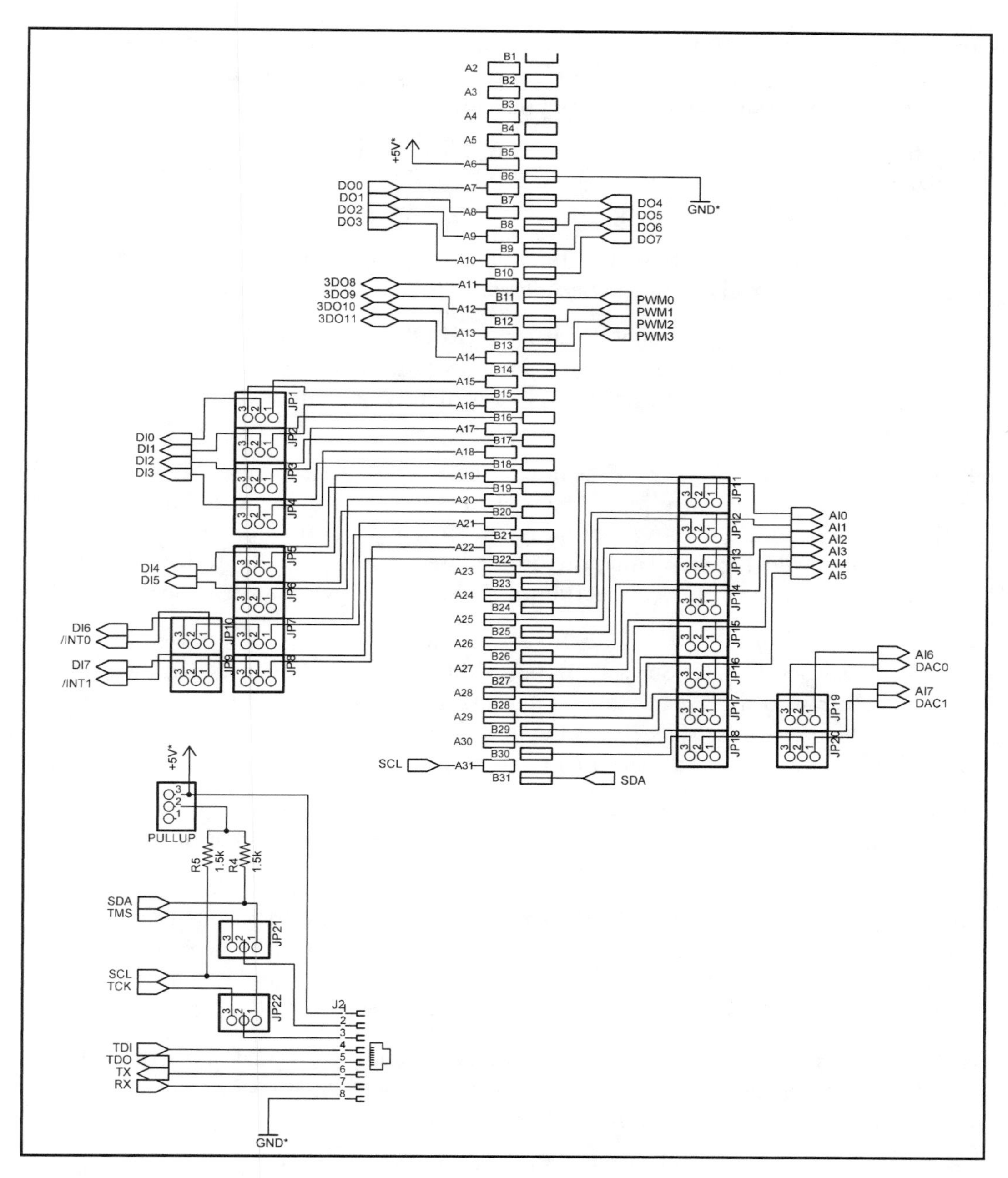

Figure 8-8 MCU (sheet 3)

Other choices for connectors include RJ11 (which your telephone uses) or RJ45 (often used for Ethernet) connectors, large (1/4") and small (1/8") stereo or mono audio plugs, and a variety of high-density computer connectors.

Figure 8-9 illustrates one form of output board. This circuit takes digital signals from the backplane and uses them to drive logic-level MOSFETs connected to Kycon's KLP42-series stacked RCA plugs.

The left edge of the circuit simply lists the call-outs that come in from the second sheet; these are the input signals from the backplane.

The MOSFETs are International Rectifier's IRLD024 logic-level MOSFET. These switch the line from the center of the RCA plug to ground. The outside of the RCA plugs are common between the stacked pairs, which is unfortunate, and hence receive the positive voltage.

Below the MOSFETs you can see a power jack. This plug is clever; it switches its output based on whether a plug is inserted or not. The inner pin is ground and is simply used to pair an incoming ground to the system ground. The outer connection, when it is not engaged, rests against the unregulated power provided to the backplane. When a plug is inserted this connection is broken and power is then sourced by the external supply.

The jack I used here is CUI Stack's PJ-102A 2mm power jack, which mates with most common wall transformers. The catch is to make sure your power source has ground in the center and positive outside, backward from most supplies. A power diode is placed inline with the output of the jack just in case things get reversed.

Figure 8-10 simply illustrates the relationship of the control signals to the edge of the card. Note the jumpers, provided to allow this card to receive input from either the A or B octet of signals.

Two of these output cards could, in theory, be put in the same backplane to manage all possible 16 digital outputs. Unfortunately for this theory the stacked RCA jacks take up two boards worth of space, making this arrangement impractical for this small backplane. Of course, a longer backplane can be made, to fit a larger box (or no box).

Figure 8-11 illustrates the connections to the backplane we are making with an input board (reversing our usual order of illustration for a moment). We will be receiving four analog and five digital signals with this board.

The four digital inputs (**Figure 8-12**), plus a pushbutton for manual access, have a few discrete components attached. These filter the inputs, reducing spurious noise, and buffer the inputs, reducing the chances of shorts and overloads. When unconnected, the digital

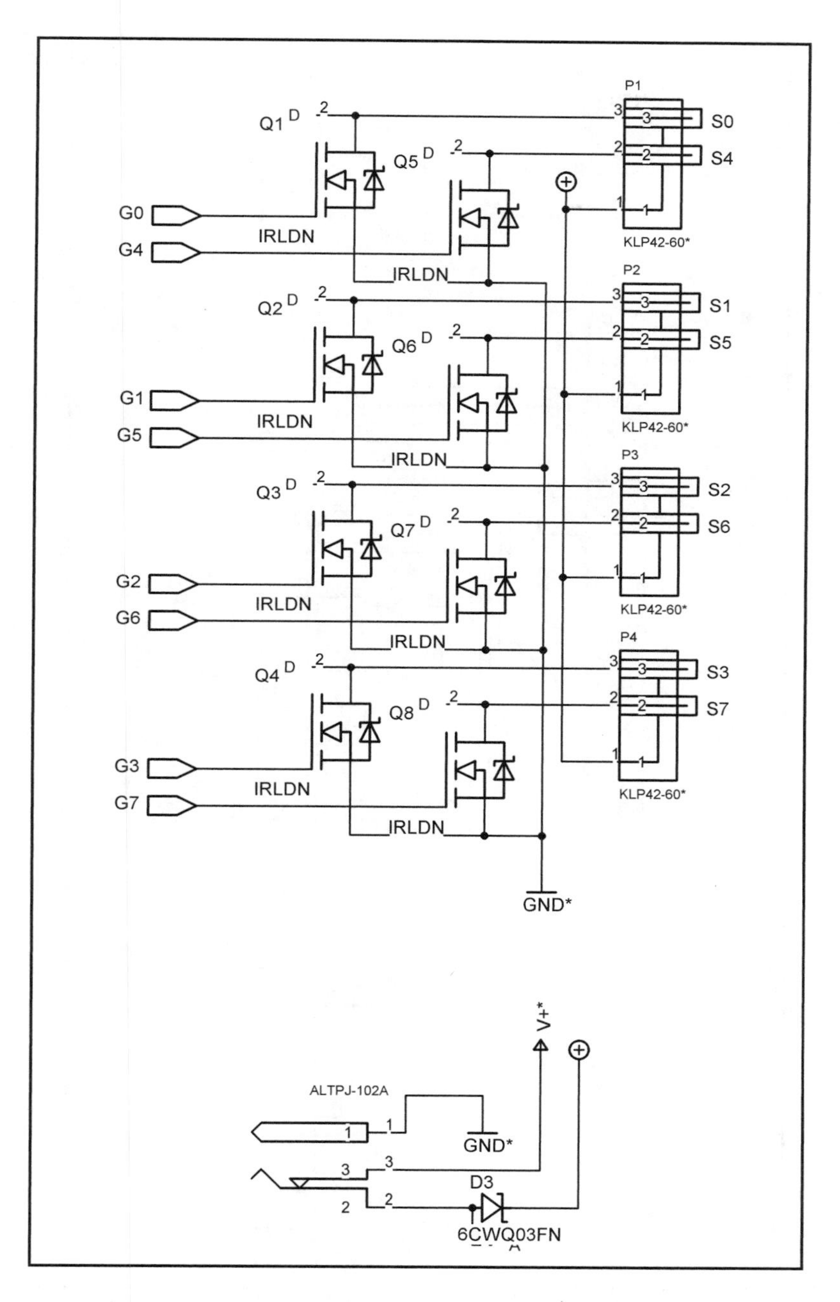

Figure 8-9 MOSFET output board (sheet 1)

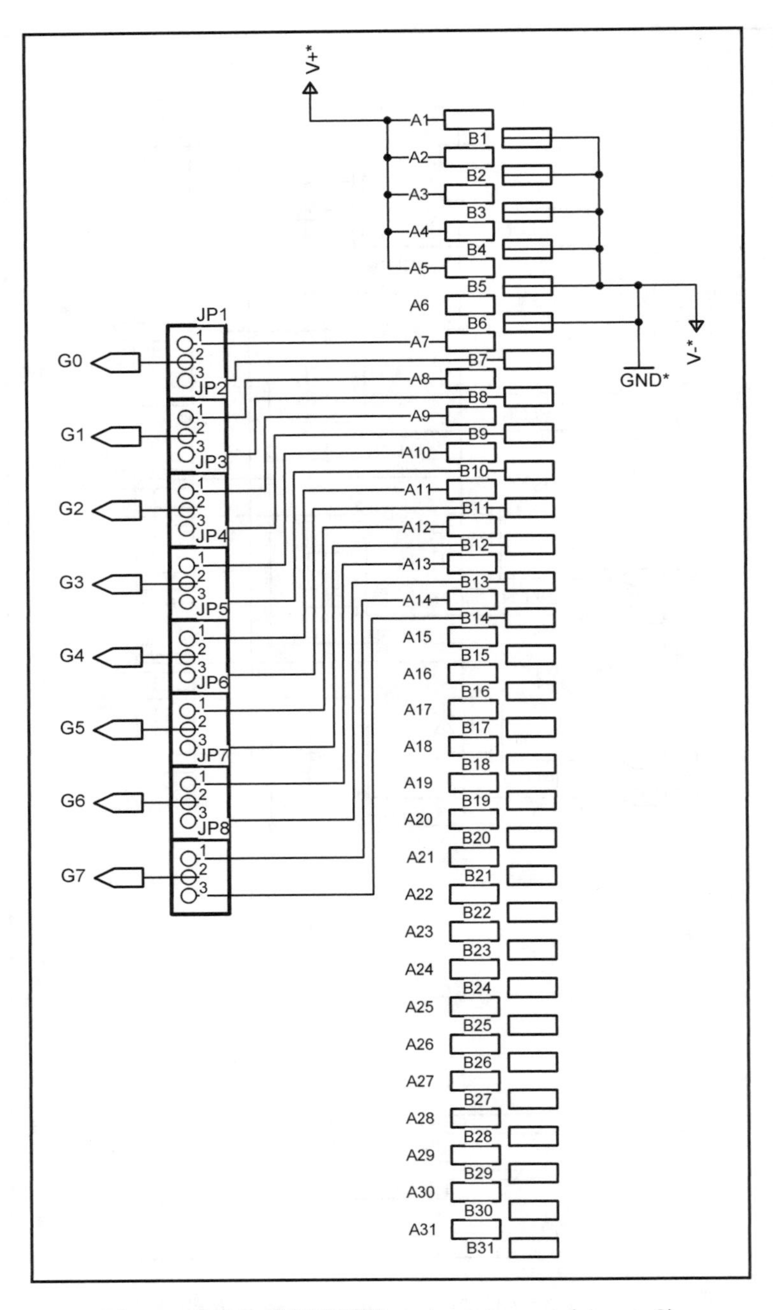

Figure 8-10 MOSFET output board (sheet 2)

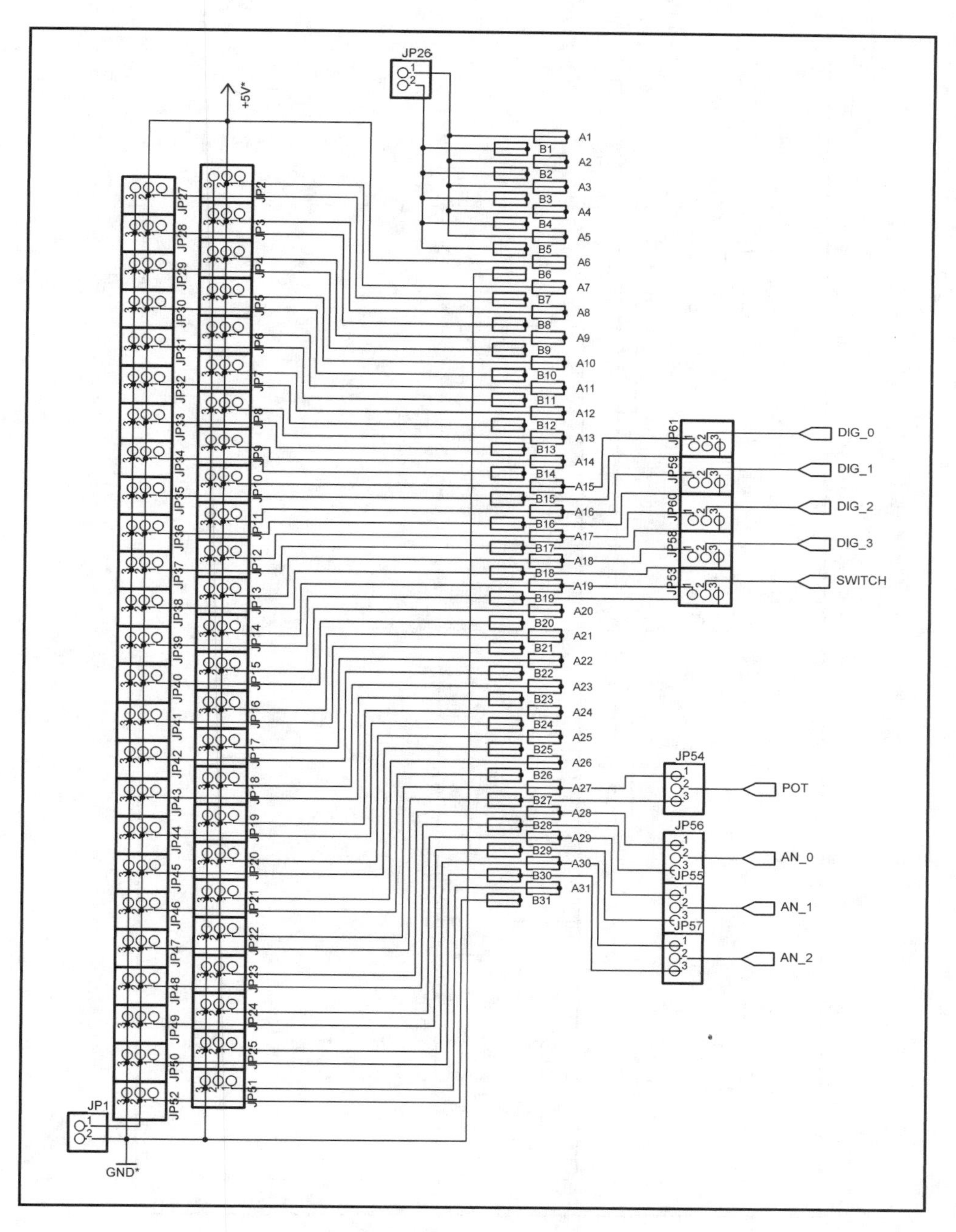

Figure 8-11 Input board (sheet 1)

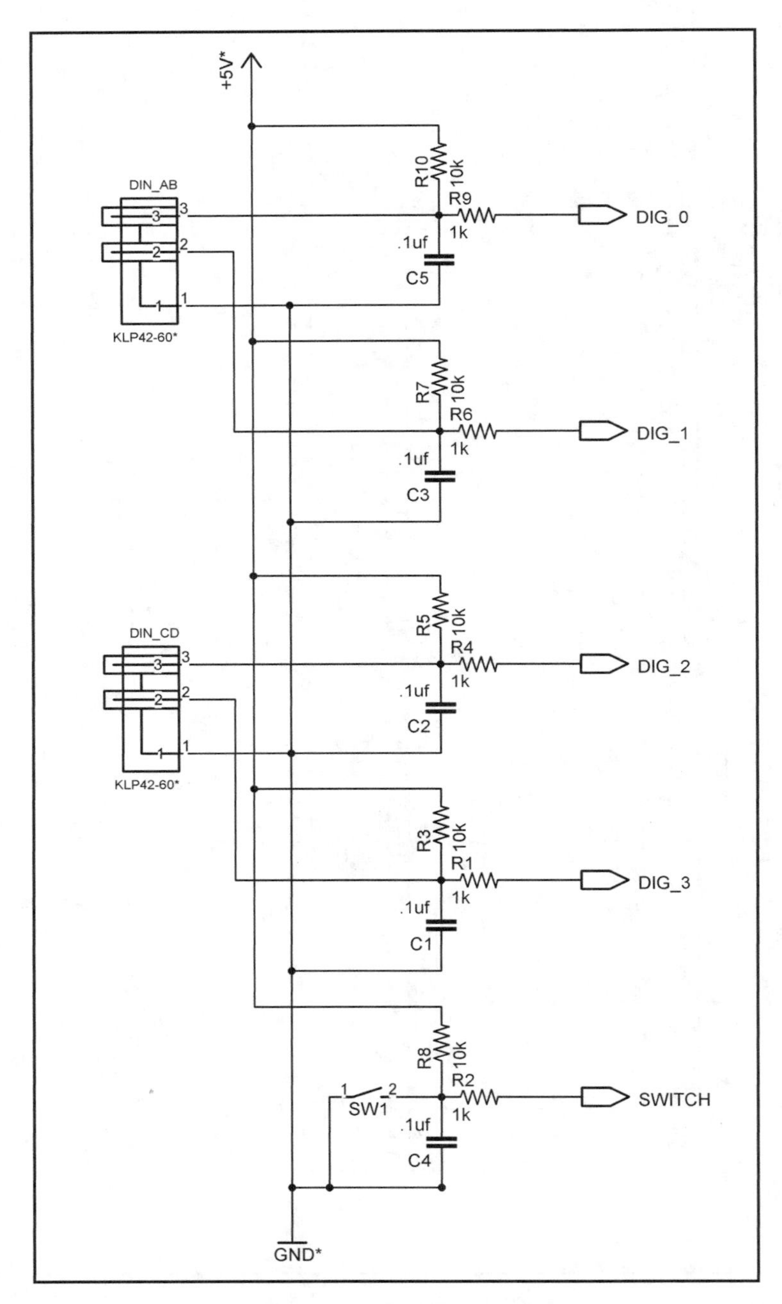

Figure 8-12 Input board (sheet 2)

inputs are pulled up to 5 volts. The intention is to have switches plugged into the jacks that will short this float down to ground.

These switch inputs, like the digital outputs, use stacked RCA jacks that, while providing an efficient footprint, do take up a second board's width on the backplane.

The four analog inputs (**Figure 8-13**) also do more than simply capture a signal. Instead of being filtered like the digital inputs (which we *could* do if we wanted to), these stereo audio jacks provide ground and 5 volts to their peripheral. Power is supplied because most analog devices need a power source to function; even the simple potentiometer. As such, the analog inputs are assumed to be attached to active devices.

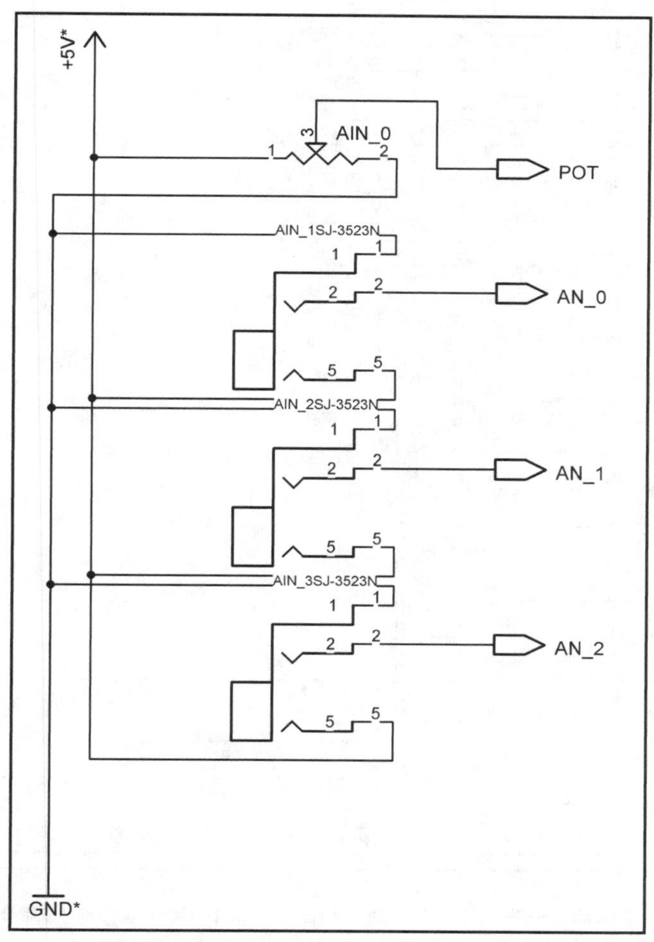

Figure 8-13 Input board (sheet 3)

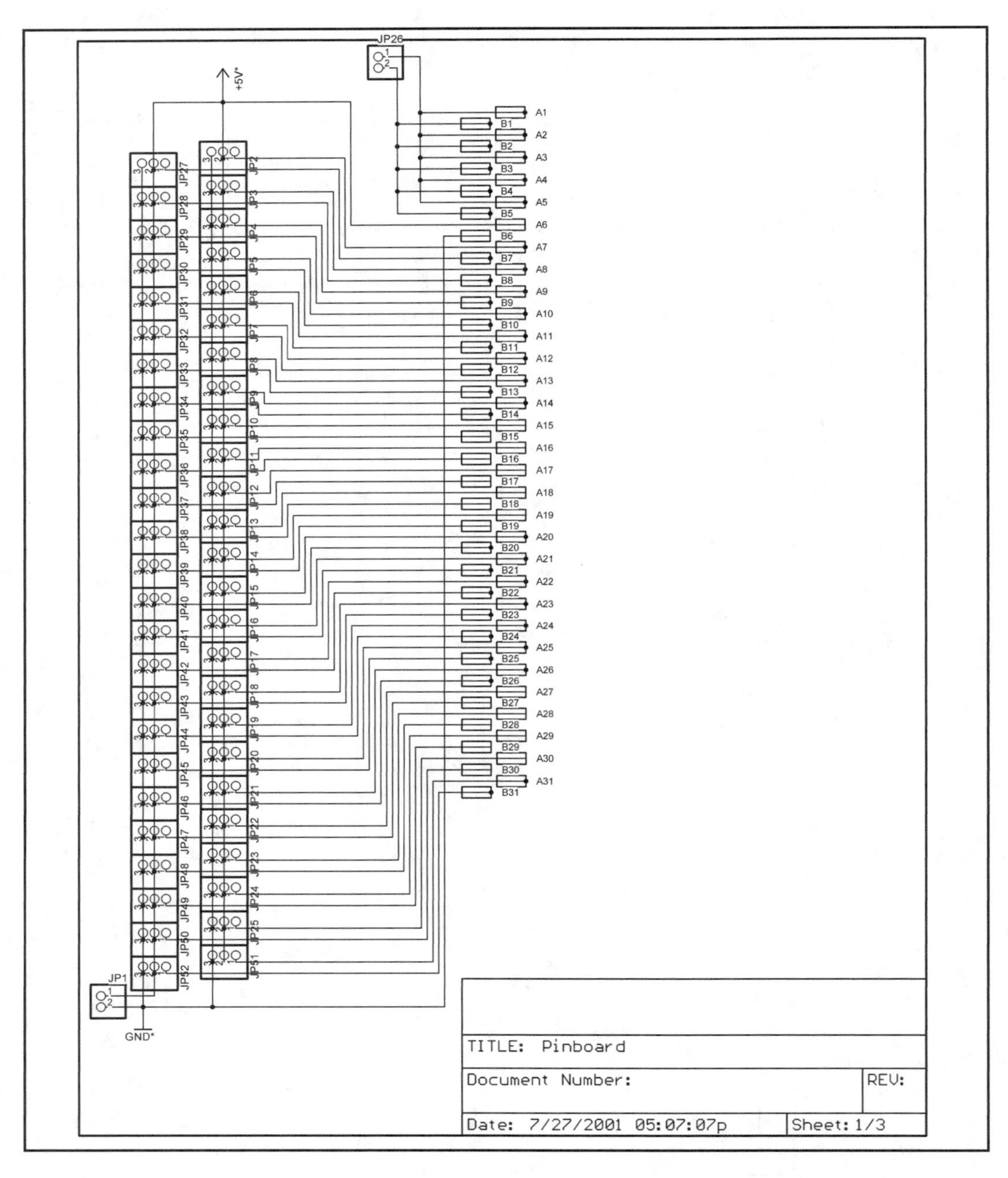

Figure 8-14 Backplane exports through pin-headers

There are many options when it comes to additional I/O boards, including a simple pin-header board where all of the backplane lines are exported for random use. A simple illustration of this is in **Figure 8-14**.

In summary, **Table 8-3** lists the various control system components and which signals in the backplane they interface with. When designing a system, this type of global checklist is necessary to verify that everything will work together.

Pin	Signal Name	Backplane	MCU	MOSFET	Input
A1	Unregulated Plus	Out		In	
A2	= A1	Out		In	
A3	= A1	Out		In	
A4	= A1	Out		In	
A5	= A1	Out		In	
A6	+5V	Out	In		In
A7	Digital Out (DO) 0		Out	(In)	
A8	DO 1		Out	(In)	
A9	DO 2		Out	(In)	
A10	DO 3		Out	(In)	
A11	DO 4		I/O 3v	(In)	
A12	DO 5		I/O 3v	(In)	
A13	DO 6		I/O 3v	(In)	
A14	DO 7		I/O 3v	(In)	
A15	Digital In (DI) 0		(In)		(Out)
A16	DI 1		(In)		(Out)
A17	DI 2		(In)		(Out)
A18	DI 3		(In)		(Out)
A19	DI 4		(In)		(Out)
A20	DI 5		(In)		
A21	DI 6		(In)		
A22	DI 7		(In)		
A23	Analog In (AI) 0		(In)		
A24	AI 1		(In)		
A25	AI 2		(In)		
A26	AI 3		(In)		
A27	AI 4		(In)		(Out)
A28	AI 5		(In)		(Out)
A29	AI 6		(In)(Out)		(Out)
A30	AI 7		(In)(Out)		(Out)
A31	SCL		I/O		

Table 8-3 Various signals

(Continued on next page)

Pin	Signal Name	Backplane	MCU	MOSFET	Input
B1	Unregulated Ground	Out		In	
B2	= B1	Out		In	
B3	= B1	Out		In	
B4	= B1	Out		In	
B5	= B1	Out		In	
B6	Ground	Out	In	In	In
B7	Digital Out (DO) 8		Out	(In)	
B8	DO 9		Out	(In)	
B9	DO 10		Out	(In)	
B10	DO 11		Out	(In)	
B11	DO 12		Out	(In)	
B12	DO 13		Out	(In)	
B13	DO 14		Out	(In)	
B14	DO 15		Out	(In)	
B15	Digital In (DI) 8		(In)		(Out)
B16	DI 9		(In)		(Out)
B17	DI 10		(In)		(Out)
B18	DI 11		(In)		(Out)
B19	DI 12		(In)		(Out)
B20	DI 13		(In)		
B21	DI 14		(In)		
B22	DI 15		(In)		
B23	Analog In (AI) 8		(In)		
B24	AI 9		(In)		
B25	AI 10		(In)		
B26	AI 11		(In)		
B27	AI 12		(In)		(Out)
B28	AI 13		(In)		(Out)
B29	AI 14		(In)(Out)		(Out)
B30	AI 15		(In)(Out)		(Out)
B31	SDA		I/O		

Table 8-3 Various signals

The first column in Table 8-3 is the backplane pin number, with the signal name next. Note that signals are named with reference to the master (as mentioned in the description for Table 8-1 and Table 8-2), so "out" is Master Out, Slave In. Finally, each board, including the backplane, is listed, with reference to whether it receives a signal (in) or sends it (out).

References in parenthesis are jumper controlled—typically choosing between the A and B versions of the signal.

A few signals from the MCU are marked "3v"—these are unbuffered connections to the MCU's ports, and as such, are optimized to drive 3-volt devices.

In general, refer to the specific device's schematic for more information on how it relates to its signals.

Note that the backplane drives only a few power lines, and it has no *active* connection to the other lines.

MCU Programmer and Interfaces

Perhaps the easiest way to program the Cygnal-based MCU controller is to use the $65 programmer from Cygnal (the so-called "Serial Adapter" that consists of a couple of interface jacks, a Cygnal MCU, and some interface logic. I suggest it is far more than a mere adapter). This programmer comes with a $99 development kit that also includes a $50 target board with an MCU on it, making it a bargain!

Using the programmer is simple, though you will need to either add a 10-pin header on the MCU board to accept the interface or create an adapter to the RJ45 jack currently designed into the MCU board.

Since I like to do things the hard way, I adapted Cygnal's cable to my RJ45 interface.

This is easy to do—simply cut the cable and cut the end off an RJ45 cable. You can also be less destructive and build up a cable that has a 2x5 pin-header on one end and an RJ45 plug on the other. The wires are then soldered together so that the relevant signals connect.

You will note from **Table 8-4** that the Cygnal interface has a number of unused wires—you can, of course, ignore these. You also need to ignore the power line—the programmer expects 3.3 volts, and the MCU board provides 5 volts. The programmer is otherwise designed to drive the IEEE 1149.1 JTAG interface ports.

1 (red wire)	3.3 Volts	6	TDO
2	Ground	7	TDI
3	N/C	8	N/C
4	TCK	9	N/C
5	TMS	10	N/C

Table 8-4 Cygnal programmer pinout

As you can also see from **Table 8-5** (based on **Figure 8-8**), there are a variety of I/O ports available on the RJ45 jack. In particular, two of the JTAG ports are jumper-selectable to represent the SMBus (e.g. I2C serial bus). For programming to work, clearly, the jumpers must be set to pass the JTAG signals instead.

1	+5V	5	TDO
2	SDA/TMS	6	TX
3	SCL/TCK	7	RX
4	TDI	8	Ground

Table 8-5 MCU board RJ45 pinout

The wires that need to be connected are listed in **Table 8-6**.

Cygnal	(name)	MCU Board
1	3.3V	N/C
2	Ground	8
3	N/C	N/C
4	TCK	3
5	TMS	2
6	TDO	5
7	TDI	4
8	N/C	N/C
9	N/C	N/C
10	N/C	N/C

Table 8-6 Cygnal programmer to MCU board crossover

There is an advantage to using a custom programming interface—you can program the MCU and then communicate to it using the RS232 serial port without changing the plug. Of course, the big disadvantage is the loss of the source-level emulator and debugger provided with the Cygnal environment.

The interface shown in **Figure 8-15** is all the hardware you need for the custom programming interface. It is shown implemented with a pair of MAX232 chips (e.g. DS14C232, and so forth; a fairly generic RS232 interface) though you could probably find a more exotic chip that would do it in one piece. This interface hardware is so simple you could put a hat on it and call it Forrest Gump. The inspiration for this design comes from Dave Everett. Thanks Dave!

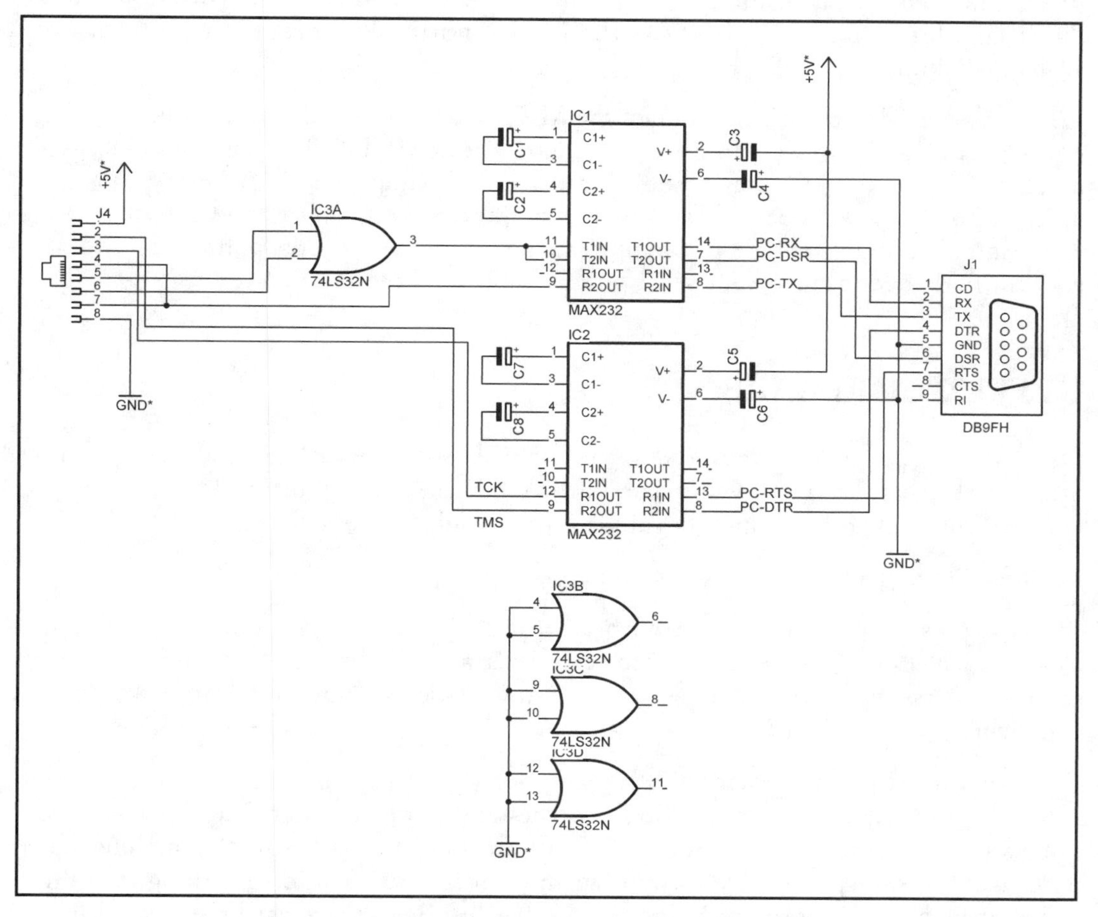

Figure 8-15 Serial interface

On the left-hand side of **Figure 8-15** is the 8-pin RJ45 plug that matches the I/O specification from Figure 8-8. The TX and TDO signals are combined by one of the four 74LS32 OR gates—because of this, you can only use one of the serial protocols (JTAG or RS232) at a time, since they will step on each other's feet.

The MAX232 chips in the middle perform the TTL to RS232 voltage level conversions.

The DB9 plug at the right hooks into your personal computer. Note that we are making use of more than just the TX and RX lines in this interface. We are using the serial port to its full capability and forcing the handshaking lines to do things they're not designed to do with the MCU.

In the "normal" serial-communications mode, the TX and RX lines communicate in the straightforward manner between the MCU and the PC, presuming you turn off all handshaking.

When you are using the PC to program the MCU via the four JTAG lines, things are a bit more complex. The software to drive this interface is not included in this book since flash programming via the JTAG protocol is not particularly simple and is of interest mostly to people who like a good puzzle. If you are one of those people, you can find the software on www.simreal.com. Even if you just want to see how to abuse the serial port until it squeaks, you can find code to drive your serial port at a low level on www.simreal.com as well.

REFLEX SOFTWARE?

Okay, this chapter may have been a bit misnamed. We have just spent a bunch of time exploring the hardware that will make up the reflexes. At this point, if the universe were fair, we would list out a bunch of software that would *be* the reflexes!

But life is cruel and this isn't going to happen.

By now you should be pounding your head against the book, crying out, "When will my robot *do something?!*" Well, stop it before you get a visible bruise and then have to face some embarrassing questions at work tomorrow. I don't know if I want that on my conscience.

First, we can't write any exciting reflex software yet because we haven't got any senses worth writing for! Sure, there are little switches and bumpers all around the robot. You can program the robot right now to drive forward until it smashes into something, then back up, turn some, and drive forward again until it smashes into something else. Preferably at high speed, until the bumpers fall off, all of your walls have been caved in, or the batteries die. That's not very elegant and I won't do it.

Anyway, reflex software is simple. Here's how it works. Each controllable aspect of the robot should consist of two elements—the actuator that moves, such as the motor on a wheel, or an arm joint, or a servo on the eye, or something; and a sensor that detects something, such as the optical encoder on the wheels, which we haven't discussed yet, or the potentiometer inside the arm joint or servo.

The software in the reflex controller has a variable somewhere that knows the value of the sensor (or, in the case of the servo, knows the value the servo has been programmed to hold—the servo itself deals with the feedback loop). Then, from somewhere, for instance, from a command sent through the serial port on the PC and read in from the serial port on the MCU, a command is sent that tells what that value *should* be.

Then the microcontroller sends control pulses to the actuator until the sensor reads the right value.

Okay, this is *grossly* oversimplified. But in truth, that is all there is to it.

There are a few complications, of course. When are there not?

For example, let's look at the optical encoders. These return a digital or analog value, explained later, that changes at a rate based on the speed the wheel is rotating. The MCU must read this changing sequence of values and from them determine *another* value that indicates the actual speed of rotation. This can be easy or hard, depending on many factors.

As another example, all mechanical systems possess a certain amount of inertia. That is, an object at rest isn't about to leap into action without fighting back a bit. Once in motion you aren't going to stop it without another fight. Because of this, if you try to move something from a dead stop to full speed, or vice versa, all at once, you have a good chance of breaking something or tipping the robot over. You need to change the motion gradually. To do this the MCU needs to ramp a motor's control signal up and down smoothly.

There are two basic techniques used to achieve this gradual change—the rather mathematical proportional-integral-derivative (PID) control formula(s) and the rather more ad hoc (yet still quite effective) fuzzy logic control. Both techniques can be used to speed up a motor slowly from zero, and slow it down gradually as it nears the commanded stop point—slowing it down *more* slowly the closer it gets, but trying not to overshoot, and if it overshoots, maybe not making a big deal out of it, depending on how you feel about such things.

Entire books have been written on both techniques, so I'm going to pass the buck here and refer you to your library. You can also do a Web search using either term and get a generous listing of sites.

Another complication may lie in the relationship of the input to the output. For example, we could have a reflex that says, "Keep the robot 6 inches away from the right-hand wall, while moving forward." The input sensor in this command might be a side-looking reflective distance sensor (such as the GP2D12 discussed later) and the output actuator may be *both* of the drive wheels. The drive wheels, in turn, have their own reflex loop that keeps the wheels turning at the rate they should be turning, regardless of bumps, inclines, battery voltage sags, and so forth. Our wall-following reflex will send commands to whatever wheel control system there is to make the robot turn slightly left or right at the same time as another control system is trying to keep the robot moving straight ahead.

At this point, we start looking at subsumption control architectures, fuzzy logic systems, and so forth—topics that don't get introduced for another couple of chapters. So let's quit while we are ahead.

CHAPTER 9

SENSE AND CONTROL: DRIVE SYSTEM

There are two elements needed to make the robot come alive—sensory input, and software to connect this input to the mechanical systems already in place.

I break away now from the project-narrative style that I prefer, where we work through projects that all tie together in one way or another. I spread before you instead a buffet of little projects that do not necessarily fit neatly into the existing work. Dim-sum for your robot.

This chapter begins with a description of what you need to drive the motor controllers from previous chapters. The companion Chapter 10 looks at a variety of input devices that could find a home on the robot.

We'll examine the hardware specifications for each device and then look at what type of MCU support might be needed for it. I include 8051 assembly code for some of the projects, implemented on the Cygnal MCU from Chapter 8. A lot of assumptions are made relating to the hardware described in Chapter 8, so you may need to flip back and forth. I would recommend putting in little colored tabs for easy access.

You also need the C8051F documentation on your desk somewhere. You can download this document from Cygnal (www.cygnal.com) as a PDF file. Print it. Use it. You may have noticed in Chapter 8 that I don't pretend to be writing a detailed tutorial on using microcontrollers. I just breezed through the MCU like it was easy or something. (It *is* easy, but only after you've made your brain sweat bullets once or twice building your first MCU. Until then, it's amazingly hard.)

This chapter continues with that blatant disregard for the new user—but with this chapter in one hand, Chapter 8 in the other hand, and the Cygnal MCU documentation held between your toes, you just *might* be able to squeeze useable knowledge out of it all.

In *Applied Robotics* I used the proprietary Fuzbol language for the programming—a controversial decision, I suppose, but great fun nonetheless. This time I am choosing to skip Fuzbol and run with raw assembly language. I apologize in advance for the horror that is assembly language—I'll intermix it with plain English descriptions as we go along so you shouldn't be presented with any long, cryptic chunks of code.

This language decision has two facets. On the lazy side, it simplifies my life in that I don't have to upgrade Fuzbol for this MCU yet...if you struggled with my implementation of Fuzbol for the Atmel AVR, you probably realize that language programming is not my central area of expertise!

On the practical side, by using the (free) assembly language system provided by Cygnal, I get to use its admirable in-circuit debugging environment.

And finally, on the third hand, since the 8051 core of the Cygnal is as old as dirt (perhaps older) more hobbyists will be able to find better support when implementing these projects. I just hope those folks who went down the Fuzbol path will eventually forgive me.

PROJECT 9-1: GENERIC PROGRAM FRAMEWORK

Okay, this is not an exciting project. The fact of the matter is, every little 8051 program in this chapter will require a certain amount of common wrapper code. This code will define the basic interrupt vectors, turn off the watchdog timer, configure the crossbar, and so forth.

To simplify all of the later chapters, we introduce this generic setup code here, now, once. This setup configures the MCU from Chapter 8 to fit into the backplane. All future programs will be extensions to this framework. If you ever get lost in these discussions, the full source code for all of these programs is available at www.simreal.com.

Note my cavalier use of assembly language without explaining it. There are good books on the subject, really, and this isn't one of them! However, for the novice, I will offer some advice.

Assembly language consists of roughly four types of instruction. First there are a group of instructions that tell the assembler what to do—the `cseg`, comments, and other various instructions that don't end up in the final code.

Then there are instructions that move data from place to place—most notably the `mov` command. There are various complications involved in moving data—where is it coming from (registers or memory or what)? Where is it going? Is there some math involved in calculating the location of the data? But mostly just moving something from one place to another.

A large number of instructions *do* things to data. A bunch of these instructions are simple things like adding or subtracting numbers. Some of them do weird computer things like shifting all of the bits left by one bit position in a register. Others do logical operations, which are just another form of math.

Finally, there are instructions that tell the MCU to start running code somewhere else. These are the jump instructions like `ljmp` (long jump), though many are branches. Branch instructions only work sometimes—they have a condition for their operation. These conditions are things like whether a particular status bit is set or not.

All languages mostly just do these four things: provide compiler control, move data, modify data using math or logic, and jump to a different bit of code. Even the fancy ones like C++ or Java.

Now, on with the show.

First, the prologue:

```
;=====================================================================
;  CYGNAL  C8051F015
;=====================================================================
$INCLUDE  (c8051F000.inc)
```

This include just sets up the various defines for this system.

```
;  ____________________
;  Data
;
   dseg
   org  0x60
   ANALOG:     ds  8
   ANALOG_IDX: ds  1
```

The data segment (`dseg`) of the code has its origin (`org`) set to the hexadecimal address 0x60. Then, eight bytes are set aside for analog input values (`ds 8`) and an additional byte is set aside for an index into these values. The data segment lives in the data memory of the MCU, which is separate from the program memory.

```
;  ____________________
;  Interrupt  vectors
;
   cseg
   org   0x00                ;  Reset
   ljmp  reset

   org   0x7b                ;  ADC  end  of  conversion
   ljmp  adc_isr
```

Each program may have a number of entries in the interrupt vector space. Interrupt vectors exist in the code segment (`cseg`) and specify the program addresses (the interrupt service routines, or ISRs) that should be called when a particular notable hardware event occurs (the interrupts).

```
;  ____________________
;     Code  space
org  0xb3
```

Finally, we prepare to write actual code. Programs should begin starting at address 0xB3 in order to be placed well after all of the possible interrupt vectors.

```
reset:
    ; Watchdog
    ;
    mov   WDTCN,  #0xDE      ; Start  disable  sequence
    mov   WDTCN,  #0xAD      ; Lock  it  in
```

The processor executes the `reset` function via the reset interrupt vector at program address 0x00. This vector is called when the processor is first turned on, and also upon other system reset events while it is running. The first thing we do in our reset is turn off the watchdog timer.

While watchdog timers are a *good* thing (they reset the MCU if too much time passes between check-ins), they add a layer of complexity that I would just as soon ignore. Read a good book on microcontrollers, or the documentation of this MCU, for more details.

```
;
; Stack  above  second  register  bank
;
mov   SP,  #0x10
```

The registers R0 through R7 normally reside at data memory locations 0x00 through 0x07. However, there are four possible sets (banks) of registers. The second bank is stored at data memory 0x08 through 0x0f. We use the second bank of registers for operations inside of interrupt service routines, and the first bank, the default, for non-interrupt code as needed.

The stack is used to preserve program addresses during function and interrupt calls. It defaults to data address 0x08, but we move it up to 0x10 so it doesn't interfere with our second bank registers.

Actual program data needs to be above the stack, at 0x60 in this case, but before 0x80 for reasons I will avoid explaining here.

```
;
; Configure the crossbar
;
mov    XBR0, #0x25; SDA, SCL, TX, RX, CEX0-CEX3
mov    XBR1, #0x14; /INT0, /INT1
mov    XBR2, #0x40      ; Enable Crossbar
```

The crossbar in the C8051F is like a telephone switchboard. It connects signals inside the various components of the MCU with physical pins that lead outside the MCU. This allows the programmer to determine which signals to use and which to ignore.

The first two commands, setting XBR0 and XBR1, connect various devices to external pins, while setting bit 6 in XBR2 enables the crossbar.

Our generic crossbar setup defines the relationships shown in **Table 9-1**.

Signal	*Port*	*Bus*
SDA	P0.0	n/a
SCL	P0.1	n/a
TX	P0.2	n/a
RX	P0.3	n/a
CEX0	P0.4	PWM0
CEX1	P0.5	PWM1
CEX2	P0.6	PWM2
CEX3	P0.7	PWM3
/INT0	P1.0	DI6/DI14
/INT1	P1.1	DI7/DI15

Table 9-1 C8051F Crossbar Configuration

The first four pins are used for communication and lead to special connectors that are off of the backplane. The last two pins are optionally connected to the backplane, replacing P2.6 and P2.7, and may be routed to DI6 and DI7, or DI14 and DI15 as desired.

All of the remaining port pins, P1.2 through P1.7 and all of ports two and three are left as general purpose I/O (GPIO) to be used as we see fit.

```
;
; Set port directions
;
mov   PRT0CF, #0xF4 ; SDA+SCL in, TX out, RX in, CEX0-3 out
mov   PRT1CF, #0xFC ; /INT0,1 in, 2-7 out
```

```
mov   PRT2CF,  #0x00  ;  0..7  in
mov   PRT3CF,  #0xFF  ;  0..7  out
```

Separately from the crossbar, we need to tell each port which direction they communicate; sending or receiving. A '1' at a given bit sets a port to output; a '0' is an input. Ports default to input.

The above code sets port zero as needed by its communication tasks; port one has the two low bits, the interrupts, as inputs with the remaining as output; port two is entirely an input; and port three entirely an output.

```
;
;  DAC
;
mov   DAC0CN,  #0x80     ;  DAC0  enabled,  H[3:0]  L[7:0]
mov   DAC1CN,  #0x80     ;  DAC1  enabled;  H[3:0]  L[7:0]
```

Though we don't use it in this book, we enable the digital-to-analog converters.

```
;
;  ADC
;
mov   AMX0CF,  #0x00     ;  ADC  all  single-ended
mov   ADC0CF,  #0x80     ;  ADC  SAR  every  16  SCL
mov   REF0CN,  #0x07     ;  A/D  internal  reference  &  bias
;  temperature  sensor  on
mov   ADC0CN,  #0x85     ;  ADC  on,  T3,  left  shifted  results
```

We definitely use the analog-to-digital converter. It is configured to trigger on overflows, when the timer reaches its largest value and resets back to zero, from Timer 3. This automates the reading of the analog inputs and frees the application from this concern. It is also possible to start the analog conversion from other sources, including Timer 2 overflows.

When the conversion is complete it triggers an interrupt so that our software can retrieve the result.

```
;
;  Timer  3  (for  ADC)
;
mov         TMR3CN,  #0x04     ;  T3  enabled,  SCL/12
mov         TMR3RLL,  #0xFA    ;  400  int/sec
mov         TMR3L,  #0xFA
mov         TMR3RLH,  #0xF2
mov         TMR3H,  #0xF2
```

This code turns on Timer 3 and sets it to get a single "tick" for every twelve system clock cycles. Then, the counter and reload words are set to 0xF2FA. With a maximum timer value of 0xFFFF, the timer will overflow once every

```
0xFFFF - 0xF2FA = 0x0D05 = 3,333
(3,333+1) * 12 = 40,008
```

system cycles. With a 16MHz system clock, that is 16,000,000 / 40,008 or about 400 overflows per second. This is easily adjusted by using a different set of numbers.

```
;
; Resets
;
mov   RSTSRC, #0x00      ; All cleared
;
; Kick it up a notch
;
mov   OSCXCN, #0x00      ; No external clock
mov   OSCICN, #0x07      ; Fast internal clock
;
; Interrupts
;
mov   EIE2, #0x02        ; Enable ADC interrupt
setb  EA                 ; Enable interrupts, in general
```

This is the last gasp of the reset setup. It clears any untidy reset sources (RSTSRC), enables the internal clock source (saving us the hassle and expense of crystals or oscillators) at its maximum speed of 16MHz, and then activates the relevant interrupts.

Note that for some applications it may be necessary to activate other interrupts—but we'll burn that bridge when we get to it.

Now let's look at our bare-bones "main" application. This will be heavily modified for later programs...mostly by the insertion of code in one of the "Insert Code Here" spots.

```
; ______________________
main:
    mov   ANALOG_IDX, #0x00
    mov   AMX0SL, #0x00     ; Select ADC register 0
    mov   ADC0CF, #0x81     ; ADC SAR every 16 SCL; scale 2

;
; Insert Setup Code Here
;
main_loop:
;
; Insert Application Code Here
;
    sjmp  main_loop
```

Very boring main program. It turns on the ADC converter, setting it to start reading the first analog input. Then it dies by going into an infinite loop.

All of the work in this naked shell occurs in an interrupt service routine—the analog-to-digital converter ISR:

```
adc_isr:
   ;
   ; capture this conversion
;
   setb RS0   ; Set to register bank 1 for ISR
   mov  R8, A
```

Note how the first thing this routine does is move the register bank and then preserve the A register. This prevents the interrupt routine from damaging the register data of any other routine. Note that ISRs cannot themselves be interrupted. Only when they return can interrupts occur again.

Note that we are *not* preserving the state flags (negative, zero, carry, etc). These flags would need to be preserved, stored in a register or something, for more complicated applications. For this preservation we need to use the stack, with code much like:

```
push     PSW
push     ACC
push     DPH
push     DPL
push     B
```

At the end of the ISR these three bytes would be restored from the stack with:

```
pop      B
pop      DPL
pop      DPH
pop      ACC
pop      PSW
```

But since we don't have any code in the main loop to be disrupted by our ISRs, it is safe to leave them with just token protection as a placeholder for better protection when the need arises.

```
   mov  A, #ANALOG
   add  A, ANALOG_IDX
   mov  R0, A
   mov  @R0, ADC0H
```

Here is where we retrieve the current analog value. The first three lines calculate the address in the analog storage array where we want to keep the value. The last line takes the value from the analog register and stuffs it into that address. The addressing form

@R0 means "The address named in register R0". Using this move instruction we can only store data into the 256 bytes of RAM—which is also shared by the stack and the registers! Very limiting. We could also put data in the larger 2048-byte RAM space using movx, but not today.

```
;
;     setup for the next conversion
;
inc   ANALOG_IDX
anl   ANALOG_IDX, #0x07
mov   AMX0SL, ANALOG_IDX
clr   ADCINT
```

Here we increment the index into the analog array, and make sure it doesn't grow out of bounds using the logical-AND anl instruction. Then we stuff that index, which is also, by design, the analog input number, into the control register and start the conversion process all over again.

```
adcisr_ret:
   mov   A, R8
   clr   RS0    ; Set back to register bank 0
   reti
```

Finally, we do the interrupt return. This turns interrupts back on and returns execution to where it left off in the main program, in this case, the infinite loop.

And, as required by the assembler, the end directive:

```
END
```

Well, I don't know about you, but I'm glad that's over with. Let's get on to the good stuff.

PROJECT 9-2: R/C MOTOR DRIVER CONTROL

While PWM is a built-in function of our hardware, the radio-control signal is not something that comes naturally to most MCUs. To synthesize this signal, actually, *eight* of these signals at once, we need to make clever use of our interrupts.

The outputs will go out P3, which is connected to the bus at DO0...DO3 and DO7...DO11. For the test, one of the signals will be based on the same potentiometer input we used in the PWM driver.

Figure 9-1 shows the form of the radio-control signal that we are trying to create. In broad terms, the period of the signal can be anything between 20mS and 40mS. The minimum

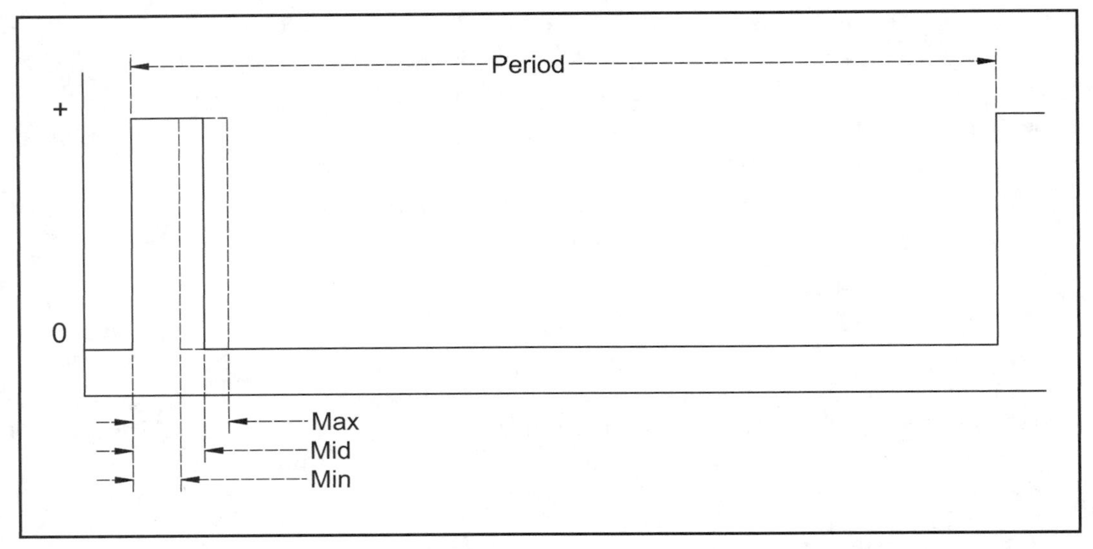

Figure 9-1 R/C control signal

pulse duration is typically described as being 1mS, middle at 1.5mS, and the maximum duration at 2.0mS. This is a 5-volt control signal (e.g. the white wire) to the servo or speed controller. Power for the servo or ESC is taken from two other wires, the black and red ones in most servos.

To be specific, the Novak ESCs I have hooked to the many motors in this robot use a pulse-width range of 0.85mS, 1.35mS, and 1.85mS. These ESCs are smart, too—they can adapt at the touch of a button (well, several touches) to the actual signal you give them. The bare minimum pulse width that they can cope with, according to their technicians, is 0.7mS, and the maximum is 3.0mS.

Depending on which standard you want to run the pulse to, we need to divide up a pulse with a variation of 1.0mS (2.0 - 1.0) to 2.3mS (3.0 - 0.7) in width.

Assuming we want to slice each pulse into 256 different sizes, for a control signal from 0x00 through 0xff, we need to control time down to 3.9µS (1mS / 255) or 9.0µS (2.3mS / 255). A control value of 0 will create a pulse of the minimum width, and a control value of 255 will create a pulse of the maximum width.

The gist of this code is to create a table of timer delays that is used to generate eight servo pulses, one on each line of the output port. Each entry in the table specifies the time until the next "event" in the ISR. The table is used to set the count of the timer that drives the ISR, and this count is updated from the table at each interrupt. At the beginning of each period, the table is recalculated.

There are two ways we can synthesize this signal, as illustrated in Figure 4-7, which showed chained pulses and spaced pulses. In the chained pulses method, the end of one pulse signifies the beginning of another pulse on the next port in the chain. Then, after all eight pulses, we have a ninth period of time that is the delay that fills out the period. In the spaced pulses method, there is a space between each pulse and the sum of all the pulses plus their spaces conveniently fills the period.

Both methods have something to recommend them; however, we will use the chained pulses method since it is somewhat more efficient.

Starting from the generic framework code, let's see what this looks like.

R/C SETUP

```
; ________________________
;         Constants
PERIOD   EQU (0xffff  -  0x9c40)
PULSE_MIN      EQU 0x03a5 ; or  0x535
PULSE_RES      EQU 0x0c   ; or  0x05
SERVO          EQU P3
POT            EQU (ANALOG+4)
```

Though the SERVO and POT constants are old-hat by now, the other magic numbers bear further explanation.

Assuming a 16MHz (16,000,000 cycles per second) clock and 12 cycles per timer tick, we define three important timer values in terms of timer ticks.

The PERIOD constant defines the number of ticks needed to create a 30mS delay. 16,000,000 / 12 gives about 1,333,333 ticks per second, or 0.75µS per tick. 30mS is 30,000µS; divided by 0.75 gives 40,000 ticks per 30mS. This explains the 0x9c40 value.

The timer counts up to 0xffff and then triggers an overflow interrupt at the next tick. Since we want 0x9c40 ticks before the overflow, we need to subtract that value from 0xffff. This gives us the value to put into the timer's counter.

PULSE_MIN is calculated the same way. Given a 0.7mS (700µS) minimum pulse, at 0.75µS per tick, that's 933 (0x3a5) ticks. The optional 0x535 described in the comment is for a 1.0mS minimum.

PULSE_RES is the resolution of the pulse. This is multiplied by a control value between 0x00 and 0xff and added to the minimum to get the complete pulse width. With a pulse variation of 2.3mS (2,300µS) divided by 255 (the maximum control value), we get a control resolution of 9µS or 12 ticks.

```
org 0x50     ; direct-addressed data above stack, below 0x80
COMMAND:     ds 8
SERVO_TIME: ds 16 ; L:H * 8
SERVO_IDX:  ds 1
PORTBIT:    ds 1
```

The data origin has been pushed down to 0x50 to allow for the extra data used in this version. In addition to the framework's ANALOG and ANALOG_IDX data, we create a few data slots for the servo code.

The COMMAND is just an array of servo positions. Each entry in this array corresponds to one output port. To change the position of a servo, you simply change its value in the command array.

The SERVO_TIME array allows two bytes (low and high halves of a word) per servo to hold the number of ticks that the servo's pulse should remain active.

SERVO_IDX is just the index into the time array and PORTBIT is used to keep track of which bit in the port is currently active (a shortcut to avoid having to calculate this each pass).

```
org   0x0b                    ; T0 overflow
ljmp  update_isr
org   0x1b                    ; T1 overflow
ljmp  rc_isr
```

We put entries in the interrupt vector section for both timers 0 and 1. T0 manages the test interface, and T1 the servo timing itself.

```
;
; Timer 0 (for update)
; Timer 1 (for R/C synthesis)
;
mov   TMOD, #0x10        ; T0 Mode 0 (13-bit) + T1 Mode
1 (16-bit)
mov   CKCON,#0x00        ; T0 through T2 use SCL/12
mov   TCON, #0x50        ; T0+T1 enabled
```

Down into the reset code, we need to set the mode for both timers. T0 is set to the same mode as described in the PWM section. T1 is set to a full-size 16-bit timer that is incremented once every 12 system clock cycles.

```
mov   IE,          #0x0A        ; Enable T0 & T1 interrupt
```

In the interrupts section we enable both of the timer interrupts.

```
mov   A, #0xff
mov   (COMMAND+0), A
```

```
    mov   (COMMAND+1), A
    mov   (COMMAND+2), A
    mov   (COMMAND+3), A
    mov   (COMMAND+4), A
    mov   (COMMAND+5), A
    mov   (COMMAND+6), A
    mov   (COMMAND+7), A
    acall rci_calculate         ; Fill the timer arrays
```

Just prior to the main loop, and just after the ADC system is initialized in main, we indulge in a bit of optional command initialization. You can set the various command slots to different values to exercise the code, if you wish.

After the initialization of the commands, we call the `rci_calculate` function to calculate the timer values for these commands.

Update ISR

Timer 0 drives an update service routine that can be used to manage any inputs. These inputs, in turn, can drive the various command values for the R/C synthesis.

```
update_isr:
    ; Adjust the first command value
    ;
    mov   A, POT
    mov   COMMAND, A
    reti
```

As you can see here, our test ISR is pitiful in its simplicity. We read the potentiometer and stuff its value directly into the first command slot. This should be sufficient to test the synthesis code.

R/C ISR

The code that manages the R/C signal synthesis is divided into two separate functions. This ISR code is called from the Timer 1 overflow interrupt.

```
rc_isr:
    setb  RS0          ; Set to register bank 1 for ISR
    ;-
    mov   A, SERVO_IDX
    cjne  A, #8, rci_do_update
    acall rci_calculate
    sjmp  rci_ret
```

```
rci_do_update:
   acall rci_update
rci_ret:
   clr   RS0
   reti
```

The ISR itself is just the manager of these two functions. The `rci_update` function is called for in eight out of nine interrupts. This function reads out the timer values from the appropriate location, updates the timer count, and manages the values in the output port.

For that vital ninth interrupt we call the `rci_calculate` function which recalculates all of the values used by the update function. The calculate function also sets the timer for the long delay that defines the period of the pulses.

Let's look at these two functions, starting with calculate.

RCI_CALCULATE

```
    clr   A
    mov   SERVO, A
```

Right off the bat, we clear the servo bits. This sets all pins of the servo port to zero.

```
    ; Some starting values
    ;
    mov   R0, #COMMAND      ; Array index to command values
    mov   R1, #SERVO_TIME   ; Array index to output times
    mov   R2, #8            ; 8 pulses
    mov   TL1, #LOW(PERIOD) ; Set period delay to full amount
    mov   TH1, #HIGH(PERIOD)
```

From there, we initialize a few registers that are used later. R0 holds the address of the command table, where we will be extracting our servo position commands. R1 is the address of the start of the time array where we will be inserting our tick counts. R2 is just a loop control counter set to the number of servos we are manipulating.

We set Timer 1 to the period of the pulses, in ticks. The timer will be counting down as we do our calculations. We will also be adding to the timer's counter, which is like *subtracting* time from it since we bring it closer to the interrupt each time, as we specify the time to be spent in each servo's pulse.

```
rcc_loop:
   ; Calculate the variable length part of the pulse
   mov   A, @R0
   inc   R0
   mov   B, #PULSE_RES
   mul   AB
```

First, we get the command value as pointed to by R0, and move R0 to the next command. This value is multiplied by the pulse resolution to get the number of timer ticks associated with it.

```
; Now add in the pulse's minimum width (in ticks)
; and store result
;
add   A, #LOW(PULSE_MIN)
mov   R5, A
mov   @R1, A
inc   R1
mov   A, B
addc  A, #HIGH(PULSE_MIN)
mov   R6, A
mov   @R1, A
inc   R1
```

The command's tick value is added to the number of ticks in the minimum pulse width and the result is stored into the timer array, via register R1. R1 is incremented each time it is used, so it is always pointing to the next blank space in the array. Note that R6:R5 store a handy copy of the current pulse width, for use in the next code block.

```
;
; Add the R6:R5 time to the timer counter, thereby
; shortening that delay appropriately.
mov   A, R5
add   A, TL1
mov   TL1, A
mov   A, R6
addc  A, TH1
mov   TH1, A
```

Since we are going to spend this time in a servo pulse later on, we can speed the timer along by the number of ticks just calculated.

Note that we are actually *in* this dead time now; the servo port has all its pins at zero, and the timer is busily counting out the dead time. Looking at a graph or oscilloscope trace, it is natural to assume that the dead time comes *after* the series of servo pulses. What is really happening is that the dead time is being spent *before* the servo pulses. The effect is ultimately the same.

```
;
; Loop to the next one!
djnz  R2, rcc_loop
```

We've calculated one pulse, and subtracted it from the dead time. Now, decrement the counter and loop until it goes to zero.

```
;
; Reset the pulse counter
mov         SERVO_IDX, #0
mov         PORTBIT, #1
ret
```

During the cleanup of this function, we reset the index to the servo timetable and prepare to set bit 1 in the servo port.

After the return, nothing happens in the `rc_isr` until the timer overflows and the interrupt is called. Then, the `rci_update` function will be called to turn on the first servo.

RCI_UPDATE

```
; First, update the output of the port
mov   A, PORTBIT
mov   SERVO, A
```

Like `rci_calculate` above, `rci_update`'s first action is to force the value of the servo port. In this case, it applies the value in `PORTBIT` to it. This value is constructed and manipulated so that it turns on one port bit at a time, in order.

```
;
; Get index into servo time array
mov   A, SERVO_IDX
rl    A           ; A = servo_idx * 2
add   A, #SERVO_TIME
mov   R0, A
```

We need to get the duration of this pulse. To do this, we index into the time array. The index value is doubled, using a roll-left command, and added to the array's base address. This is then put into R0 for later use.

```
;
; Move ~time into timer 1 count registers
mov   A, @R0
cpl   A
mov   TL1, A
inc   R0
mov   A, @R0
cpl   A
mov   TH1, A
```

The timer count is extracted from the array, via R0, complemented, flipping each bit from 0 to 1, or vice versa, which is like subtracting it from 0xffff and stuffed into the timer's counter registers. All the hard work of determining *what* the timer count should be happened earlier.

```
;
; Increment index
mov   A, SERVO_IDX
inc   A
mov   SERVO_IDX, A
```

Now that we've dealt with this pulse, we move the index on to the next one.

```
;
; Adjust port bit
mov   A, PORTBIT
rl    A
mov   PORTBIT, A
```

Likewise, we go from this bit to the next; 0x01 is rolled to 0x02, to 0x04, and so forth.

```
ret
```

PROJECT 9-3: WHEEL ENCODERS

With the two output systems just defined, PWM and R/C control signal, you can get the MCU to drive your robot around. You've just reviewed the code that lets you remove the radio control receiver from your robot! It has the power to be autonomous.

There is *one* inconvenience still, a fly in the ointment of computer-controlled motors. The problem is that each drive motor will react differently to the control signal you send to it. Even if the ESC or other motor driver you are talking to creates *identical* motor control outputs, the mechanical nature of the motors will prevent them each from responding in precisely the same way.

One way around this problem is to use a stepping motor. But since we don't use these expensive devices in this robot, we need to take the other path around the problem. Close the loop and add feedback to the motor so we can tell how fast it is turning and adjust our control signal in response.

Some motors come with an encoder built in. A rotary encoder is like several switches on one shaft—as the encoder's shaft is rotated, the switches close in different patterns. By watching the patterns of activation, the controller can determine both the direction and speed of rotation.

Some encoders are optical—the pattern is switched by optical sensors "reading" a moving disk. Some encoders are mechanical, with actual switches. Others use magnets.

Some encoders are relative. You can't tell what position the encoder is at, but you can tell the direction of motion. The information from a relative encoder is usually coded in two lines using "quadrature" encoding (**Figure 9-2**). This is two pulses that are 90 degrees out of phase, like the sine and cosine functions, only digital. Looking at **Figure 9-2**, try to figure out how you can determine the direction of rotation from the two signals A and B.

Other encoders are absolute, like a digital potentiometer—you can read the actual position of the encoder from its output. Absolute encoders can have binary or Gray-code data. Look to the data-sheets of any given encoder to see what information is available and for hints on how to decode it.

A device related to the encoder is the resolver. This is an analog position sensor, not unlike a rotating transformer or two-phase generator. The resolver provides two outputs whose voltage levels are essentially the sine and cosine of the angle of the shaft.

For the most part, both encoders and resolvers are expensive devices, resolvers more so, with their analog control overhead. It is possible, though, to get a few low-resolution encoders through companies like DigiKey. These tend to have just a few (4 or 16, for example) output pulses per revolution.

It is possible to create your own optical encoder using inexpensive components and a printer. This section discusses the creation of such an encoder and some of the issues involved in using it. Of course, it's usually cheaper and easier to *buy* something than build it, but building is so much more educational!

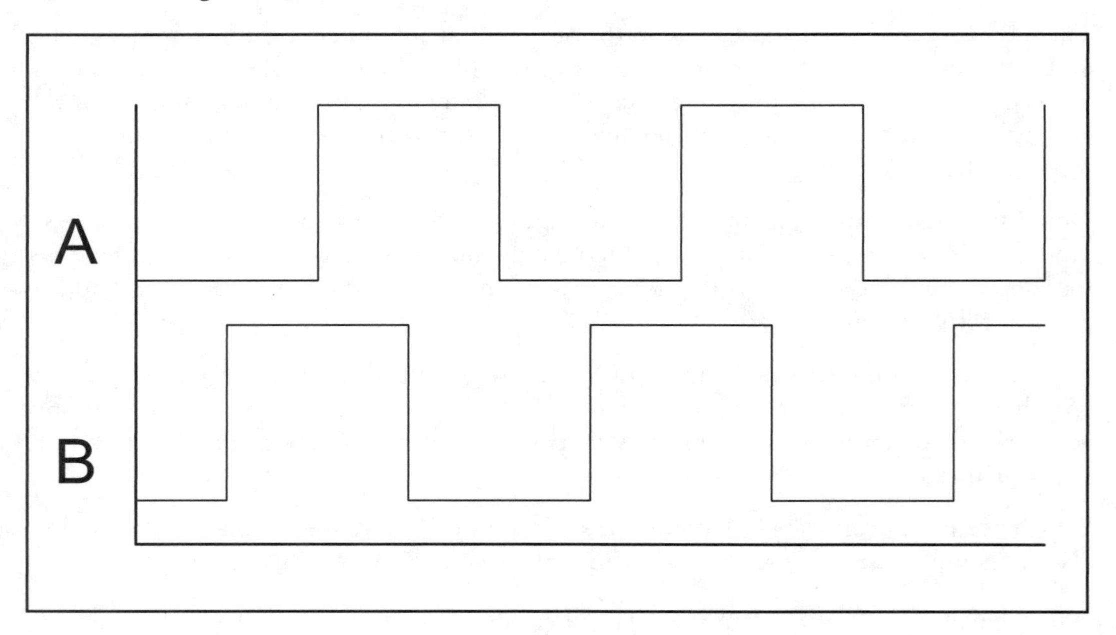

Figure 9-2 Quadrature encoding

WHERE TO ENCODE?

There are roughly three places you can stick an encoder on a geared-down drive train. One is on the motor itself. Another is at the output of the gear train, or in the middle of the gearing between the output of the gearmotor but before the chain to the wheel's sprocket. The third is at the output device, such as at the wheel's drive sprocket.

For a low-resolution encoder, such as a 4-pulses-per-revolution device, it is best to attach to the back of the motor. This way, you get a better *relative* resolution. Since the encoder is counting turns before the drive train, the pulse per revolution (PPR) of the encoder relative to the final output is much higher. It is scaled up by the same amount as the motor's rotation is scaled down at the wheel.

A high-resolution encoder will also work at the motor, though you'll need higher-speed logic to read it.

Since we are building our own encoder from parts, it is easier for us to attach something to the sprocket on the wheel. Because this is the slowest-moving part of the drive assembly we want to make the encoder as high-resolution as we can.

ENCODER DISK

The wheels used on this platform have built-in sprockets. These sprockets have a nice space between the mounting bolts and the teeth, where we can paint or glue on a reflective encoder disk. If you recall back in the construction of the frame, we provided a mounting hole for the sensor. The encoder pattern must be placed near this mounting point, or, conversely, the sensor mount must be placed near the encoder pattern.

A simple encoder pattern is shown in **Figure 9-3**. These can be easy to create, depending on what drawing package you have. This one has thirty-six sections, eighteen white and eighteen black that were created by drawing black-filled polygons at the intersections of the inner ring, outer ring, and the thirty-six lines that were created using a simple copy-rotate function.

How many segments you put in your ring depends on two factors. The first is the diameter of the pattern at the point your sensor will be reading it. The second factor is the minimum resolution, or width, that your sensor can "see."

For example, say the sensor is reading the pattern 2.5" out from the axis, giving a 5" diamter. And, for example, the sensor can discern a 0.2" diameter dot. This means that, at 2.5" from center, the pattern's section should be about 0.2" wide. The math to determine the number of sections is simply the circumference of the circle at that diameter divided by the section width of 0.2":

$$c = 2\pi r$$
$$c = 5\pi$$
$$c = 15.708''$$
$$n = c / 0.2''$$
$$n = 78.5$$

So we should have about seventy-eight sections. To make the math in the microcontroller simpler, we can adjust that to the nearest power of two, or sixty-four sections. This gives us a nice tidy encoder signal from the sensor.

Figure 9-3 Encoder pattern

Now for the cheat: if we put in 128 sections instead (or perhaps better, 160, even though it complicates our counting) the signal from the sensor will never see a "clean" segment of the pattern. The amount of light and dark will vary continuously as the wheel turns, but it will never be *just* white or black. In fact, it will look a lot like a sine wave, or the output of one signal from a resolver.

So pop your wheel off of the robot, take a few measurements, and make your own encoder wheel pattern. You could even make several different types, to see how they work. Take your time, I can wait.

Once you have the patterns printed out, cut them out with a sharp knife or scissors.

You have several choices on how to stick them to the sprocket. Regardless of how you want to proceed, clean the sprocket thoroughly with alcohol or window cleaner to get all of the grease and dirt off.

Now take a ruler and a felt-tip pin and measure from the center of the wheel's axis out to the radius of the inside (or outside) of your sensor pattern. Mark this distance several times around the wheel so you can use these marks to center your pattern.

Then, using an adhesive of your choice, glue your sensor ring to the sprocket. I used an evil spray adhesive, 3M's Super 77, which will glue almost anything to anything. You could also print your pattern onto a full-sheet adhesive label such as Avery's 5265.

Once the pattern is in place, cut a ring of clear laminating plastic that is somewhat wider. Stick this over the pattern so that it also sticks to metal on the inside and outside of the paper ring. Burnish it down nicely so there are no air bubbles, and your pattern is now protected from the elements.

Though we are using a reflective sensor, you can do essentially the same thing with an interrupt-type sensor. Print the pattern on clear acetate and stick a same-sized lamination sheet to the toner side of the sheet to protect it from being scraped off.

Encoder Sensor

The sensor we are using for this example is the Fairchild Reflective Object Sensor number QRB1134. This sensor consists of two parts—an infrared LED and a phototransistor whose base is driven by the light reflected off the pattern.

Running both the LED and the phototransistor at about 25mA at 5V gives the schematic shown in **Figure 9-4**.

The precise values of the two resistors are not terribly important.

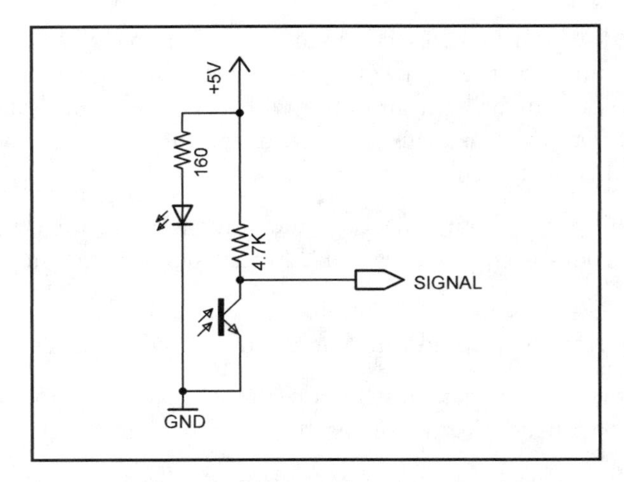

Figure 9-4 Reflective photosensor

When it is all plugged in, the light from the LED bounces off the pattern and varying amounts of this light bounce back into the phototransistor, which in turn creates a varying signal at its output.

If you want to try and use this system as an analog resolver you can feed the signal directly to one of the analog inputs. Otherwise, you can feed it through a Schmidt trigger or comparator, which will convert it to a digital pulse to be read by one of the digital inputs.

For the next section, we assume a digital-output encoder like that shown in **Figure 9-5**.

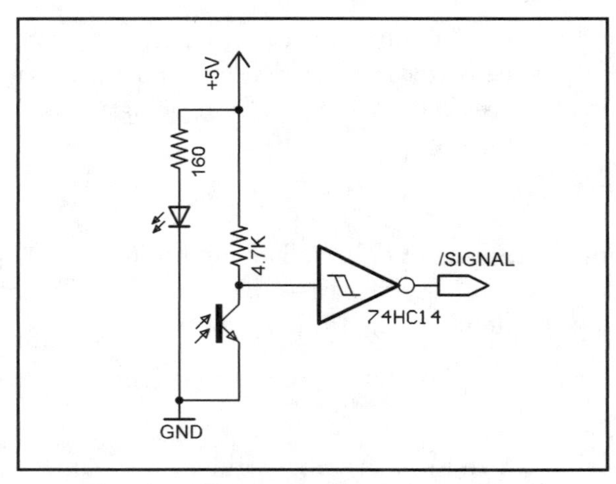

Figure 9-5 Digital optical encoder

ENCODER USAGE

First off, we aren't going to write any definitive encoder software here. What we are going to do is discuss the issues, and then outline what the software will look like. This is the first of several steps away from practicality and into the realm of theory. Painful, but necessary if we are going to fit everything into this book!

QUADRATURE

There are two benefits that a quadrature encoder gives that we need to compensate for in our single-output encoder, unless you want to put a second sensor on the frame that is one-half of a section removed from the first one. The first benefit is that quadrature gives us a sense of direction, and the second is that quadrature eliminates the problems associated with jitter. These benefits are related.

Looking back at **Figure 9-2**, let's imagine what happens as the wheel turns and these waveforms are presented to the microcontroller, or dedicated electronics. Assume that the A signal determines the direction of a counter and the B signal is the clock that drives the counter. We watch the *level* of A and the *transition* of B (that is, B is *edge triggered*—it signals a change only at that time when it goes from low to high).

When A is low and B changes from low to high, the count is incremented. You can see that this combination of A high, B rising only occurs as you read the pattern from left to right. The reverse condition holds when you read the diagram from right to left—B goes from low to high only when A is high, so this combination is used to *decrement* the counter.

Now, to make it both faster and more robust, we want to trigger on the falling edge of B as well. This time, when A is *high* and B is falling we increment. A low and B falling decrements.

The two signals clearly give us the direction of motion.

With this scheme you only get one timing pulse per full cycle of B, and A is essentially ignored with respect to timing. If you want to get fancy, you can calculate the timing with twice as many pulses by using the levels of both A and B in combination with both the rising and falling edges of both A and B. This is left as an exercise to the reader.

JITTER

Now, what about jitter? First, we need to define what jitter *is*. When a sensor is right at the edge of a transition, noise and/or mechanical vibration can cause that sensor to turn on and off rapidly—even though the object it is sensing isn't really moving at all. Now imagine that you have a single sensor, like our reflective encoder, and you are watching its signal oscillate wildly. From this, you would be led to assume that the wheel (or whatever) is spinning madly, when in reality it is just sitting there on the edge of a pattern wedge.

Now imagine how this looks in quadrature. Since the A signal is solidly on or off when B is at the edge, the oscillations only occur in B. With A high, for example, the first fall in B will increment the counter by one. But then, because of jitter, B will rise again, incrementing the counter. This can repeat forever and, since it looks like the wheel is only turning forward and backward a minor amount, it will have no detriment to the wheel rotation counter except for a slight wobble.

Using our one-sensor encoder we have to be more clever, plus hope for a bit of luck in our environment.

First, we know what we are *telling* the motor to do; the known command sent to the motor driver. We know if we think we are moving forward, and we know if we think we are moving backwards. We also know if we are telling the motor to go fast or slow.

For the most part, the motor will be doing what we tell it to. There are two notable exceptions. The first exception has to do with momentum. If the motor is at a dead stop and we suddenly ask it to spin full speed, it won't do that right away—it will ramp up to that speed. The heavier the load on the motor, the longer it will take to reach full speed. If the load is too heavy it may never even *make* it to the speed we requested. Likewise, if the motor is moving quickly and we ask it to suddenly stop (or worse, reverse), it won't.

The second issue relates to forces outside of the robot. If the robot is being acted on by forces that are sufficient to override the motor's current state, such as the robot is being pushed, or is on a steep slope and begins rolling, the sensors will be reporting a sense of motion that is, in fact, true and not jitter at all.

USING THE SINGLE ENCODER

To use the encoder you need to do four things: read the sensor, time the width of the pulses, manage errors, and correlate the encoder pulses to the motor driver output.

Reading the sensor. While you can read the sensor using any of the digital input ports, this requires that you sit and monitor those inputs waiting for the input to change. Since these encoder pulses happen regularly and fairly quickly, and since we want to get accurate timing from them, that would mean watching the ports a lot. If you want to watch a bunch of dynamic I/O, you don't have much choice but to drop that code into the main loop. But if you are only watching one or two signals, it is easier to attach them to one of the two external interrupt pins, to one of the two comparators, and write an ISR to deal with them. In fact, pins 4 through 7 of port 1 (which are exported to the bus without buffers, as 3DO8 through 3DO11, that is, on pins A11 through A14) can be configured as inputs which trigger interrupts on the falling edge of any signal presented to them.

Some of the interrupt inputs can be set to trigger on a rising edge, a falling edge, or both (IE0 and IE1); others only respond to the falling edge (IE4 through IE7). Depending on various factors, the width of encoder pulse when it is high, when reading a white sector, may be different from the width in the low half, when reading a dark sector. Because of this possible asymmetry, it is better to measure the width across a full cycle of pattern—white plus black—so you only need to interrupt on one edge. Of course you *could* measure the two halves of the cycle separately and average them or something.

Timing the Pulse. Once you have the sensor triggering an interrupt, you need to have a timer running to measure the time between interrupts, that is, the amount of time it has taken for the wheel to move two sections of the encoder pattern past the sensor. Though there are four timers on this MCU, not to mention the PCA array, it takes some thinking to determine which one to use.

First, you need to decide how you want do your time accounting. Do you want to count the number of encoder interrupts that occur during a given amount of time, or do you want to measure the amount of time that elapses between interrupt pulses? We talk about the latter method here.

Now, assuming that we are using an R/C signal to drive the motor controllers, Timer 1 is already being used to manage that. Timer 0, of course, is ticking away at about 162 overflows per second to manage the user inputs. Timer 3 is busy at 400 overflows per second to drive the analog inputs. That leaves Timer 2 free—except, as we'll note later in the book, it is needed to time the UART for external communications! We need to overload one of these other timer's interrupt routines so it can be used for our encoder timing.

There are only two timers with a regular pulse; Timer 0 and Timer 3. Assuming that our encoder will be creating pulse widths between 20mS to 200mS (looking at my oscilloscope), Timer 3 looks like a good one to use, with an interrupt once every 2.5mS or so.

To use Timer 3 for this purpose, all you have to do is add one 8-bit counter for each encoder signal and increment each counter every time Timer 3's interrupt service routine is called. If one of the encoder's counters overflows it means the wheel isn't moving or is moving too slowly for us to manage it, so the ISR may want to make note of that.

At each execution of an encoder's ISR, it needs to read the appropriate counter and then clear it. The value from the counter is the amount of time that has elapsed since the last ISR execution. Of course, if the counter has overflowed, the ISR would want to make note of that and discount the actual counter value.

Encoder errors. There are three types of error you need to watch for in the encoder ISR. The first is counter overflow. This condition can be noted and flagged within the timer ISR itself, and just noted in the encoder ISR.

Then there is the jitter problem. You can reduce jitter issues by noting how short your fastest encoder pulse is and simply reject any values less than this—assume the wheel is jittering and call it stopped.

Finally, there is the issue of determining whether the encoder value is tracking the motor control signal correctly. This is a bit more complicated. It also leads us to the fourth task of the encoder software.

Correlating the encoder input with the motor output. This is the hard task. The encoder's timing value does not directly mirror the command value that drives the motors. You could probably do some math to make them both fit the same scale, but MCUs are weak in the mathematics department.

One of the purposes of adding the encoder feedback in the first place is to help the robot drive in a straight line, by compensating for differences in motor response to their inputs. To do this you don't need to match the encoder's input with the motor output—you need only compare the two encoders' values and see how they differ. Then, you can adjust the motor command signal until they match. This will give you straight lines.

By thinking about what you are trying to accomplish with the encoder input, you might be able to find other simplified shortcuts to using them.

CHAPTER 10

SENSE AND CONTROL: OTHER SENSES

There are many other senses you can add to the robot, and to fit them into this chapter I'll have to be brief. But first, it is necessary to provide a window into the working of the MCU to make it easier to explore these sensory inputs.

PROJECT 10-1: RS232 OUTPUT

Testing sensory input devices can be fun, but unless you're an especially gifted psychic (or able to discern high-speed binary data by brail) it helps to have human-readable output so you can see what the senses are doing. These are the times when your microcontroller needs to communicate to your desktop computer. And the easiest way to do this is through the RS232 communications port in that computer.

If you are really interested in the RS232 protocols, you would be well served to run out and purchase a book dedicated to the subject such as Jan Axelson's *Serial Port Complete*. This book is full of both useful and trivial information, such as the detail that the 9-pin D-sub connector everyone calls DB-9 is actually *DE-9*, where the E indicates the shell size. Only the 25-pin is size *B*.

Chapter 8 referenced JTAG programming software (which can be found on www.simreal.com); that code does some serious bit-banging to achieve its goal. Compared to that interface, the serial port code in this chapter is a cakewalk.

You should be able to use any straight-through serial port cable for this project (not a null-modem cable). If the PC, for some reason, insists on handshaking protocols, you can make a special loop-back cable. A loop-back cable is a simple rewiring of the serial port that turns the hardware handshaking lines back in on themselves. The standard form for this cable for a DB9 connector is shown in **Figure 10-1**.

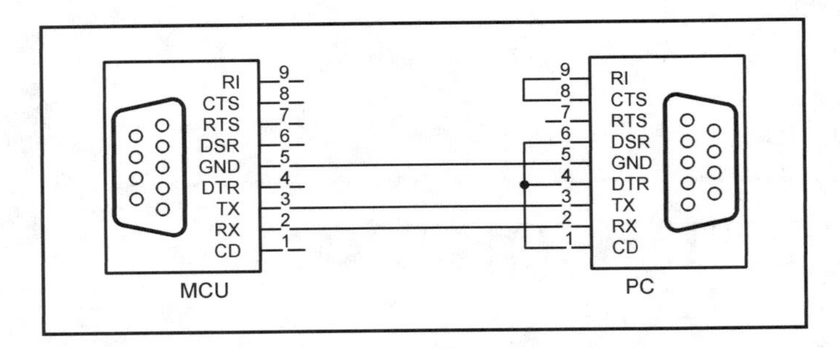

Figure 10-1 DB9 loop-back wiring

The only wires necessary for an RS232 interface, assuming no handshaking, are 2 (RX), 3 (TX), and 5 (Ground). On the PC side of the cable you wire together lines 1, 4, and 6, and lines 7 and 8. With this configuration, if the PC attempts any handshaking it will be, essentially, shaking its own hand.

The software on the PC side of the cable can be the standard-issue HyperTerminal found in the Accessories/Communications folder of a Windows PC. Of course, you are also free to adapt the CSerial class from www.simreal.com. On the MCU side things are a little more complex, as usual.

MCU Serial Software

Most microcontrollers have a built-in UART (Universal Asynchronous Receiver/Transmitter) and the Cygnal 8051 used so far is no exception.

Like most of the software throughout these chapters, the code presented here is unique to this particular MCU, but the concepts can be generalized to many others. Once you know what to *expect* from a particular peripheral, you know what to look for in its different uses and variations.

Serial communications, such as via the UART in the MCU, are performed by sending information one bit at a time across a single wire (more or less; you need at least two wires, one of which carries the reference voltage or a complementary voltage). These pulses of data are all sent with respect to the constant ticking of a clock.

In synchronous interfaces this clock is transmitted right along with the data signal. We don't have that luxury here, since we are using an asynchronous protocol. In this case, both sides must agree in advance what frequency clock to use to time the bits of data.

Starting off the asynchronous communication is a start bit. This is simply a transition from the normal "resting" state of the data line to an "on" state. This bit lasts one tick of the communications clock. After the start, the full set of data bits are presented one at a time, each lasting one tick of the communications clock. When all of the data is sent there is a rest period, the so-called "stop" bit, that lasts, usually, one tick of time. Of course, the space between bytes of data can be as long as forever, but the stop bit time is the *minimum* allowed.

The receiving device sits around watching the data line and, when it changes state for the start bit, it starts paying close attention. Since the receiver knows the communication frequency, having agreed to it before hand, it waits for the duration of one and a half bits and then checks for the state of the data line. This delay skips over the start bit, the sole purpose of which is to wake-up the receiver, and puts the test right in the center of the first data bit.

The receiver then waits a tick and reads the next bit, and so on. When it's done, it may test the state of the "stop" bit—and if it isn't in the correct rest state the UART might signal a framing error, indicating that something is not quite right.

Anyway, the key to all of this is the shared communications frequency.

TIMING

The UART mode we are going to use here is mode 1: 8-bit asynchronous variable baud rate, driven by Timer 2.

The timer is driven by the system clock divided by 2, divided again downstream by 16. So the baud rate is ultimately based on the system clock divided by 32; this clock/32 advances Timer 2. Each communicated bit is, in turn, processed during the Timer 2 overflow.

The baud rate (bit rate) is calculated by:

$$baud = \frac{SCL}{32 * (65536 - T2)}$$

where SCL is the system clock rate and *T2* is the value plugged into the timer's reload registers. *T2* is subtracted from 65536 (0x10000) since it is counted up to that value for an overflow.

Turning things around a bit, we can determine the value to plug into *T2*:

$$T2 = 65536 - \frac{SCL}{(32 * baud)}$$

Then, for a 16MHz internal clock desiring a 9,600 baud communications, the reload value is:

$$= 65536 - \frac{16{,}000{,}000}{32 * 9{,}600}$$

$$= 65536 - 52$$

Which gives us a reload of 0xffcc.

The problem is that the internal clock is *not* guaranteed to be 16MHz. According to the documentation, it could vary anywhere from 12.8MHz to 19.2MHz. This gives us timer reload values anywhere from 0xffd7 to 0xffc1.

There are two ways to tackle this problem. The cheapest way is to simply tune your reload values until your PC recognizes the communications from the MCU. In my test circuit a reload of 0xffd0 worked fine, for what it's worth.

The *reliable* way to manage the problem is to spend a few extra dollars and make a subtle change to the circuit, adding a crystal oscillator to the MCU, as shown in **Figure 10-2**.

I've used both crystals, with their associated capacitors, and oscillators to drive MCUs, and I must say I prefer oscillators. Though more expensive, they are also more reliable and easier to use.

In addition to this hardware change, you need to load the oscillator control register OSCXCN with 0x20 instead of 0x00 to indicate its presence.

If you intend to do much communication, the selection of oscillator frequency bears special thought, especially if you want to run the UART in the higher baud rates. For example, a system clock rate of 18,432,000 (18.432MHz) makes the math work out very nicely—check it and see for yourself.

SETUP

With all of these preliminaries in place, let's look at the actual code changes to the framework:

```
DATA equ   R0 ; Assign to any convenient registers
UART_TEMP   equ   R1
```

We have the traditional beginning of assigning various magic numbers to names. In this case, we also create aliases for certain registers we use. This indirection allows us to change our register use easily as needed, to prevent conflicts between different sets of code.

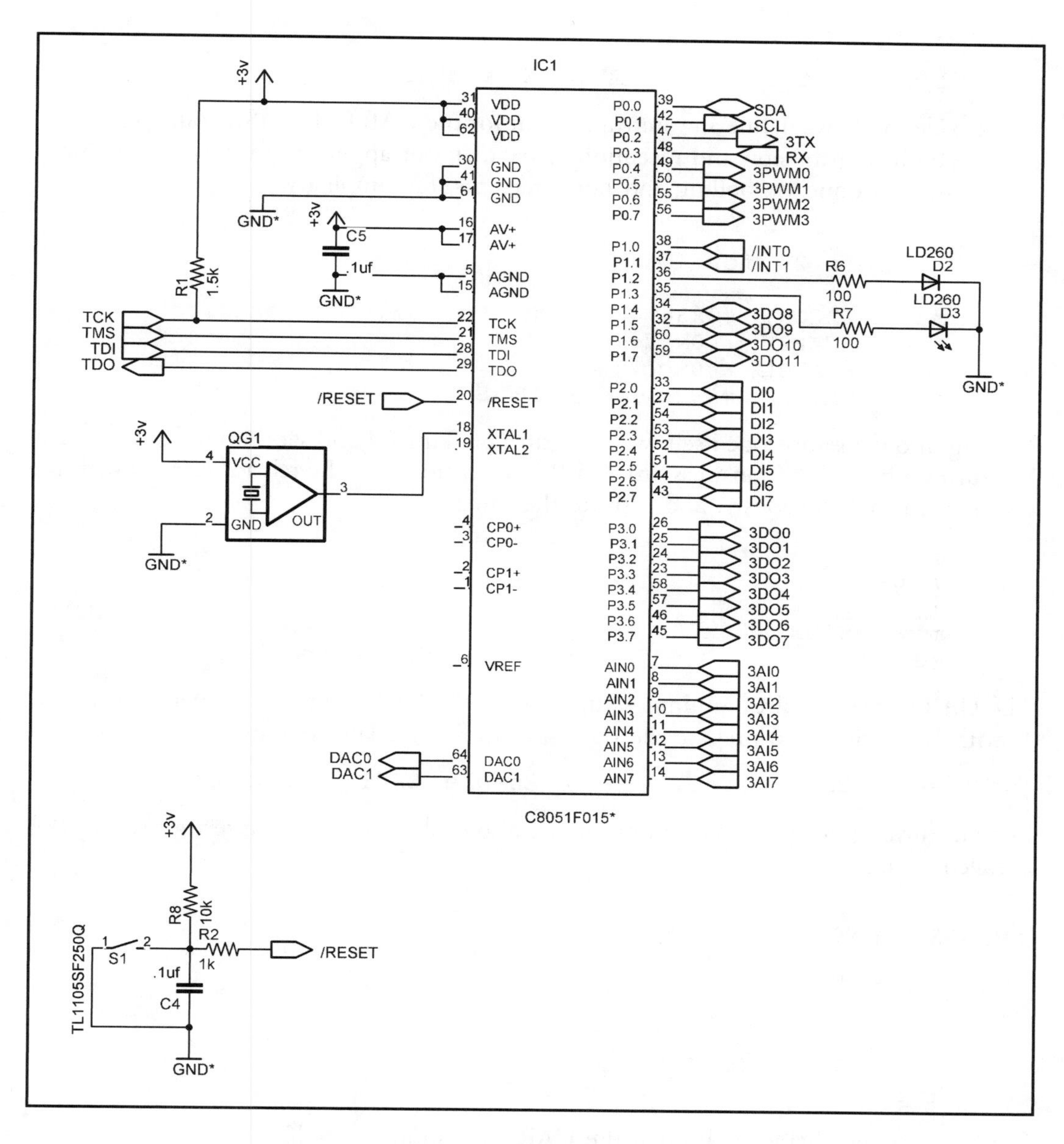

Figure 10-2 MCU with oscillator

```
org   0x23
ljmp  uart_isr    ; UART RX and TX
```

In the code space, we define an interrupt vector for the UART. This ISR could, in theory, manage both transmission and reception of data. In our application it just receives data; transmission is done via polling and not interrupts, for simplicity.

```
;
; Timer 2 (for UART)
;
mov   T2CON, #0x34      ; T2 for RX, TX, enabled
mov   RCAP2H, #0xff
mov   RCAP2L, #0xcc
mov   CKCON, #0x20      ; T2 SCL
```

Moving into the setup code itself, we must turn on Timer 2 and set it up for UART duty. We turn on both TX (transmission) and RX (reception), set the reload registers with the values calculated previously, and activate the timer.

```
;
; UART
;
mov   SCON, #0x50       ; Mode 1, RX enabled
setb  TI                ; Force TX to be ready
```

The UART itself requires minimal setup. We set the mode we want to operate in (mode 1, with RX active) and prepare the transmission flag for later polling.

```
setb  ES                ; Enable UART interrupt
```

Prior to turning on the system interrupts we also enable the UART interrupt so our ISR is called as needed.

TRANSMISSION

```
tx_char:
   jnb   TI, tx_char
   clr   TI
   mov   SBUF, UART_DATA
   ret
```

Sending a single character through the UART is simplicity itself.

The first line waits until the data buffer is ready to receive a character. This loop simply waits until the transmit ready bit (TI) is set, wasting clock cycles until any previous character has been cleared out.

We then clear this bit, dump our data into the buffer, and leave.

The TI bit remains cleared until the character is completely transmitted and then TI is reset by the hardware.

Sending a string of data is only marginally more complicated. First, the calling program must define a string in the code segment. For example:

```
string: db "This is a test of the emergency UART
system."
        db 0x00
```

The last byte of the string must be zero.

The address of this string is put into the 16-bit data pointer register, via its two halves DPH (the high byte) and DPL (the low byte).

```
   mov   DPH, #HIGH(string)
   mov   DPL, #LOW(string)
   call  tx_str
```

Finally, call the tx_str function, shown here:

```
tx_str:
   mov   UART_TEMP, #0x00 ; index counter
txs_loop:
   mov   A, UART_TEMP
   movc  A, @A+DPTR        ; String in code space, in
DPH:DPL
   jz    txs_exit          ; if zero, exit
   mov   UART_DATA, A
   call  tx_char
   inc   UART_TEMP
   sjmp  txs_loop
txs_exit:
   ret
```

This code loops through the string until it hits a zero byte. For each non-zero character found, it calls `tx_char` to send it.

RECEPTION

After each byte received, and after each byte sent, the UART calls the appropriate interrupt vector—if it is turned on, of course. In our case, this vector jumps to `uart_isr`:

```
uart_isr:
   jnb   RI, uarti_exit
   clr   RI
   mov   A, SBUF
uarti_exit:
   reti
```

First, the ISR tests the RI bit to see if we received any data. If not, we exit. Otherwise, it loads the received byte into the accumulator. What you do with it here is up to you; we don't practice reception of data from the PC in our tests.

If you wanted to, you could also test the TI bit here and manage the transmission of data. As you notice above, we poll the TI bit directly in the simple `tx_char` function.

PROJECT 10-2: ANALOG SENSES

There is a wide array of sensors that output an analog voltage. This voltage may need to be amplified, filtered, or otherwise cleaned up prior to feeding it into the MCU.

Once you *have* an analog voltage, it is merely a matter of plugging it into the appropriate port in the system and then reading the corresponding entry in the analog input array. We've been reading these signals and building this array of values since the very first framework code, so using analog values should be simplicity itself.

GP2D12 OPTICAL DISTANCE SENSOR

Sharp's GP2D12 optical distance sensor is a particular favorite of mine (see **Figure 10-3**). It measures distance to an object through parallax and is reasonably immune to variations in surface color and texture. Internally, it consists of an IR LED and a position-sensitive detector (PSD).

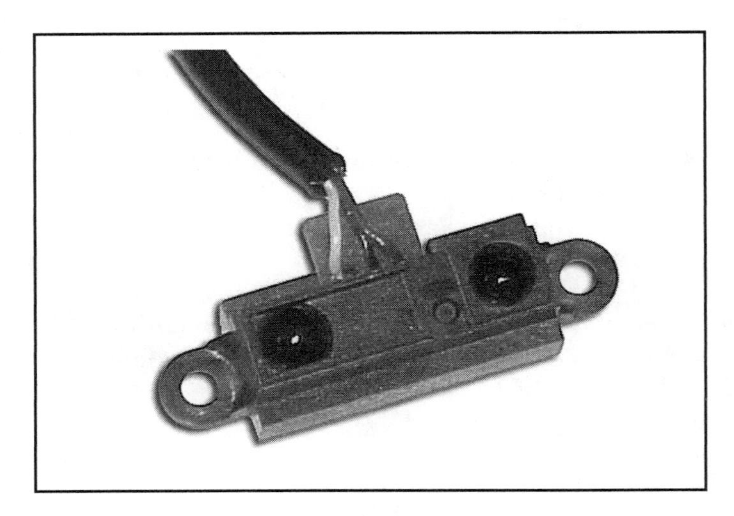

Figure 10-3 GP2D12 sensor

The PSD is a strip of silicon that varies its output signal based on the centroid of the light falling on it. These tend to be expensive devices when purchased on their own, not to mention the hassle of arranging the optics to make them work reliably.

The optics in the GP2D12 focus the light from the LED's reflection onto the sensor. This device has a theoretical sensory range of 10 to 80 cm (3.9 to 31.5 inches). For a while, it was hard to find this sensor on the market. Today, several distributors carry it. Try Acroname, www.acroname.com, and HVW Technologies (no Web site, but see the Appendix for complete address), among others.

Sharp has several sensors in this family, including the GP2D02, which provides a digital serial output, but with only half the sampling speed, the GP2D05, which acts as a Boolean switch with adjustable trigger distance, and the GP2D120, which has a range of 4 to 30 cm.

The output of the GP sensor is not linear, but follows the curve illustrated in **Figure 10-4**. The output from the sensor is updated roughly every 40mS, or about twenty-five times a second, according to the specifications, though it may update more frequently in practice.

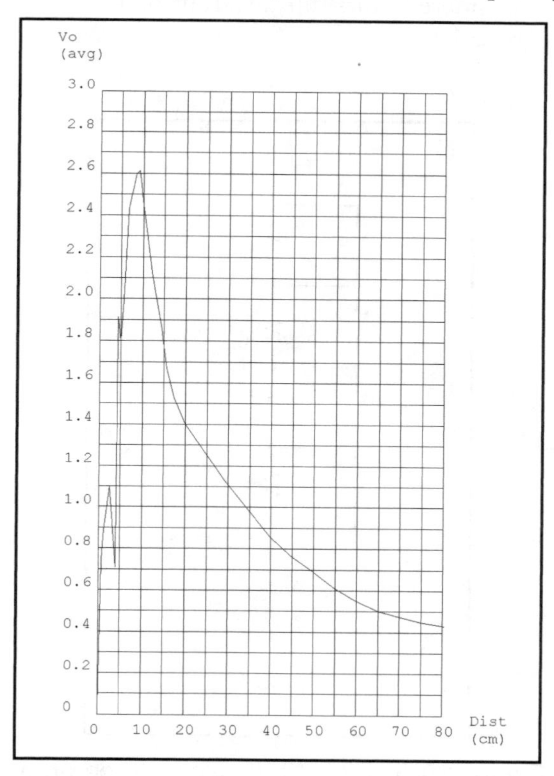

Figure 10-4 GP2D12 voltage output

If you have difficulty with noisy signals from the device there are several remedies you can take. Sharp's application notes indicate that you should ground the conductive-plastic case. You can also place a capacitor across the power leads into the sensor, near the device, to provide a stable power supply. Finally, you can filter the output line with a small capacitor to ground.

Because of the minimum sensor distance of 10cm you should set the sensor back into the robot if you want to see things at the edge of the robot. Alternatively, you can point the sensor across the robot's body.

BUMPER ARRAY

Any given bumper switch is a digital device. But considering that there are eight switches shown on the sample robot, and your machine may have more...who wants to take up all of the digital input ports with these?

These switches can be considered "bits" in an array of resistors making up an analog "nibble" as shown in **Figure 10-5.**

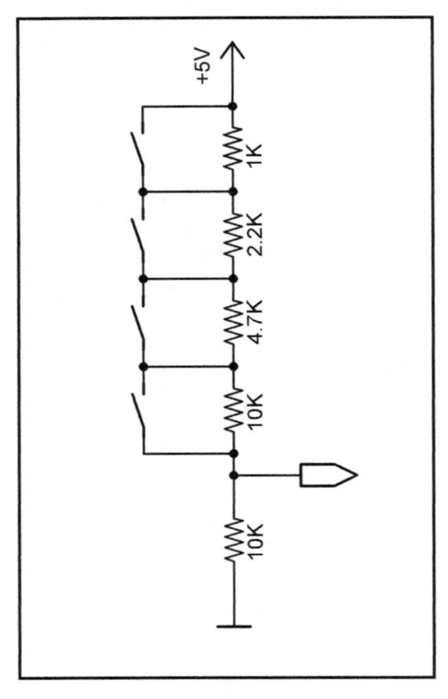

Figure 10-5 Digital-to-analog switch array

This is an extension of the classic resistor bridge, with all of the resistors above the signal tap being R1, and the 10K resistor to ground being R2. The output voltage is described by the familiar math:

$$Vout = Vin * \frac{R2}{R1 + R2}$$

Different on/off configurations of the switches set different, unique, resistances for R1. The resistances of the various sub-resistors in R1 should follow a power-of-two progression. The value of R2 can be chosen to select different ranges of output.

Reading the switch arrangement is done through an analog input port and decoded to determine what switch was pressed.

Using the very common values available at any Radio Shack (as shown in **Table 10-1**), we have sixteen possible R1 resistances. With R2 of 10K and a positive voltage of 5V, we can easily calculate the output signal.

S1	S2	S3	S4	R1	Vout
1K	2.2K	4.7K	10K	17.9K	1.79
	2.2K	4.7K	10K	16.9K	1.86
1K		4.7K	10K	15.7K	1.95
		4.7K	10K	14.7K	2.02
1K	2.2K		10K	13.2K	2.16
	2.2K		10K	12.2K	2.25
1K			10K	11K	2.38
			10K	10K	2.50
1K	2.2K	4.7K		7.9K	2.79
	2.2K	4.7K		6.9K	2.96
1K		4.7K		5.7K	3.18
		4.7K		4.7K	3.40
1K	2.2K			3.2K	3.79
	2.2K			2.2K	4.10
1K				1K	4.55
				0	5.0

Table 10-1 Switch array values

You can make up a similar array of values for whatever resistors you actually use.

In theory, with an eight-bit A/D converter you can have eight switches in the array. Due to noise in the system and imperfect resistor values, it is better to keep the switch count down from this maximum.

TEMPERATURE

There are many choices available for temperature sensors—a quick search of DigiKey's Web site, www.digikey.com, brings up sixty different variations. Casting a quick eye through the list we pick one to illustrate here—National Semiconductor's LM50.

This sensor is almost embarrassingly simple. A three-pin device, you apply power and ground to two of the pins and it feeds you a temperature-related voltage out the third; not unlike a temperature-driven potentiometer.

This device doesn't create a strong signal. The output varies linearly from 100mV at -40° C to 1.75V at 125° C. The formula for the voltage output in millivolts for a given temperature T is:

$$Vout = (0.010 * T) + 0.5$$

This voltage can be fed directly to an analog-to-digital converter. Depending on what voltage scale your A/D converter is using, you may want to amplify the signal first.

The single-ended amplifier shown in **Figure 10-6** should suffice for most applications.

The amplification of these circuits is given by the formula:

$$Vout = Vin * (1 + R2 / R1)$$

It should be a simple matter to find values for R1 and R2 to give the desired output.

A longer treatment on sensor amplification was given in *Applied Robotics*, Chapter 5, or you can find details on the subject in any number of other sources.

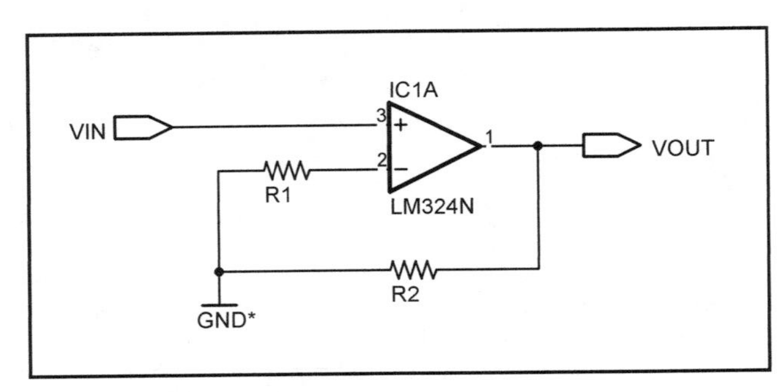

Figure 10-6 Simple OpAmp

PIR (PASSIVE INFRARED)

A popular sensor in the small-robotics field is the passive infrared, or pyroelectric, detector. These sensors respond to variations in emitted IR, that is, heat.

PIR is used for fire-fighting robots, for competitions, or for security patrols. They are also used as people detectors for burglar alarms and motion-activated lights in all of the consumer detectors. One way to get a PIR sensor is to hack one of these mass-market systems.

You can also find PIR sensors from such diverse places as QuickKits (www.qkits.com) and Acroname (www.acroname.com).

Other than the relative difficulty in finding a PIR sensor, these are much like any other analog sensor in application.

FORCE, SOUND, LIGHT, VOLTAGE...

There are any number of sensors that return a variable resistance or voltage. Most of these sensors need some form of signal enhancement in the form of a resistance bridge (either a simple two-resistor bridge, or a four-resistor Wheatstone bridge) or an amplifier.

One new sensor is the Force Sensitive Resistor (FSR) from Interlink Electronics (www.interlinkelec.com), which was discussed in Chapter 7. This is a pressure-variable resistor and can be used like any other resistive sensor.

SERIAL SENSES

Some peripherals come with convenient data streams built in. In this section we look at a few protocols and peripherals that provide a data stream rather than a voltage level.

I^2C (CLOCK/CALENDAR)

Why not give your robot a sense of time? The chip we look at here, the PCF8583, has its own clock at a frequency well suited to timekeeping, and some memory that the microcontroller can use for other purposes. Philips Semiconductor's PCF8583 is advertised as a clock/calendar. As part of this function it also has a programmable alarm. The 240 extra bytes of memory are a bonus if the MCU you are using has limited available RAM.

The main thrust of this section, though, is not the peripheral itself, though I do feel that a robot should have some sense of time. The clock/calendar is used as an example with which to explore the wonders of the Inter-IC (I2C or, more properly, I^2C) interface, which is useful in so many ways.

There is a growing family of devices designed for the I2C protocol, not only the various peripheral devices such as those available from Philips Semiconductor (www.philips.com), but also microcontroller systems like the OOPic (www.oopic.com). You can find many different peripherals that use I2C such as I/O buses, clocks and calendars, EEPROMs, and so forth.

I2C is a simple two-wire protocol developed by Philips. Each of the two lines in the I2C interface is pulled up to Vcc by resistors somewhere on the bus. Each device on the interface can allow the line to float, or it can pull it back down to ground. There are four types of devices on the I2C bus. The Master Transmitter drives both the clock (SCL) and data (SDL) lines. A Master Receiver drives the clock but listens to the data. Slave devices listen to the clock and, depending on whether they are sending or receiving, drive or listen to the data line.

I2C is a multimaster protocol. There can be any number of "master" devices on the bus. There is a well-defined arbitration system for determining which master wins use of the bus in the case of conflicts. Most of this hassle is managed by an I2C-friendly MCU.

A great deal of information about the I2C specification can be found on the Philips Web site, www.philips.com. In this section, we focus on the specific details needed to get our Cygnal 8051 to talk to the PCF8583 clock/calendar.

The PCF8583's operation is managed via control registers, accessed as memory addresses through the I2C communications bus. The chip itself has the single input pin A0 to augment its internal address; with these, the clock can have an I2C bus address of 0xA0 or 0xA2. The least significant bit is not part of the address.

Device addresses in I2C consist of several distinct parts. The most significant four bits of any I2C device address are fixed and are used to indicate the type of device in use, though the four-bit values of 0x0 and 0xf are reserved for special cases. In PCF8583's case the group nibble is 0xA.

The next three bits are programmable address bits so there can be up to eight instances of any given device or device "group" on the bus at one time. For the clock/calendar in hand, only the lowest bit is adjustable with the two other bits fixed at zero.

Finally, the least significant bit is a Read/Write control flag; a zero indicates you are writing to the device and a one that you are reading it.

For more complicated environments I2C also supports 10-bit addressing, expanding on the 7-bit default addresses. In this case, the first five bits of the address must be 11110

(0xf0). The next two bits are the most significant bits of the address. The last bit of this first byte is still the Read/Write control flag. Finally, a second byte contains the remaining eight bits of the address. This odd separation of the address bits allows both 7- and 10-bit devices to operate in the same system.

The schematic for the sample clock/calendar circuit (**Figure 10-7**) reflects both the insane simplicity of this device as well as my preference to use oscillators rather than crystals for my circuits. Of course, a 32.768kHz crystal plus trim capacitor to improve the accuracy of the clock, can be used instead. Look to the PCF8583 data sheet for more details on clocking.

It is important to connect the clock and data pins from the PCF8583 to the appropriate SDA and SCL ports on the MCU. If you are building this as an external circuit you can plug it into the MCU communications port (as shown in Figure 8-8), taking care to set the jumpers correctly. If you want, you could also build this circuit onto a separate card and communicate to it through the backplane, leaving the MCU's communication port free for other tasks.

In addition to the I2C bus, the PCF8583 provides an interrupt output that can be used to signal the MCU of significant events, such as alarm events or, by default, a 1Hz pulse train. This output is normally a floating open-drain output. When set, though, it pulls the line to ground. Because of this passive/floating "off" condition the signal must be pulled up to register a high value. Though the Cygnal 8051 ports allow for a weak internal pull-up, the interface buffers do not. The pull-up resistor, R1, can be anything handy from 1K

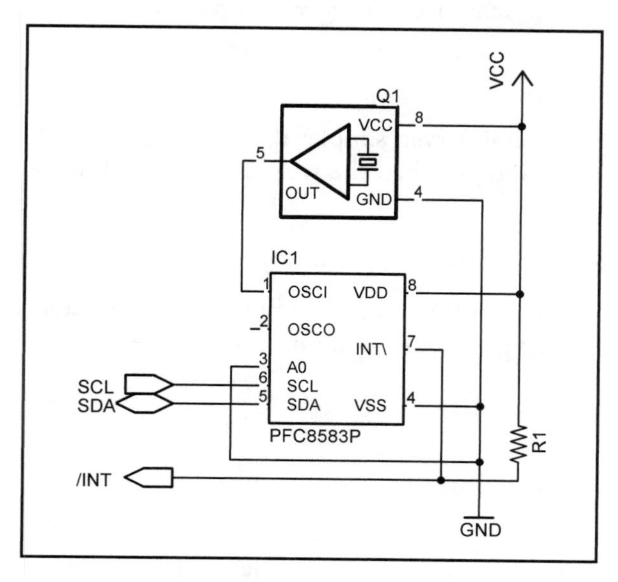

Figure 10-7 I2C clock/calendar

to 10K. Lower values, of course, increase current consumption when the interrupt is pulled low; higher values of R1 are more susceptible to noise.

From a master device's point of view there are two ways to communicate to an I2C peripheral. The master can be sending data to the peripheral, or the master can be requesting data from the peripheral.

Debugging the I2C code can be a bit trickier than something like the UART code or PWM system. First, the Cygnal sample code is run as a state-machine, driven by the I2C status byte. Beyond this structural oddness, though, is the fact that the I2C coordinates all activity through a sequence of Start, Stop, and Acknowledgment signals, plus a timeout system. If anything takes too long (because, for example, you are watching it in a debugger), the I2C drivers time out and go back to a waiting state.

Requesting Data

The process of requesting data from an I2C peripheral is fairly simple. First, the master (the MCU) triggers a start condition and sends the slave device address, with the R/W bit cleared, for writing. Then it sends a byte that specifies the address within the peripheral it wants information from.

The master then sends the slave address again, this time with the R/W bit set, for reading. The data byte is then received.

Once finished reading, the master can trigger a stop state.

Sending Data

Sending data to the peripheral is even simpler than reading data. The sequence starts out the same, with the device address, then the word address. Following this, any number of data bytes can be sent.

Once finished writing, the master triggers a stop state.

Additional I2C interfacing information, including sample source code, can be found at Cygnal's Web site, www.cygnal.com, in application note AN013.

SPI (Compass)

Motorola's Serial Peripheral Interface (SPI) is like I2C in that it is a serial interface supported by many microcontrollers.

Unlike I2C, SPI is defined for a single master device (the MCU) communicating with multiple slaves (peripherals). It is possible to have a multimaster SPI bus, but it may require extra work. The Cygnal C8051 MCU supports multimaster SPI systems. In SPI you have three communications lines plus the slave-select line for each slave, which is used instead of the device addresses we saw in I2C. The communication lines are the serial clock (SCK), Master In/Slave Out (MISO), and Slave In/Master Out (MOSI). With its dedicated (but well-named) transmit and receive lines, SPI is like a synchronous RS232 bus.

One of the goofier aspects of SPI is the relationship of the clock to the data. Different SPI devices may require different phases and/or polarities of the clock. The polarity of the clock determines if it is active high or active low. The phase determines whether the data is latched on the rising or falling edge of the clock—for active low; reversed for active high.

You can find the same range of devices for SPI that you see for I2C—I/O ports, analog-to-digital converters, serial EEPROMs, and the like. One special device that is worth noting, however, is the Vector Electronic compass from Precision Navigation, www.precisionnav.com.

While this is a lovely little peripheral, measuring about an inch and a half square, there are a few caveats in its use.

The basic V2X vector compass (as shown in **Figure 10-8**) needs to be mounted level to gravity to work correctly. Tipping the sensor will introduce errors into the heading. You can buy a gimbaled version of this, the 2XG, for more money.

Figure 10-8 Electronic compass

In the basic operating mode the vector compass sends out a heading value in degrees (0 through 359) in 16-bit binary mode, or 16 bits of binary-coded decimal (BCD). This heading information can be correlated with the calculated heading you get from the robot's odometry. It is important to correlate multiple senses to get a final position and/or heading value—each of the senses alone is inaccurate, but together you can hope for a reasonable answer.

You can also use the compass in "raw" mode, where it returns 32 bits of data (two times 16-bit binary values) describing the field strength in each sensor coil. The application notes from Precision Navigation tell how to convert this field strength to a heading, if you want. But more interesting could be the uses of these strengths directly—giving the robot a sense of magnetic flux. Not only would the robot know roughly which direction it was facing but it could "sense" the presence of metal and magnetism. Such a sensitive magnetic device as the vector compass is going to be affected by any large body of metal, such as the robot's body, as well as static or dynamic magnetic fields, like the robot's motors and power lines. Because of this, it is prudent to mount the compass on a pole made of fiberglass or some other nonmetallic material.

In addition to the three wires needed for SPI communication, the vector compass uses an array of other inputs to control its behavior. Some of these, like Calibration, Raw, BCD/Binary, and Master/Slave can be hardwired or, at worst, attached to a manual pushbutton for intermittent use. Others, like Poll/Continuous or Slave Select need to be hooked to a port on the MCU for good operation.

Look to the compass's manual for further details on its use.

Microwire

Microwire is another 3- or 4-wire interface, developed by National Semiconductor. It is a subset of SPI (essentially SPI with phase zero, polarity zero). There are lots of devices that use Microwire, such as EEPROMs, temperature sensors, and so forth.

What more can I say? One can only talk so long about serial interfaces before it stops being fun. If you want the lowdown scoop on all of these interfaces, I would recommend the book *Serial PIC'n* by Roger Stevens. Though that book was written for the PIC microcontroller, much of the information is still valid for the 8051 or, for that matter, any MCU.

Dallas 1-Wire

Also worth mentioning is the 1-wire bus from Maxim/Dallas. This is a clever interface that uses just one wire, as the name suggests. A related system is the Dallas iButton. You can find the usual range of suspects using this interface.

A 1-wire bus is not unlike the familiar RS232 serial bus but with shared Rx and Tx lines. One difference, though, is in the timing of the communication. There is some special synchronization done by the master, and the bits themselves are not transmitted with respect to a clock but have different widths (a '1' is a pulse of 15μS or less and a '0' is 60μS or so). Okay, so it's not much like RS232 after all. You can find devices and information on this protocol from Maxim/Dallas, www.maxim-ic.com.

iButtons can add special services such as serial numbers, passwords, and other identification-oriented features. The iButton is designed to be an external device that is attached (briefly) to a system in order to activate it or otherwise enhance its operation. It raises some interesting possibilities for the mobile robot.

One kind of iButton provides a unique, read-only, serial number. You can key your robot with this so it only functions, or only performs certain functions, when you authorize it. Another iButton provides 4K of read/write memory, or up to 64K of write-once memory. This could be used to take snapshots of the robot's state or to transfer information from the robot to a special iButton port on your PC, also available from Maxim.

BIT-BANGING SENSES

There are a number of sensory devices with digital outputs that will not fit onto a tidy I2C, UART, SPI, or Microwire interface. These devices need to be managed the hard way, by reading and setting bits on the MCU's I/O ports.

If your MCU doesn't support a particular protocol you can emulate it on the ports of the controller with nothing more than a timer interrupt and your wits.

Two senses that you may want on your robot that fit into the bit-banging category include an ability to recognize TV remote control signals and the ability to decode the signal from a radio-control receiver.

REMOTE CONTROL

In *Applied Robotics* I touched on the coding systems and receiver modules needed for a robot to receive TV remote control signals. Alas, that's all I'm going to do in this book, too—touch on the subject, but from a slightly different direction.

HARDWARE

The signal sent out by your average remote control unit consists of numerous pulses of modulated light.

The information being transmitted is the control signal (as shown in **Figure 10-9**). Internal to the remote control, this control signal is used to modulate a carrier frequency. What is flashed out the front of the unit, in infrared light, is the modulated signal.

The hardware aspect of this project converts the modulated signal back into a control signal. It is this control signal that the MCU then tries to decode.

The easiest way to decode the control signal is by purchasing a remote control receiver module tuned to the appropriate carrier frequency. These modules can be purchased for many different frequencies—I've seen them in 32.75kHz, 33kHz, 36kHz, 40kHz, 48kHz, 56.8kHz, and others. The actual frequency you buy isn't vital since they tend to be fairly lenient on what they actually accept. However, the closer your module is to the frequency used by the remote the better it will work.

A popular module that Radio Shack used to carry is Sharp's GP1U52X. The GP1U series is obsolete now (and Radio Shack has stopped carrying it), but it has been replaced by the GP1UD and GP1UM series from Sharp (www.sharpsma.com). These do the same job in a smaller package, if you can find them. Lite-On also used to have a nice series of modules, the LTM-97 series. Unfortunately, these have also been discontinued since I last purchased them. Finally, Vishay (www.vishay.com) has its TSOP12 series modules.

If you can find one, buy a module. They are easy to use. You send power in, they send a control signal out. Nothing could be easier. You may want to filter both the power supply to the module and the signal coming out of it (as shown in **Figure 10-10**).

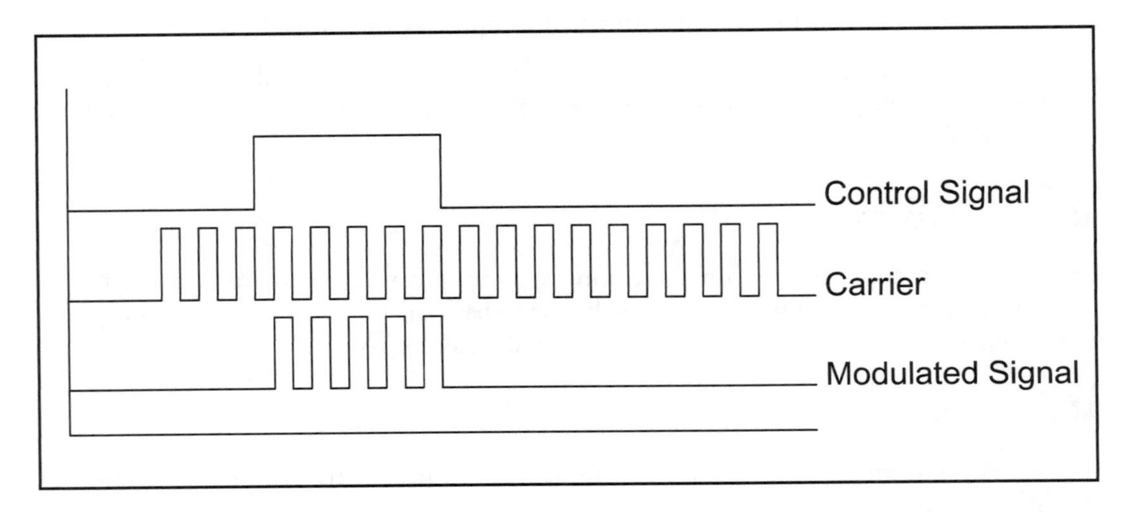

Figure 10-9 Remote control signal

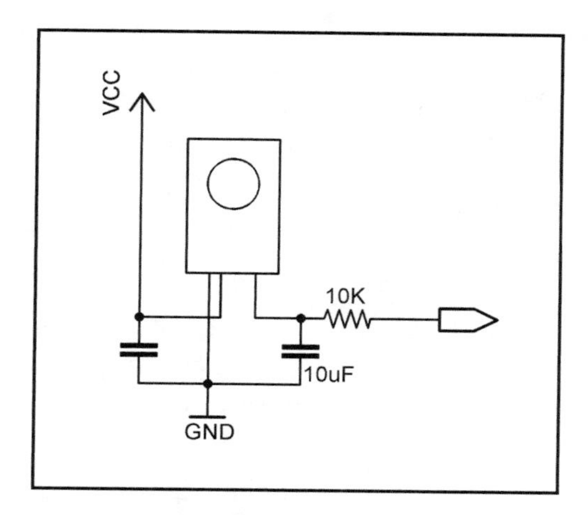

Figure 10-10 Remote control module

If you can't get a module you *could* build your own, though as usual I don't recommend making what you can buy.

SIGNAL CODING

Now that you have a nice digital pulse stream from the receiver hardware, you need to decode it inside the MCU. This is a lot like the problem of reading rapidly changing wheel encoders from Chapter 9. Unfortunately, we have a limited number of external interrupt sources—there are the external interrupts IE0 and IE1, then the four high bits in Port 1 (IE4 through IE7; there doesn't seem to be an IE2 or IE3 in the C8051). Though they aren't hooked up in the bus shown in Chapter 8, the comparator inputs can also trigger interrupts.

Once the signal is driving an interrupt, you need to time the distance between pulses, again, like you did for the wheel encoder. Depending on the remote control, the pulses may vary from 0.5mS to 2.0mS in duration. At least they will be fairly predictable.

Everybody has a hand in the consumer remote control protocol business, and this leads to a stack of different protocols, such as RC-5, RC-6, RECS-80, NEC, RCA, HP-SIR, and Sharp-IR protocols. In general, though, there are two classes of coding—pulse-width and pulse-phase coding.

The easiest code to read is the generic pulse coding signal, shown in the top of **Figure 10-11**. Each pulse is either short (coding a zero bit) or long (for a one bit). The long pulse width will typically be some multiple of the short pulse width; something like three times the duration of the short pulse. This broad disparity in widths should make these pulses easy to distinguish from each other. This coding is used by, for example, Sony.

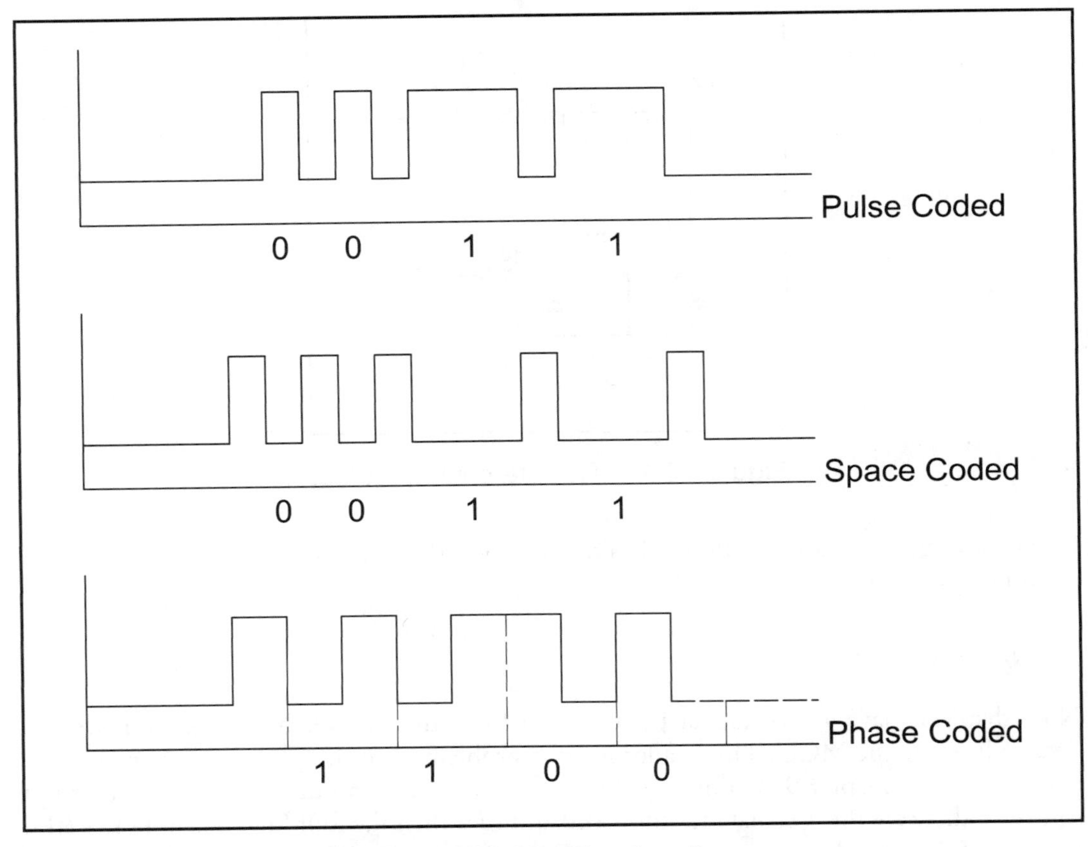

Figure 10-11 Signal coding

Related to pulse coding is space coding. In this method each pulse is the same width. The width of the *spaces* between the pulses is what defines the bits. The spaces will be long or short and should be fairly easy to distinguish. This is the RECS-80 coding scheme.

If you record the combined time of the pulse and the space after it you can decode pulse and space coding systems with equal ease.

The last coding style, phase coding (or shift coding) is different. Once the signal has started each bit occurs in a fixed-length time slot. The lead-in to the signal will typically be a pair of '1' bits to wake up the receiver and define the time slot duration. After this, the receiver must watch for the transition in the middle of each time slot. If the time slot has length *t*, then the receiver can check the signal level at time *t/4* and time *(3*t)/4* and compare the levels. A high-to-low transition is a zero bit, and a low-to-high transition is a one bit. This is the RC-5 coding scheme used by, for example, Philips.

You can find detailed information on the (many) various coding methods on the Internet. One good place to find this data is in the specifications of the ICs that *generate* these signals—such as the SAA3010 remote control transmitter IC from Philips.

R/C Input

Now that you are warmed up to the thought of measuring pulse widths using an external interrupt and a timer, a concept we've visited several times in this book, you may want to try your hand at decoding the signals coming out of your R/C receiver. These were shown in Chapter 4, in various ways, by Figure 4-5 and Figure 4-7.

The first problem is that there are a bunch of separate wires carrying one R/C signal each. It would be a nuisance to use up that many interrupts, even if you had them.

The first step is going to be to combine the signals, as shown in **Figure 10-12**. The easiest way to do this (assuming you don't have simultaneous output signals as shown in **Figure** 4-7, which you probably won't) is with a multiway OR gate like the 4-input 4072. You can also pair together both 4-input gates of the 4072 with the 74HCT32. You can, of

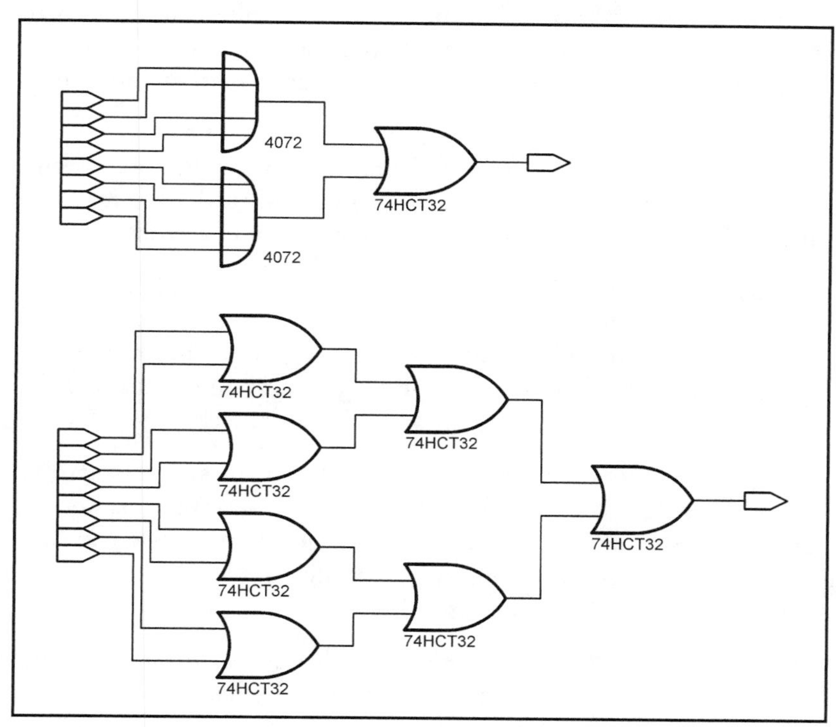

Figure 10-12 Multiplexing RC cignals

course, cascade any number of 74HCT32 gates as well. One nifty device that can be used in several different ways is the 8-input multifunction logic gate 4048. You could probably even smack an 8-channel multiplexer like the 74HCT251 into service, though the MCU would have to drive the channel select to choose the next signal to decode.

Once you have a single stream of pulses you need to measure them. Measurement is complicated by the detail that you need to make timing distinctions down to a few microseconds, which is not very many MCU clock cycles.

VIRTUAL SENSES

We've looked at several types of physical senses so far. Some senses help locate and orient the robot in time and space, like the clock/calendar, wheel encoders, and compass. Other senses provide a sense of space around the robot, such as the GP2D12 distance sensors, sonar if you go that route, and bumpers. And still others give a sense of the robot's state, measuring things like temperature, battery power, light, sound, and so forth.

Using the location and space-sensing inputs, you can, in theory at least, create a map of the environment. This is, like everything in robotics, easier said than done—but it can be done. Or if you particularly enjoy tedium, you can create a map of your environment for the robot to start with.

Once you have a map you can use the robot's senses to align the robot within the map. But the fun begins when you (a) have a map in hand, and (b) have a reasonable sense of where you are in the map. This is when virtual senses can come online.

A virtual sense is just a bit of software that looks into the map and reports what is there. Given an assumed robot position and orientation, the virtual senses can provide information about what *should* be around the robot, even though it is out of range of the robot's physical sensors. These virtual senses don't have to be "point" senses, either, but can accumulate or average information over an area of map space.

If you have map path-planning software in the computer, you can give the robot an "urge" to follow a path. Given a position and a desired direction to travel, this homing sense can nudge the robot along the path. If your robot has some practical task, such as vacuuming or delivering stuff, the high-level planners can create simple directional urges that the hardware reflexes can fulfill. These are also virtual senses.

Ultimately, the robot's control system needs to synthesize all of the competing impulses, inputs, and information in order to make a decision on what to do next.

CHAPTER 11

PC Control

Robots are complex and difficult machines, as you have probably noticed by now. And if you want your robot to do anything other than the most basic of reflexive tasks, such as follow a line, follow a wall, avoid light, etc., you will find yourself wanting more computer power than the basic MCU can provide.

The easiest choice for a computing upgrade is the personal computer. Your modern 1.6GHz PC is running 100 times as fast as your basic 16MHz MCU, plus it will have a math co-processor for high-speed calculations. And since used laptop prices are very reasonable you can now afford to put a powerful computer on the robot itself.

Of course, an off-the-shelf PC is not the only solution to the need for greater computing power. You can purchase embedded computers that run off of versions of the same CPUs that your PC uses, except these computers are *designed* for use in a robot-like environment. Embedded options include PC/104, EBX, and ETX-sized systems, plus more standard-sized motherboards designed for an embedded environment.

As you might expect, there are tradeoffs.

Your garden-variety laptop computer is going to be the least expensive solution for any given system configuration simply because it is a mass-market product. You can also find used and reconditioned laptops for excellent prices at various places on the Internet, such as eBay.com, uBid.com, and Pricewatch.com, as well as through local sources. You can find all manner of development software, toolkits, libraries, and other support systems for your portable PC. The biggest downfall of a PC, however, is in its operating system. Unless you move into Linux territory, which we don't explore in this book, your machine will be running some flavor of Microsoft Windows (or MacOS)—which will then proceed to eat many of your CPU cycles, a lot of memory, and so forth. Of course, these operating systems also provide device drivers, utility libraries, and other software resources to make your programming job easier.

TRADEOFFS

Embedded machines come in many different capabilities and sizes and you can find them from many different vendors, for example, Ampro (www.ampro.com), EMJ (www.emjembedded.com), and ZF MicroDevices (www.zfmicro.com) among others. Embedded computers come in a variety of form factors, differentiated mostly by size. One of the most popular is PC/104, which is 3.6" by 3.8" in size. Others include EBX (5.75" by 8.00") and ETX (about 3.9" by 4.5"). Though embedded computers are not going to come with the same advanced speeds as laptops, you can get some very respectable systems on these tiny boards. And, as for all things miniature, you will be paying a premium. There will be less overhead consumed by an operating system (none, in some cases), as well as less support for your software task. Also on the plus side, these computers will be easier to integrate into the power and space provided by the robot.

You don't necessarily need to put all of your computing power on the robot anyway—you can communicate with stationary brainpower using a wireless data link.

WIRELESS COMMUNICATIONS

Radio modems have been on the market for quite some time and are inexpensive and reliable—though they are also fairly slow and usually no faster than 19,200 baud. Most radio modems operate in the very busy 900MHz radio band, along with many cordless phones and other wireless devices. There is no easy solution to the radio interference you get between the robot's radio modem and these other devices.

If you want to get all masochistic and build your own wireless communications device, you can. There are a number of wireless communications modules available from manufacturers such as Abacom (www.abacom-tech.com) and Radiometrix (www.radiometrix.co.uk). You can also purchase radio transceiver ICs for some hardcore suffering. Of course, these all tend to operate in the same crowded radio space.

Right now the best approach to creating a radio link between two computers would be 802.11b wireless Ethernet. This is a technology with a rapidly growing user base and devices are reasonably priced, with base station prices under $200 and cards for your PC or laptop costing less than $100. 802.11b can provide network speeds of up to 10Mbit/sec, given the right hardware and a clean signal. The advanced 802.11a standard is promising over 50Mbit on the 5.4GHz band—but at the expense of reduced range and increased power consumption.

The range of a generic 802.11b link is about 100 feet or less. The system was designed to work inside a building with the computers and antennas arranged for optimum reception. A wandering robot can easily fall into blind spots where it loses communication with the base station. You can increase the range with a high-gain antenna such as a dish antenna, but then you add the problem of keeping it pointed in the right direction, since a dish is highly directional. You could, in theory, kick up the range by increasing the power of the transmitters, but the FCC would definitely disapprove. You have to have a special amateur radio, or "Ham," license to operate a higher-power transmitter.

Wireless Ethernet operates in the 2.4GHz radio band, which used to be fairly free of clutter. Unfortunately, wireless phones and other devices are beginning to escape the 900MHz band and move into 2.4GHz territory to escape the 900MHz clutter. Of course, this isn't so much helping the 900MHz band as making a mess of 2.4GHz.

A fairly new kid on the wireless communications block is the oddly named Bluetooth standard. Unfortunately, as of this writing, Bluetooth offers mostly promises with little consumer equipment to back them up. Bluetooth was designed to replace short-run cables, such as between your printer or keyboard and the desktop computer. With a range of 30 feet, it would provide little help for the mobile robot.

Finally, if you are serious about going wireless, you should consider amateur radio. An amateur radio license is the only gateway to the high-power transmitters that are so useful for long-range communication. The new "no code" Technician license allows you to operate on many different frequencies and with significant power. Packet radio in the amateur bands provides a convenient (and time-tested, showing up on the amateur radio scene in 1978) method of data communication. Packet radio protocols allow two computers to communicate across the radio as if they were connected directly together, across a range of many miles. Unfortunately, data speeds only range from 1,200 baud to 9,600 baud. Amateur TV can be used to supplement this data link, however, providing for standard 30-frames-per-second video data sent across long distances.

Not only can you run data links between a computer on the robot and a computer on your desk, you can run video links from a video camera on the robot to a video capture card on your computer. Some video links, such as the X10 video transmitter, suffer from some of the same antenna problems as the basic 802.11b Ethernet, such as directionality, low range, and dead spots. But scout around, you can find a bunch of interesting systems in the spy camera business, among others.

Brain Cluster

Assuming you have a high-speed link between the robot's onboard computer and a desktop computer, there is no end to what you can do. For one, you are not limited to doing your computations on a single machine. You can use a whole network of computers to do the thinking. The one PC that communicates with the robot can parcel out robot data to any number of other machines on a network. It can then integrate the control data from these same machines and send commands back out to the robot.

Computers have become so cheap, and technology has progressed so quickly, that older computers are tossed in the trash or sold for pennies on the dollar after only a few years of use. These older machines are still very fast, especially compared to MCUs, and can be pressed into service as part of your distributed robotic brain.

While you could probably do some good work with ad hoc networking, there has been a bunch of recent interest in building supercomputers from stock hardware. One project that leaps to mind is the Beowulf Cluster project, which has numerous Web sites devoted to it, as well as a number of books. Check out the References list at the end of this book for several that I have found useful. A search for "PC Cluster" and "Beowulf Cluster" on the Internet should turn up a wide variety of references and resources.

A cluster computer, such as Beowulf, uses special software to tie the various machines that compose it into a single synchronized virtual computer. This type of system is best applied to problems where one algorithm is applied repeatedly to different pieces of data—such as subregions of a video image. Clusters can be assembled to tackle extremely processor-intensive applications like image processing and tied in to other computers with more traditional networking to coordinate other tasks.

A picture of what is possible for your robot's brain is beginning to emerge. First, like the animal brain, the robot brain is going to be constructed out of a hierarchy of devices. At the lowest level you will have any number of reflex controllers—the microcontrollers that read and drive the actual hardware. These reflexes take simple, primitive commands and execute them. They also report back on the raw (or, perhaps, slightly cooked) data they receive from the senses. While the MCU is in constant communication with the hardware, its communication to the higher-level PC is limited. Commands are simple and short, and data is only sent up when it changes or is requested, and the data itself might be filtered and adjusted to simplify it further.

One step up from these reflexes will be a single MCU or smaller embedded computer that coordinates them. This machine is like the spinal cord, perhaps. It doesn't do much in the way of work itself, but it coordinates the efforts of the outlying systems on the reflex network, which may be I2C, SPI, or whatever. On the robot itself, I've found that the RS485 hardware interface chips provide a very reliable communication platform in the noisy environment of the robot. Regardless of the reflex network you end up using, this spinal cord controller provides a single point of communication with the next higher level.

The next level above the reflex network is also where we will see the first use of a laptop or more powerful embedded computer. This could be like the brainstem—it performs more advanced processing but it is still no rocket scientist (or even a rodent, for that matter). The tasks that this computer performs will depend in large part on the model you are using for the robot intelligence. Perhaps this computer is the sum total of your robot's higher brain, performing all mapping, planning, and so forth. For your first robot or three this will likely be the case. And you can do a *lot* in one computer these days.

However, if you are motivated, have lots of time, and deep pockets, you will want to think about the next step. This is the high-speed link from the robot to one or more computers that make up the robot's higher brain functions. Once you have an architecture in place for this network, you can slowly add to it over time until, with any luck, your machine moves out of the house, gets a good job, and begins to support you in the manner in which you would like to become accustomed. Well, maybe not. But we can always dream, right?

(Much of the information about wireless communications and clustered computing in this section is from Glenn Currie, a founding member of the Robot Group in Austin, Texas. Thanks, Glenn).

COMMUNICATION

There are two levels of communication to look at here. The first is between the MCU "spinal cord" command router and the outlying MCU "reflexes." The second is between the onboard PC and the spinal cord.

REFLEX NET

Assume, for a moment, that you have more than one reflex microcontroller (you may not, but you might), and you need to connect these MCUs together so that they can communicate with a gateway MCU, which in turn will communicate to the PC using plain-vanilla RS232 serial communications. So, how do you tie all of the reflexes together? You have choices.

Regardless of the choice of protocol you use, you should do everything you can to protect the physical wiring from noise. Robots are very noisy environments. The motors and high-current wiring throw all sorts of electromagnetic signals through the air and mess with the voltage levels in the power system. This noise will not only disrupt communications between MCUs, in extreme cases it can reset the MCU. You can begin to protect yourself by keeping wires away from these noise generators, building well-shielded circuit boards with nice ground planes when possible, and using shielded wire for communications. When applicable, use twisted pairs of wire with a differential signal, such as RS485, for communication.

UART NET

Though it wasn't really designed for this, the UART on your MCU, traditionally used for simple RS232 communication, can be used to drive a reflex network. This network must be a single-master system, but that fits our current needs anyway. The gateway MCU is the master and each of the reflex MCUs would be a slave (see **Figure 11-1**).

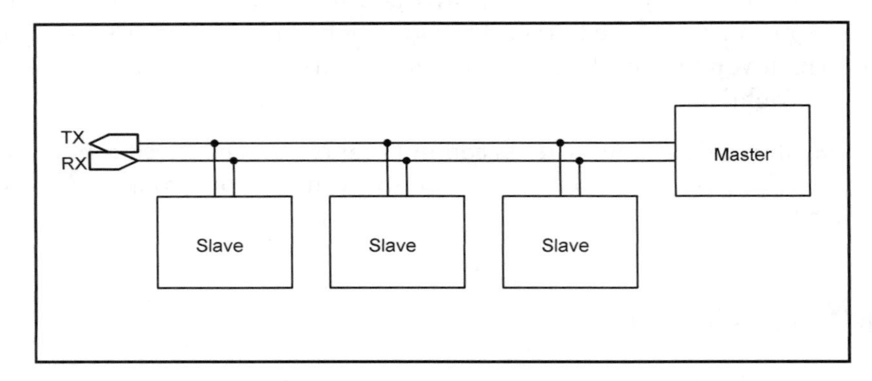

Figure 11-1 UART net

Anything sent by the master MCU will be heard by all slaves so it is important to be able to give each slave a unique address. This can be done with hardware, that is switches or jumpers on the MCU, or, though more cumbersome, through software, for instance, you turn on one slave, give it an address; turn on the next, and so forth.

Since all of the slaves can, in theory, send information to the master at any time, it is important that they speak only when spoken to—that is, the only time they wake up their communications to write to the master's Rx line is when they have been asked to by the Master. At all other times the slave's transmit circuit should be disconnected from the line. This circuit is shown in **Figure 11-2**.

The output from the MCU's UART will normally be a 3-volt or 5-volt signal, depending on the MCU. Though this could be used directly, when communicating between MCUs at least, this signal is easily corrupted by noise and won't travel very far (though this last factor may not matter much unless you have a very large robot).

The first obvious option is to boost the signal using the cheap and familiar RS232 level converters. While this works very well it is still prone to noise in some robots. I originally used RS232 levels in my Boris robot, but with all of the noise from the many pneumatic valves, each a little inductive transmitter of noise, (ouch) I had problems.

My favorite communications physical layer, because of its extreme toughness in the face of adversity, is the RS485 driver (for more information, you can read about the RS485 and RS422 standards and products online).

Figure 11-2 shows a sample RS485 circuit. This one is using the MAX489 chip, though there are many others available, such as the LTC1481, or the related 3.3-volt LTC1480. The wiring can be done with standard Ethernet cables plugged into RJ45 sockets. Though not shown in the figure, be sure to include ground in the wiring.

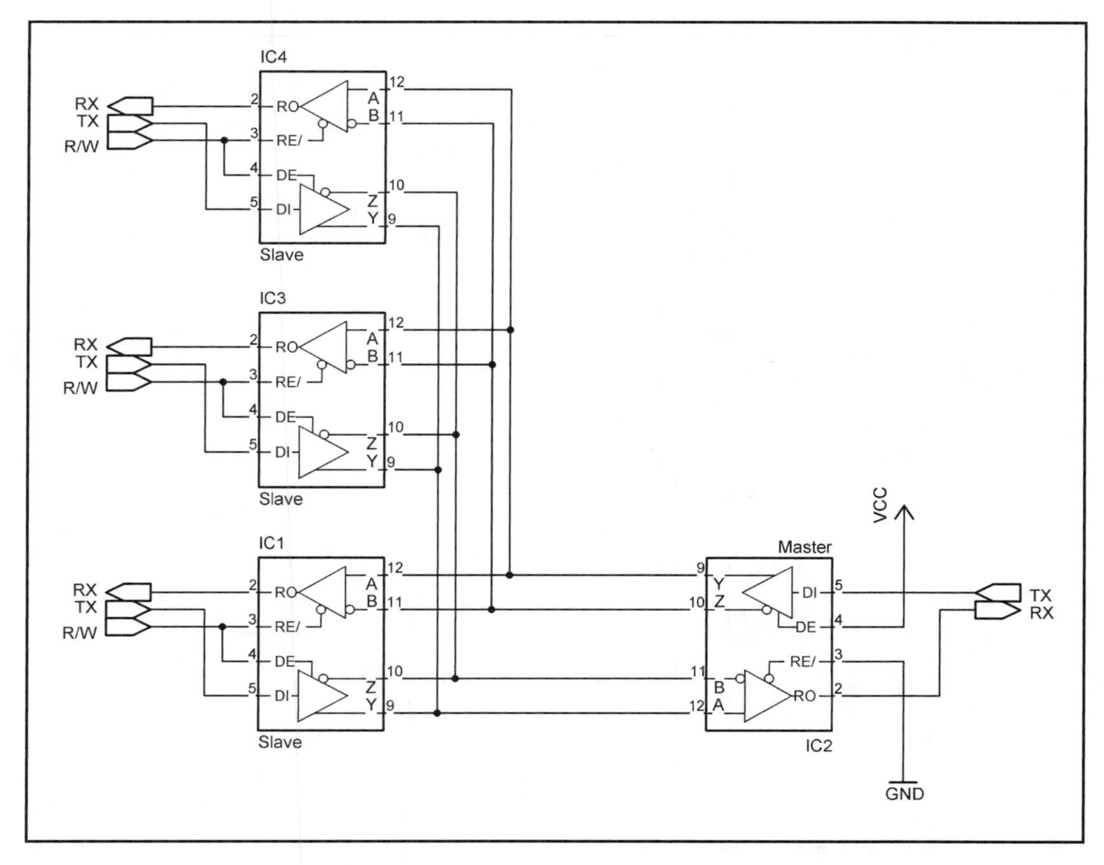

Figure 11-2 RS485 physical layer

Be aware that if you want to communicate for long distances across a RS485 physical layer, such as the theoretical 4,000-ft. maximum, or at very high speeds, you need to pay careful attention to line termination and other technical factors relating to the wire you use. These concerns are addressed by most RS485 IC specification sheets and don't affect us anyway for our short distances and relatively slow network speeds.

In a differential driver, each communication direction uses two wires. The bit status is taken from the difference between the wires and not just a stand-alone voltage level. Since noise should affect both wires roughly equally, especially if they are closely spaced, or twisted together, the information carried on the pair should survive even in a very noisy environment.

The circuit in Figure 11-2 is full duplex. You can also use a half-duplex circuit as shown in **Figure 11-3**, using the read/write enable line to avoid conflict on the wires.

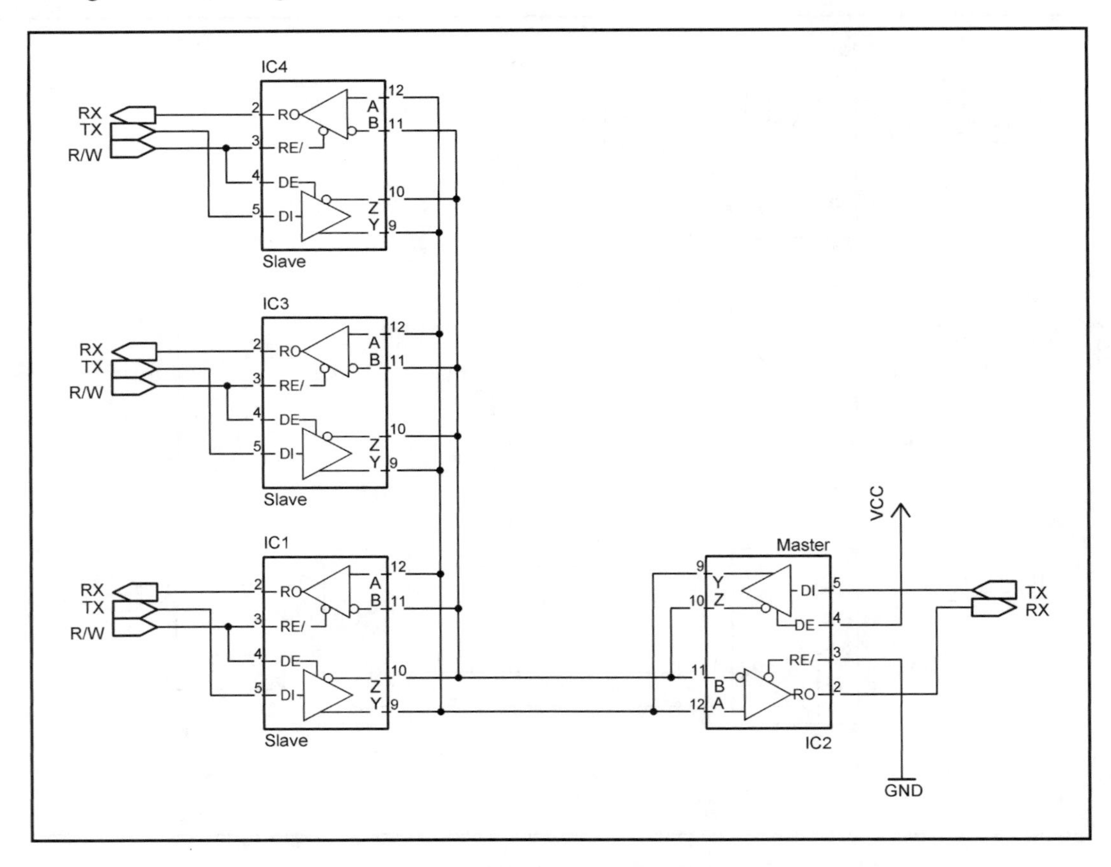

Figure 11-3 Half-duplex RS485

If you wanted to get really clever you could hook up an RS485 driver to an RS232 driver and tie the reflex net directly to a laptop or PC, instead of communicating through an intermediary MCU. This might be best, in fact, since not all MCUs have the two UARTs needed to act as a gateway.

If you do use an intermediate MCU to tie this network to a PC, it would need to either switch between two drivers (an RS485 to the network and another RS232 to the PC, with the inactive drive set to standby mode) or simulate a second UART in software. (You can find examples of this for most MCUs in their application notes.) It might be easiest to use the hardware UART to talk to the PC and use a different software communications protocol to create the reflex net.

SPI NET

Since the Serial Peripheral Interface (SPI) is designed to work as a single-master network (see **Figure 11-4**), it can be connected in the same ways as the UART net described above. There are two complications. The first is the need for a third master transmit driver for the clock, and the second is in the slave-select line.

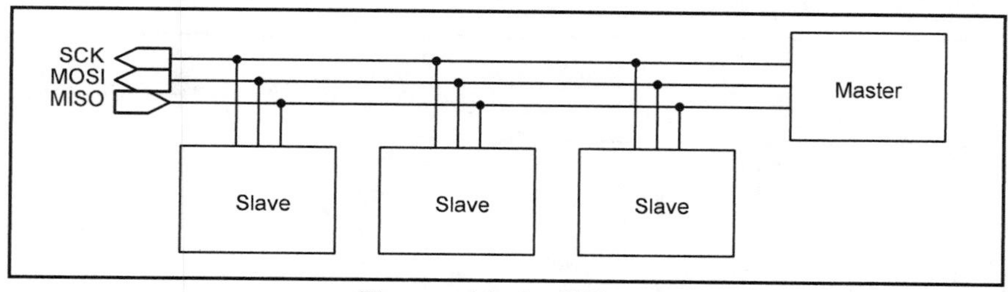

Figure 11-4 SPI net

The extra driver can be handled by buying a different IC, with two drivers and a receiver for the master, and with two receivers and a driver for the slaves. Of course, this is not such a common requirement, so you will probably end up using a pair of RS485 chips per MCU.

The issue of slave select can be handled by virtue of ignoring it—and using an addressing scheme to identify the slave explicitly.

I²C NET

The I2C interface does not lend itself to the RS485 solution like the UART and SPI did. I2C, you recall, is a multimaster interface where each line, SCL and SDA, is normally left floating at VCC and is pulled down by each device as needed (see **Figure 11-5**). This prevents the overt conflicts that would arise from more than one device trying to positively drive the signals at the same time. Of course, I2C was designed as a local protocol for use on a single circuit board between ICs (hence the name *Inter-IC*) where the native voltages are perfectly acceptable. But for long-distance use in a noisy environment we need to do something different.

If you are using I2C to connect to an OOPic device, you won't be doing anything fancy—simply connect SDA and SCL from the MCU to these same signals on the OOPic. Of course, the communication protocol you use to talk over these lines is between you and your version of the OOPic.

One solution to boosting I2C distance or reliability is the fairly obscure 4-wire version of the interface. This, however, is not that well supported.

The official solution is to drop a Philips 82B715 current driver onto each node of the net. With proper application this chip can give you I2C communications for up to a mile. Before you start stringing up your neighborhood, though, you should visit Philips Semiconductor (www.semiconductors.philips.com) and read up on this chip and its uses.

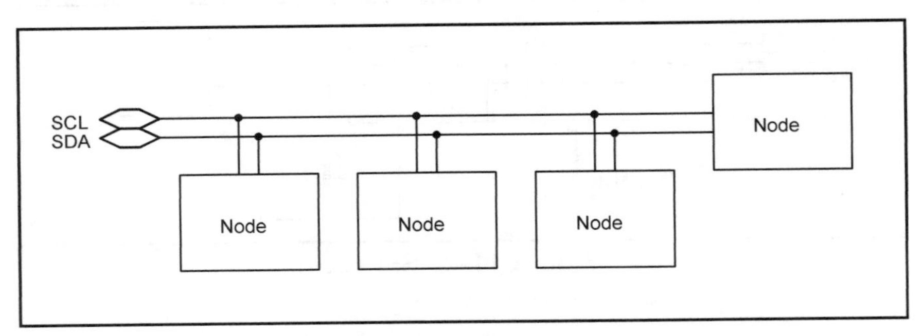

Figure 11-5 I^{2}C net

MCU TO PC

There are two ways to go here. One way is to eliminate the gateway MCU (the "spinal cord") entirely and just hook the PC up to the reflex net. The other way is to hook the gateway MCU to the PC using RS232 communications through the UART while talking to the reflex net with a different protocol, such as SPI, or with a bit-banged UART.

A simple bridge circuit can be used to eliminate the gateway MCU, as shown in **Figure 11-6**. In this case, though, the reflex net must be running on standard UARTs so that the

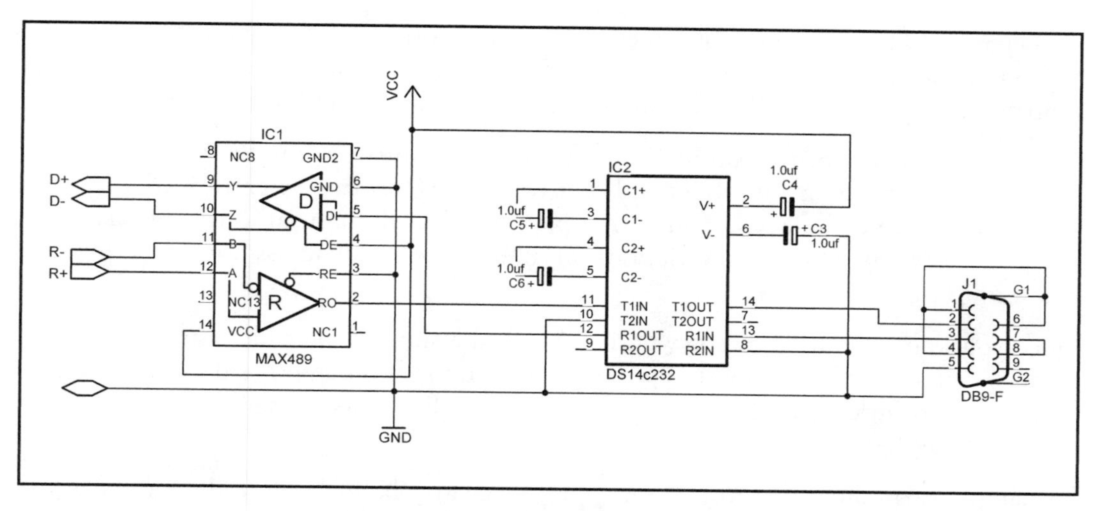

Figure 11-6 Gateway eliminator

serial port on the PC will be talking the same protocol. In theory, you can use advanced serial port manipulations to emulate, for example, the SPI protocol, but that is just making extra work for yourself.

The fairly obvious layout of the PC, master gateway MCU, and slave reflex MCUs is shown in **Figure 11**-7, in this case, using an SPI reflex net.

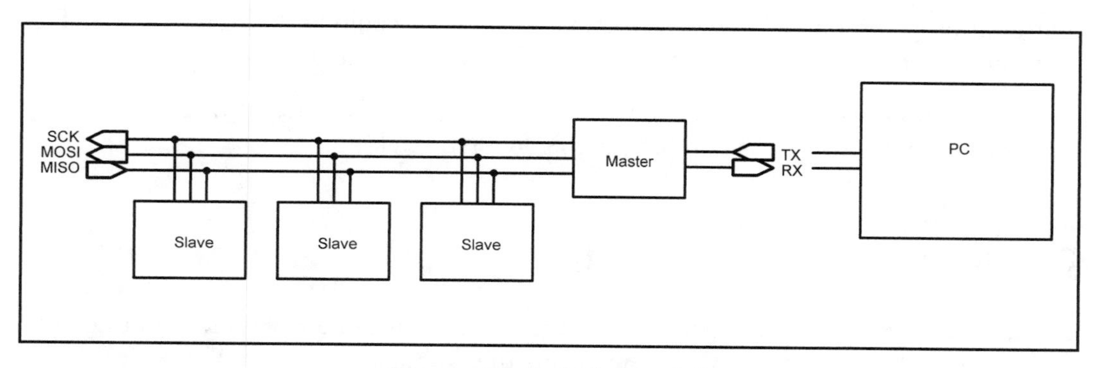

Figure 11-7 MCU to PC

Command Protocol

Once you get past the hardware details, it is time to think about what you will be saying across the hardware.

To begin with, for debugging purposes, you should work with just two devices—the PC and one MCU for example, or two MCUs. Start from the easiest to debug on platform, probably the PC, and work your way down to the deepest interior of the system.

The first step is to create a simple echo. A character sent from the PC is, for example, incremented by one on the MCU and sent back. This test can be performed with very simple software on the MCU and with a plain-vanilla telecommunications package on the PC. This same test, or variations of it, can be used across each layer in the communications hierarchy.

Once you know the hardware is working, using the simplest possible tests, you can get fancy. Probably the most robust approach to communication is to use data packets. The packet itself is little more than a wrapper around application-specific information, whatever that may be.

For simple two-point communications, the packet may be little more than a size and a checksum:

0:	Data size (0..255)
1..N:	Application data
N+1:	Verification code

This checksum or verification code at the end provides a way to determine that the data arrived safely. Upon receipt of this packet the recipient could reply with an acknowledge or error byte.

For multidrop communications, the destination address would be added to the mix. This way, the packet would be ignored by all but the intended recipient.

Communication between the processors can be as simple as ASCII commands that can be typed from a terminal, or as complex as the TCP/IP stack.

PROJECT 11-1: CAMERA EYE

The sense of sight is normally out of the range of microcontrollers (with few, but notable, exceptions such as the $100 CMUcam from the Robotics Group at Carnegie Mellon University). After all, even given a fairly small video frame size of 320 by 240 grayscale bytes at 30 frames a second, you are processing over two million bytes of data per second! With color cameras, there is even more.

With a PC at the helm of the robot a camera becomes more practical. This project describes the process of getting the images out of the camera and into our application.

When you want to attach a camera to your computer, there are several ways you can go. Traditionally you had to stick a digitizing card into one of the PC's slots and from there you could hook up any number of different cameras.

One specialized camera system of note is the SRI Small Vision System (www.ai.sri.com), marketed by Videre (www.videredesign.com). This system takes data from a pair of cameras and sends out its own video signal where shades of gray indicate distance from the camera. If you are interested in stereo vision, check these guys out. Intel, in late-breaking news, has a stereo vision shareware library...so bop over to Intel.com and see what it has cooking!

These days you can attach inexpensive "web" cameras into various computer ports. Though these inexpensive (many less than $50) webcams are widely available and easy to use, they won't give you the same video quality or frame rate as a digitized video signal.

Once the camera hardware is attached to the computer, driver software is needed to interpret the signals. Fortunately, these drivers will be shipped with your webcam or digitizer.

Finally, you access this driver to get the video data into your application. This is where it becomes complicated.

If you are using a Microsoft Windows operating system on your computer you have two basic choices for video driver: Video for Windows and DirectShow, as a part of the DirectX 8.0 system. Video for Windows (VFW) has been around for ages so you will find drivers for most cameras for it. Since VFW has been around as long as it has, technology has done its best to pass it by—so DirectShow (XDS) was created to manage the modern video device.

If you are using any other operating system, you are on your own. There are solutions available, of course, I just haven't had the opportunity to use them.

Microsoft Research (research.microsoft.com) has gone a step further and simplified your ability to access video data with its free Vision SDK. This package provides a bunch of libraries, templates, MFC wizards, and sample programs to aid in video processing. The Vision SDK also interfaces with the Intel Performance Libraries (developer.intel.com) and the ImageMagick (www.imagemagick.org) libraries. These libraries add optimized image and graphic file processing to the Video SDK.

Vision SDK

This project is based on the Microsoft Research Vision SDK. You need this package, and you can find it on the MSR Web site, research.microsoft.com. The full source and libraries makes for a zip file that is almost seven megabytes long, so get a head start on this.

I, for one, am very happy that Microsoft spends some of its cash flow on research projects and provides the results to us, the developing audience, for free. I only wish it would keep its documentation up to date. The help file is, as of this writing, for version 1.2 of the SDK, two years out of date, which makes this interface all that much more frustrating to work with.

Once you have downloaded and installed the Vision SDK, follow the instructions in the included documentation to register your camera. Now compile and exercise the test programs. Try making a new project using the Vision Wizard.

Pretty nifty stuff.

CAPTURE PROGRAM

Now we can look at a simple program that provides real-time video access. There are three parts to this program: the MFC Application that is the entry point to the system, a dialog that acts as our main loop and gives us access to various controls, and the vision class that wraps up various Vision SDK calls into a tidy package.

Since there is a lot of code involved in even the most basic Windows application, I'm afraid this section might get a bit tedious. I'll try to breeze by the framework and get to the good stuff, the vision class, quickly.

CCAPTUREAPP

The Capture Application is the starting point for the application. It is a very basic MFC application framework, as shown dissected here:

```
class CCaptureApp : public CWinApp
{
public:
    CCaptureApp();
    //{{AFX_VIRTUAL(CCaptureApp)
    virtual BOOL InitInstance();
    //}}AFX_VIRTUAL
    //{{AFX_MSG(CCaptureApp)
    //}}AFX_MSG
    DECLARE_MESSAGE_MAP()
};
```

The only function that *does* anything is `InitInstance`, and all that does is start up the Capture Dialog:

```
BOOL
CCaptureApp::InitInstance()
{
   Enable3dControls();
   CCaptureDlg dlg;
   dlg.DoModal();
   return FALSE;
}
```

Hopefully you can reconstruct the actual MFC application from these nuggets.

CCaptureDlg

The Capture Dialog is a little bit more interesting; it is also rather a bit longer, so I will only reproduce the most relevant pieces here. Both the Capture Application (above) and the Capture Dialog (here) were originally generated using the MFC AppWizard (exe) to create a Dialog Based application.

Since this is a dialog, you will want to give it the look shown in **Figure 11-8**.

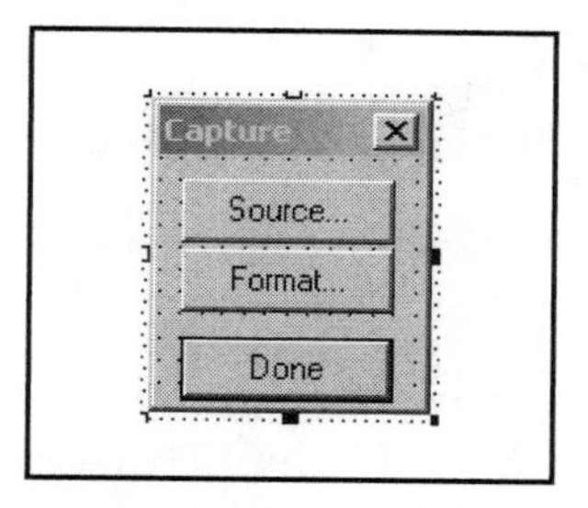

Figure 11-8 Capture dialog

The IDs associated with the buttons are `IDC_SOURCE`, `IDC_FORMAT`, and `IDOK` respectively. You can use the Class Wizard to hook up the `BN_CLICKED` messages for the first two buttons. `IDOK` is handled automatically for you.

The Capture Dialog class definition is very simple, as shown here:

```
class CCaptureDlg : public CDialog
{
public:
   CCaptureDlg(CWnd* pParent = NULL);     // standard
constructor
   virtual ~CCaptureDlg();
   //{{AFX_DATA(CCaptureDlg)
   enum { IDD = IDD_CAPTURE_DIALOG };
```

```
    //}}AFX_DATA
    //{{AFX_VIRTUAL(CCaptureDlg)
    protected:
    //}}AFX_VIRTUAL
protected:
    CVision     m_vision;
    HICON       m_hIcon;
    //{{AFX_MSG(CCaptureDlg)
    virtual  BOOL OnInitDialog();
    afx_msg  void OnSource();
    afx_msg  void OnFormat();
    //}}AFX_MSG
    LONG OnVisionUpdate(UINT, LONG );
    DECLARE_MESSAGE_MAP()
};
```

Note the `CVision` member object; this is where the work is going to happen. `OnVisionUpdate()` is going to be the main loop of the application.

Here are a few select methods from the source file:

```
BEGIN_MESSAGE_MAP(CCaptureDlg, CDialog)
    //{{AFX_MSG_MAP(CCaptureDlg)
    ON_BN_CLICKED(IDC_SOURCE, OnSource)
    ON_BN_CLICKED(IDC_FORMAT, OnFormat)
    //}}AFX_MSG_MAP
    ON_MESSAGE((WM_VISIONUPDATE), OnVisionUpdate)
END_MESSAGE_MAP()
```

The message map is fairly important. Most of it is automatically generated by the ClassWizard. However, you need to add the `WM_VISIONUPDATE` message by hand. This message is sent by the camera drivers when a new image is ready to be captured.

```
BOOL
CCaptureDlg::OnInitDialog()
{
    CDialog::OnInitDialog();
    if (!m_vision.AttachToSource())
    {
          AfxMessageBox("Unable to attach to a camera
source");
          return FALSE;
    }
    m_vision.StartCapture( m_hWnd );
    SetIcon(m_hIcon, TRUE);                  // Set big icon
    SetIcon(m_hIcon, FALSE);            // Set small icon

    return TRUE;
}
```

The initialization method does little more than attach the vision class to a data source and then turn it on.

```
void CCaptureDlg::OnSource()
{
   m_vision.SourceDlg();
}
void CCaptureDlg::OnFormat()
{
   m_vision.FormatDlg();
}
```

These methods (OnSource() and OnFormat()) are used to call up dialogs to change the camera source and capture format. They can only be used once a capture source has been started, so you can't put calls to the dialogs in the initialization method. The AttachToSource() call should ask you for a camera source if one has not be previously set.

```
LONG
CCaptureDlg::OnVisionUpdate(UINT, LONG)
{
   m_vision.Capture();
   return 0;
}
```

The main loop is painfully simple; it calls the vision Capture() method, which does all the work.

CVISION

The vision class is where the fun happens. We will review this class in detail. It is listed in its entirety here—but bear in mind that we are only touching the surface of the Vision SDK. You will want to spend some quality time with the Microsoft documentation and header files for the Vision SDK.

We start with the header file, Vision.h:

```
#if !defined(_VISION_H)
#define _VISION_H
#pragma once
//
=================================================================
//      Vision
//
// Vision wrapper class – implements a version of
Microsoft's
// Vision SDK.
//
```

```
// Might work under DirectShow (XDS), however it has only
been
// tested under Video for Windows (VFW).
//
// Copyright 2001, Simulated Reality Systems, LLC
//
//
==============================================================
```

Okay, so the header block isn't very exciting.

```
#include <VisImSrc.h>
#include <VisDisplay.h>
```

Of course, you need to include a variety of Vision SDK headers. There are others, but we don't delve into their code today.

```
//
==============================================================
//      Type Definitions
//
// Convenient shorthand for various classes and templates
// Provides for quick and easy data-type changes.
//
#define COLOR_PIXEL 0           // 0 for B&W, 1 for Color
#if (COLOR_PIXEL)
typedef CVisRGBABytePixel       tPixel;
#else
typedef CVisBytePixel           tPixel;
#endif
typedef CVisImage<tPixel>       tImage;
typedef CVisSequence<tPixel>    tSequence;
```

This vision class is (compile-time) switch-settable between color and monochrome. To make this transparent throughout the rest of the code, we define a few `typedefs`.

```
//
==============================================================
// Various definitions
//
#define WM_VISIONUPDATE  (WM_USER+1)
#define PANES_X     3
#define PANES_Y     1
#define BLUE        0
#define GREEN       1
#define RED         2
```

A few constants for later use.

```
// ================================================================
    class CVision
    {
    public:
        CVision();
        virtual ~CVision();
        bool AttachToSource( void );
        // Source Control
        void SourceDlg( void );
        void FormatDlg( void );
        // Capture control
        bool StartCapture( HWND parent );
        bool StopCapture( void );
        bool IsCapture( void );
        bool Capture( void );
    private:
        void process_image( void );
        void set_image(tImage& refimage);
        void calculate( tImage& src1, tImage& src2, tImage& dst );
        void manage_dlg( EVisVidDlg dlg );
        // ——
        CVisImageSource                     m_imagesource;
        CVisImageHandler<tPixel>     m_imagehandler;
        CVisPaneArray                            m_panearray;
        CVisCritSect                             m_critsect;
        tImage*     m_prev_image;
        tImage*     m_calc_image;
        HWND        m_parent;
    };
    #endif
```

Finally, the list of public and private methods and data. This is a bare-bones wrapper class.

The bare-minimum sequence of operation is (1) `AttachToSource`, (2) `StartCapture`, followed by repeated calls to (3) `Capture`. Everything else is ornamentation.

Let's see how it all happens in Vision.cpp:

```
//
==================================================================
//      Vision
//
// Vision wrapper class — implements a version of
Microsoft's
// Vision SDK
//
// Copyright 2001, Simulated Reality Systems, LLC
```

```
//
//
==============================================================
#include "stdafx.h"
#include "Vision.h"
```

Again, the header section and includes.

```
//
==============================================================
CVision::CVision()
   : m_imagehandler( true, WM_VISIONUPDATE ),
     m_panearray(PANES_X,PANES_Y)
{
   m_parent = NULL;
   m_prev_image = NULL;
   m_calc_image = NULL;
}
CVision::~CVision()
{
   m_imagehandler.KillThread();
   if (m_prev_image) delete m_prev_image;
   if (m_calc_image) delete m_calc_image;
}
```

The constructor does three important things. First, it starts the image handler class with the relevant parameters. Note especially the `WM_VISIONUPDATE` message value. Second, it starts the pane array which will later display all of our video images. Finally, it clears a few internal pointers.

The destructor turns off the image handler and deletes any memory laying about.

```
//
==============================================================
bool
CVision::AttachToSource( void )
{
   m_imagesource = VisFindImageSource(NULL, false);
```

Try to find a default image (video) source.

```
   if (!m_imagesource.IsValid())
         m_imagesource = VisFindImageSource(NULL, true);
```

If that failed, prompt the user for an image source.

```
   if (m_imagesource.IsValid())
   {
         m_imagehandler.ConnectToSource(m_imagesource);
```

```
m_imagehandler.SetThreadPriority(THREAD_PRIORITY_BELOW_NORMAL);
          return true;
    }
    return false;
}
```

If we finally get an image source, we connect the image handler to it and turn it down a bit so it won't hog the computer.

```
//
==============================================================
//        SourceDlg
//        FormatDlg
//
// Built-in VFW dialogs to setup camera information.
// Have to be attached to a source already for these to
work.
 //       You must also have captured some video first.
//
void
CVision::SourceDlg( void )
{
   manage_dlg( evisviddlgVFWSource );
}
void
CVision::FormatDlg( void )
{
   manage_dlg( evisviddlgVFWFormat );
}
```

These front-end methods simply call a generic setup dialog manager:

```
void
CVision::manage_dlg( EVisVidDlg dlg )
{
   bool running = IsCapture();
   if (running)
   {
         StopCapture();
         // Give it time to wind down, otherwise this
fails
         Sleep( 500 );
   }
```

If the system is currently capturing video we stop it and wait for any pending operations to complete. Without an appropriately long `Sleep()`, the dialog will simply refuse to appear.

```
        m_imagesource.DoDialog( dlg, m_parent );
        if (running)
                StartCapture( m_parent );
    }
```

After running the appropriate dialog, the capture is restarted (if it was running).

```
    //
    ==================================================================
    bool
    CVision::StartCapture(
        HWND parent )
    {
        m_parent = parent;
        m_imagehandler.SetNotifyWindow( m_parent );
        m_imagehandler.Run();
        return true;
    }
```

The `StartCapture()` method hooks the image handler up to the application window and turns the handler on. Once it has been activated, the image handler will send `WM_VISIONUPDATE` messages to this window at regular intervals.

```
    bool
    CVision::StopCapture( void )
    {
        m_imagehandler.Pause();
        return true;
    }
```

This method pauses the image handler, stopping the flood of `WM_VISIONUPDATE` messages.

```
    bool
    CVision::IsCapture( void )
    {
        return m_imagehandler.IsRunning();
    }
```

Test the status of the capture system.

```
    //
    ==================================================================
    bool
    CVision::Capture( void )
    {
        // Some tests
        //
        if (!IsCapture())
                return false;
```

This method (and its supporting methods) does all of the work. But it only does it if we are actually capturing video.

```
    tImage new_image = m_imagehandler.Image();
    if (!new_image.IsValid())
        return false;
```

These lines get a reference to the current video image and exits if it is bogus.

```
    //
    // Ensure we have a window up to display the capture
    //
    if (!m_panearray.HasWnd())
    {
        m_panearray.CreateWnd( evispaneDefault &
~evispaneAutoTitle
                | evispaneAutoDestroy);
        m_panearray.SetSizePaneInside(new_image.Size());
        m_panearray.SetTitle( AfxGetAppName() );
        m_panearray[0].SetTitle( "Live Video" );
        m_panearray[1].SetTitle( "Previous Image" );
        m_panearray[2].SetTitle( "Calculated Image" );
    }
```

This is as good a place as any to ensure we have a valid display window. If we don't, we create one. Each of the three internal panes in the pane array is given an appropriate title.

```
    if (!m_panearray.HasWnd())
        return false;
```

If, after all that, we still don't have a display window, we punt.

```
    //
    // Now process and display this new image
    //
    m_critsect.Enter();
```

We are entering the chunk of code that manipulates image data. To mark this auspicious occasion, we inform the critical section handler.

```
    {
        m_panearray[0].Display(new_image);
```

Draw the live video.

```
        if (m_prev_image)
        {
            if ( (new_image.Width() == m_prev_image-
>Width())
                && (new_image.Height() ==
```

```
m_prev_image->Height())  )
                        {
                                m_panearray[1].Display(*m_prev_image);
                        }
                        else
                        {
                                delete m_prev_image; m_prev_image =
NULL;
                                delete m_calc_image; m_calc_image =
NULL;
                        }
                }
```

If we have something from the previous video frame, we make sure it is still the same dimension as the current video frame. This can *change* by calling the `FormatDlg()` method, so this is an important test.

If it succeeds, we draw the previous image in the correct display pane. Otherwise, we delete the bogus previous image as well as the (assumed) bogus calculated image.

```
                if (m_calc_image)
                {
                        calculate( new_image, *m_prev_image,
*m_calc_image );
                        m_panearray[2].Display(*m_calc_image);
                }
```

If we have a container for the calculated image to go into, we perform our calculation and display the results. What that calculation *is* is defined later.

```
                //
                // Now, preserve the current image.
                // (and create image buffers if need be)
                //
                if (!m_prev_image)
                {
                        m_prev_image = (tImage*)new_image.Clone();
                        m_calc_image = (tImage*)new_image.Clone();
                }
                else
                        new_image.CopyPixelsTo( *m_prev_image );
```

If we don't have containers for the non-live video images, clone our current image. This creates all of the relevant data in these containers and allocates the appropriate amount of memory.

If we already have these containers, we copy the current image to the data space where it will be the previous image in the next pass.

```
    }
    m_critsect.Leave();
    return true;
}
```

Tidy up and return happy.

```
//
=================================================================
//      calculate
//
// Do any image processing here.
//
void
CVision::calculate(
   tImage& src1,
   tImage& src2,
   tImage& dst )
{
   ASSERT( src1.Width() == src2.Width() );
   ASSERT( dst.Width() == src1.Width() );
   ASSERT( src1.Height() == src2.Height() );
   ASSERT( dst.Height() == src1.Height() );
```

Biological vision is not a static process; it relies heavily on dynamic changes in its input. To this end, our calculation is based on two sequential frames of video. For simplicity, it also requires that all image buffers be the same size.

```
      register BYTE* src1_ptr;
      register BYTE* src2_ptr;
      register BYTE* dst_ptr;
```

For speed, we keep pointers to the current picture element (pixel) in each of the three images. As a hint to the compiler, we specify that we want these pointers in registers. What the compiler does from there is a mystery.

```
      for (int row_idx=0; row_idx<src1.Height(); row_idx++)
      {
            src1_ptr = src1.PbFirstPixelInRow( row_idx );
            src2_ptr = src2.PbFirstPixelInRow( row_idx );
            dst_ptr = dst.PbFirstPixelInRow( row_idx );
```

The images are composed of multiple rows of data. Each row has multiple columns, the pixels themselves. For each row, we get the pointer to the start of that row.

```
            BYTE prev_delta = 0;
```

The `prev_delta` value is only used for the color test. We start it with zero.

```
        for (int col_idx=0; col_idx<src1.Width();
col_idx++)
        {
```

For each column in the row, we look at the relevant pixels and perform our calculation. Note that the following code is just an *example* of how to access the data. Real-life vision calculations tend to take more into account than the current pixel, and will normally involve more complicated access across the image buffers.

```
#if (COLOR_PIXEL)
                // Highlight red and red motion
                register BYTE delta;
                BYTE red1 = *(src1_ptr+RED);
```

Let's highlight red for this example. Red plus image motion. Yeah.

```
                if (red1 > *(src2_ptr+RED))
                    delta = (red1 - *(src2_ptr+RED))>>1;
                else
                    delta = (*(src2_ptr+RED) - red1)>>1;
```

This is a simple split calculation to get the absolute difference in values between the current and the previous value of this pixel. We are only watching the red component for now. Oh, and we divide the value by two, a shift-right by one ">>1" is the moral equivalent of divide by two for non-floating-point values. The resulting delta value will be between zero and 127 inclusive.

```
                if (delta > 64)
                    delta = 128;
                else
                    delta = 0;
```

Here we perform a simple threshold operation.

```
                red1 = (red1/3)<<1;
                if ( (red1 > *(src1_ptr+BLUE))
                    && (red1 > *(src1_ptr+GREEN)) )
                    delta = 255;
```

Finally, if the red channel is significantly stronger than both the blue and green channels, we flag it.

```
                *(dst_ptr+BLUE) = delta&prev_delta;
                *(dst_ptr+GREEN) = delta&prev_delta;
                *(dst_ptr+RED) = delta&prev_delta;
```

Since this is a color image, we need to put our results into all three color channels for the results to appear in grayscale. Note that we also combine the current value with the previous value using a binary AND operation; this has the effect of filtering some of the speckles out.

```
prev_delta = delta;
src1_ptr+=sizeof(tPixel);
src2_ptr+=sizeof(tPixel);
dst_ptr+=sizeof(tPixel);
```

Preserve the delta, increment our pointers, and go on to the next iteration.

```
#else
                    // Highlight motion. A nifty form of edge-
finding!
```

Here we perform the calculations for the noncolor version (i.e. monochrome).

```
if (*src1_ptr > *src2_ptr)
     *dst_ptr = (*src1_ptr - *src2_ptr);
else
     *dst_ptr = (*src2_ptr - *src1_ptr);
```

Note that the calculation to get the difference between the two frames is greatly simplified from the color version.

```
src1_ptr++;
src2_ptr++;
dst_ptr++;
```

So is the increment!

```
#endif
          }
     }
}
```

This simple difference between frames calculation has the effect of highlighting the edges of moving objects. It's really effective and extremely simple; it is also only one of literally hundreds of possible transformations or calculations available for vision processing.

One possible next step to integrate the vision sense to the robot's behavior would be to reduce the amount of data being processed. One way to do this is to break up the video field into just a few "zones," and reduce (through accumulating, averaging, or other methods) the many pixels in that zone into a single datum for later use.

Entire books have been written on the subject of computer vision. There are also many code libraries on the Internet to support the eager researcher. With this video framework in hand, you should be able to make use of many of the resources available online.

CHAPTER 12

SIMULATED INTELLIGENCE

Once you have the body of your robot in hand, wired, sensed, and hooked into reflexes, your work really begins. The purpose of a robot is to *do*, and in order to *do*, the robot must have some kind of controlling intelligence.

For a remote-controlled robot, such as seen in RobotWars, BattleBots, or various projects in telepresence, that controlling intelligence is the man behind the curtain—you. Even if there are many reflexive support computers on the robot end of the joystick, all of the really hard control work still happens between your ears.

When you want to give your own personal intelligence a rest, you have to turn the controls over to some form of simulated intelligence. This is another one of those things that falls under the category of easier said than done.

There are two intertwined prongs to creating a simulated intelligence. On one side we have the task to perform. On the other side we have a bunch of different technologies that we can use. Each side of this equation depends on the other rather intimately, and the choices you make up front will limit the types of things you can do for a long time down the road.

The task we want to perform is to create a brain. Unfortunately for us, this is currently impossible for two reasons. The first has to do with the mind-numbing complexity of, for example, the human brain, and how many computers it would take to emulate it to any reasonable degree. The second reason is a bit more fundamental in that we don't really know how a brain works anyway. But, in this field, there is no room for pessimism!

The technologies we have on hand to create this brain (or brain subset) are numerous and often contradictory. At least the pantheon of existing technologies gives us something to work with until computers catch up with neurons.

THE TASK: CREATING A BRAIN

What is a brain? Since each of us has a nice squishy one behind our eyes, perhaps we can be forgiven for getting anthropomorphic when we think about brains. But brains are surprisingly hard to pin down.

Any sufficiently complex cluster of neurons can qualify, and goodness knows there are plenty of examples from human brains on down to the small wad of neural tissue that makes slugs all sluggy. But even single-cell creatures show some aspects of brain-driven behavior. Some protistas avoid hazards, approach food, eat, reproduce, and generally act like an animal, all with only one cell. Clearly they don't have a brain in any traditional neural sense, but that doesn't keep them from performing their complex behaviors anyway.

A brain is that part of the animal that answers the question, "What do I do now?" Whether the brain is a bundle of neurons, or something simpler inside an amoeba, the product of that brain is, ultimately, behavior. Our personal experience is likely to include a rich tapestry of imagination, memory, and emotion in addition to our observable behavior; these come from our brain as well. Of these, behavior is the one thing we can put our finger on, quantify, and otherwise analyze.

In an absurdly reductionary example the brain can be considered an opaque "black box," as shown in **Figure 12-1**. Information (e.g. sensory stimulation, emotions, and whatnot) feeds into the brain, which ultimately, observably, sends signals out that make the body "do stuff."

While this model may be technically true, it is also completely useless.

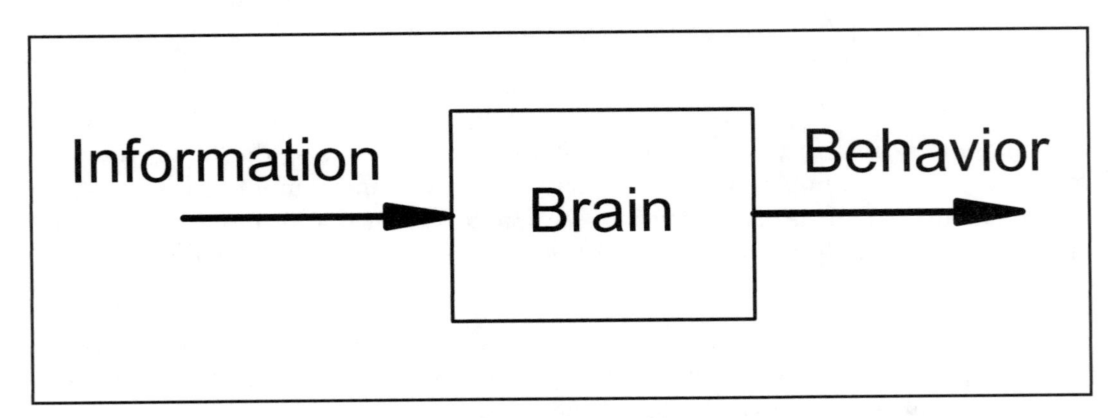

Figure 12-1 Black-box model

Usefulness isn't even well represented by **Figure 12-2**, where the brain is broken down into subparts until there are perhaps thousands of little boxes with information swirling all around them in a flurry of bits and neurotransmitters.

The problem is that even if we could put happy labels on all of the boxes in Figure 12-2 to correspond to the tasks they perform, and if we had a good representation of how the boxes communicate with each other, we still don't have more than a toehold on the problem of how to implement it inside our robot. And of course, a review of the literature today will show that we don't even have all of these happy labels yet.

Let's try to get a sense of what we *can* do.

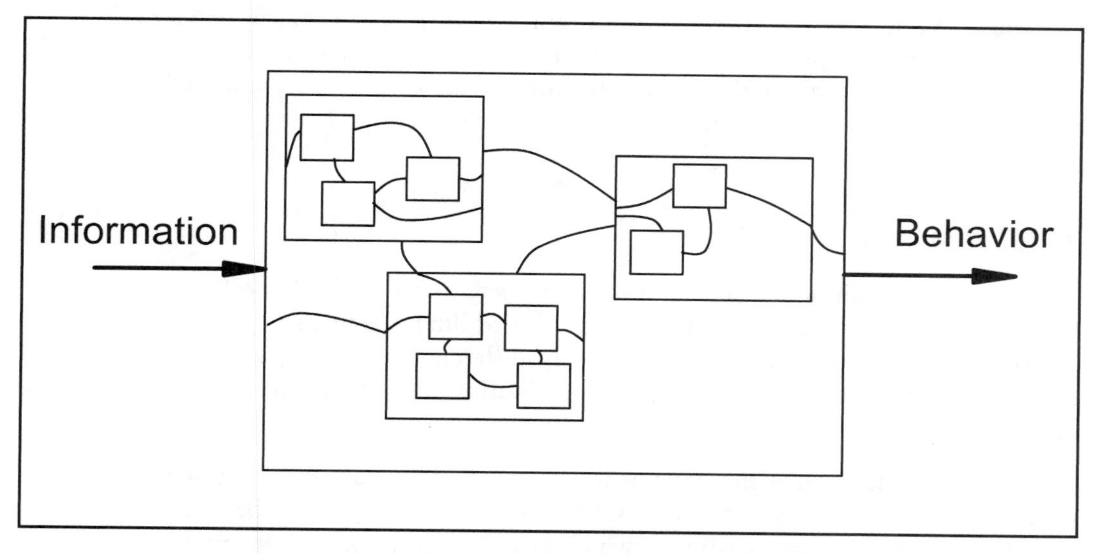

Figure 12-2 Hierarchical black box

A QUESTION OF SCALE

Let's look at the immensity of the problem we are approaching. For example, observe the human brain. (Hey, put away that bone saw!)

In that three-pound lump there are between 10^{11} and 10^{12} neurons, though I've seen it counted as a mere 10^{7} neurons, and if you were a nematode you would have only 302 neurons. This is in the vicinity of a trillion individual neurons. Each neuron, in turn, can have thousands of synapses, giving your brain more synapses than there are stars in the galaxy. And each synapse, axon, and neural body functions because of complex chemical processes involving untold numbers of molecules.

Though each neuron processes information very slowly, with processing turnaround on the order of 1/1,000 to 5/1,000 of a second, this still puts us to the task of performing about 10^{15} neural computations per second.

Even your modern gigahertz computer is only ticking over its clock at 10^9 cycles per second. So if, for example, that computer could perform one full neural computation per cycle (it can't), it would take it one million seconds, over eleven days, to come up with the results of one second's worth of your brain's output.

So we are not going to be simulating a human brain any time soon. Or even an octopus brain, with 10^8 neurons, or probably even a leech brain, with only 10^4 neurons.

Of course, we still want some kind of a brain for our robot. What we have to do is get a bit more clever about how we build it than simply copying our own brains. It still doesn't hurt to look at modern biological brains to see how they cope with things—nature has had a very long time to work on the problem and may have a few insights up her sleeve for us.

Your Brain

At first glance a brain looks like a wrinkly, undifferentiated blob of goo (see **Figure 12-3**). On closer inspection, and after many years of meddling, scientists began to discover that the brain appears to be composed of many different patches of neurons working together—with each patch performing or contributing toward a particular function. Brain modules, if you will.

Our early clues as to what the parts of the brain do came from studying people who had suffered some brain injury or trauma. Modern information is rapidly being filled in by the various scanning techniques, some of which give detailed information about brain activity while it happens.

A detailed description of the many parts of the brain, what they do, and how they interconnect, is far beyond the scope of this book (as so many things are). I recommend that you do some reading on the subject as inspiration for your robot's brain. There is a lot of information available. One good introduction is *Neurons and Networks: An Introduction to Neuroscience* by John E. Dowling.

Technologies

For the purposes of creating a robotic brain, it might be useful to look at things from a slightly different direction. Since we are building from scratch, perhaps we should look at

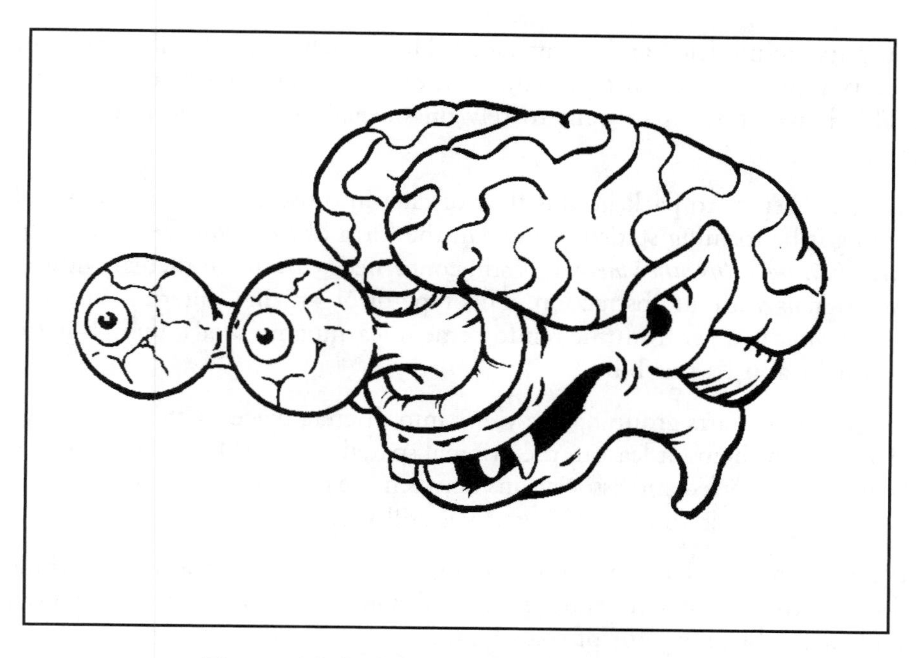

Figure 12-3 Brain (art by Shawn Sharp)

what we are attempting to accomplish in terms of different levels of processing detail—each level, perhaps, with its own appropriate technology. Note that this is just a *brief* tour of things that I have found interesting and have had time to look into at this time, a small snapshot into my large collection of notes. There are literally hundreds of papers and projects out there that I still want to explore, and hopefully I can find the time to do that once I finish this book!

Most books and theses on intelligence and the brain are concerned with *what* the brain is doing; however, as experimenters we are more interested in *how* to reproduce the effects ourselves.

THE BRAIN AS SEMANTICS

Though the "spatial semantic hierarchy" (see the technical report of the same name by Dr. Benjamin Kuipers) is proposed for the management of maps in a spatial environment (hence the name), it might provide insight to the different levels of processing needed by a brain in general. Dr. Kuipers proposes four levels of processing:

The *control level* has no knowledge of places, things, or abstract labels of any kind. It exists as a continuous, egocentric flow of data. This level is where the sensory inputs and motor outputs are modeled and/or managed. Here is where we need a simple mapping from sensory input to internal representation, or from higher-level command to motor pulses. This is where we may want to rely on specialized symbolic programming and fuzzy logic.

The *causal level* begins to package this flow of data into useful chunks or views. Though this level might be forming statistical links in the form of *from this view I do that and get (more often than not) this other view*, it is still concerned with the immediate flow of senses and motor signals without abstraction. This type of organizing, filtering, and predicting is not unlike what you get from the reinforcement learning methods, though self-organizing maps might also fit this bill.

The *topological level* starts grouping the views into discrete places with their many regions plus paths between them (at least in the original spatial sense of the model). Here is where things begin to get labeled and associations between the greater things are formed. Though we are beginning to get conceptual here, it is still very qualitative.

Metrical level. This level extends the topology (places and connections; symbols and associations) with quantitative values such as distance, direction, and shape (and color, flavor, and so on, breaking out of space again).

Who knows if these four levels have any "real" meaning in our brains? Regardless, they are one way to try and cut the problem space down into smaller, more manageable pieces.

THE BRAIN AS MAPS

The brain seems to be structured as a large set of interconnected topological *maps*, where a point in a map corresponds to a "point" in what it represents, such as, for instance, an input nerve on the hand, a receptor in the eye, or a concept. These maps are topologically structured so that a nearby point on a map corresponds to a closely related thing, such as a neighboring input nerve on a finger, a nearby receptor in the eye, or a related concept. The toe nerves are mapping next to the foot nerves, the foot nerves are mapping next to the leg nerves, and so forth.

The earliest neural maps were found for sensory inputs and motor control. In the case of touch, it is as though there is a "little person" laid out flat along a strip of the brain and, when you poke this representation in the brain (don't try this at home), you feel the effect in the relevant part of your body. With more research, similar topologically structured maps are being found for things as diverse as taste, smell, and even abstract concepts. It makes sense that one type of map is found all over the brain; nature likes to reuse things. This is not to say that the brain is only composed of one form of neural map.

Symbolic Programming

The traditional field of artificial intelligence was originally concerned with the manipulation of symbols by sets of transformation rules. Information coming in would be manipulated by these rules, associated with internal data, shaken, stirred, and spindled, and ultimately some response, generated by more transformation rules, would come out.

Expert systems are the most visible result of this type of AI, and though some feel that symbolic AI has been discredited, some significant projects, such as the grand and interesting Cyc project by Cycorp, are under way in the field.

Another area where you can see this type of AI at work is in the chatterbot (or chatbot), most of which are extensions of the old Eliza wheeze.

Fuzzy Logic

Fuzzy logic is a way, like most of these technologies, of specifying transformations from input data to output results. Fuzzy systems have made a very fine showing in the realm of motor and process control, areas previously managed by proportional and proportional integrated differential (PID) control systems. Of course, we aren't tolling the death of PID systems yet.

The input to a fuzzy system can be a single value, such as you might get from a sensor, such as *temperature*, *illumination*, or *velocity*, or it can be multiple values, such as one or more sensor values, plus additional information such as a command value like *desired velocity*.

Each input value is evaluated for membership in one or more categories. In traditional Boolean logic, an input can be evaluated as being a member of a category (is the light *bright*?) with only two answers possible: yes and no (true and false). In fuzzy logic, a value's membership in the category is a matter of degree (is the light *bright*?) with a range of answers from zero percent (no, the light is not bright) through a range of *sort of bright* answers, up to 100% (yes, the light is bright).

Each of these categories carries with it one or more response values. These responses determine the action the system would take if that category had full control.

The various responses from all rules are then combined to create one final answer for the system.

There are subtleties to fuzzy logic, with many choices to make along the way. What shape are the rules or categories? How are the membership results combined? How is the final answer crystallized out of this?

Let's follow an example through one type of fuzzy system and see what happens to it. Usually, fuzzy tutorials have you checking whether some temperature is hot or not. Let's pick an example, instead, from robotics. Let's try to move (or hold) an angular sensor to a specified position by issuing commands to a motor controller.

PREPROCESSING THE INPUT

Sometimes (though rarely) the values we get from a sensor or subsystem can plug directly into a fuzzy system. Other times, you need to massage them to get them into the range you desire. In this example we have two values and we want to combine them to get our desired value, *position relative to the desired position*, or more succinctly, P_R. We are given the actual position P_A from one system, and the commanded position P_C from some different system. Let's assume that the inputs are in the range of 0 to 1 (though in fact, the range will more likely be 0 through 255, or some other natural range).

$$P_R = P_A - P_C$$

Other transforms are possible. In this one, our relative position P_R varies from +1 (actual greater than commanded) to -1 (actual less than commanded) at opposite extremes, with a relative position of zero when the sensor exactly matches the command.

In this example, the commanded position is 0.75 and the actual position is 0.43. This gives a relative position somewhat less than the commanded position:

$$P_R = 0.43 - 0.75$$
$$P_R = -0.32$$

FUZZY RULES

Before we can apply our fuzzy rules to this input, we need to define them. Each rule consists of a membership function (shape), plus a result associated with the rule.

This sample system has nine rules, each with a different output recommendation. The output is a command to a motor, from -1 for full reverse to +1 for full forward.

The membership function or, more accurately, membership diagram, is shown in **Figure 12-4**. Though there are many different shapes this could take, this one is both useful and easy to calculate.

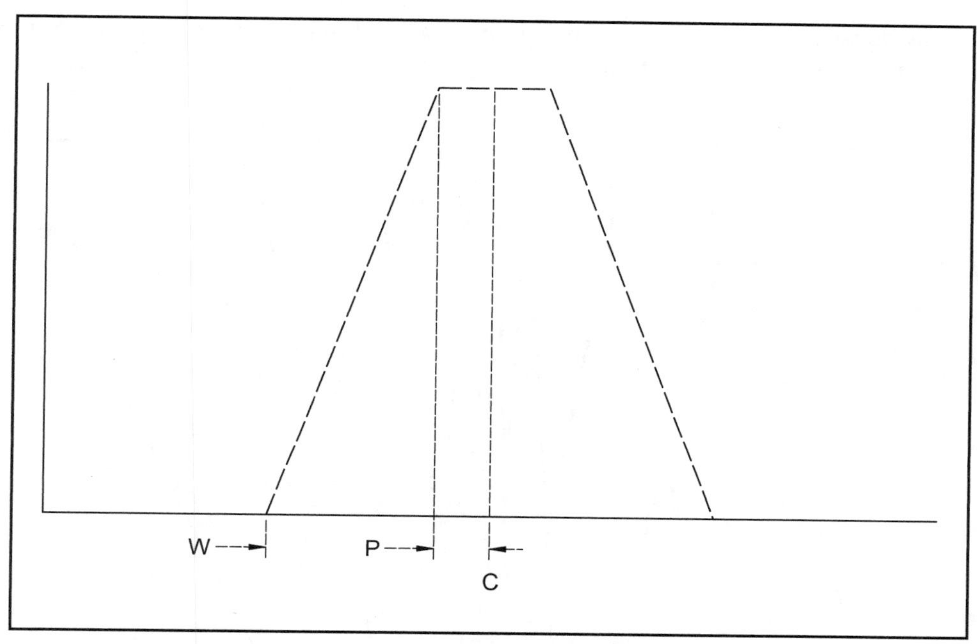

Figure 12-4 Membership diagram

The shape is fully defined by three values: its overall width *W*, the plateau width *P*, and the center position *C*. The motor command value is the result *R*. The rule is represented by these four values. The nine rules in our system, shown in **Table 12-1**, are:

#	C	P	W	R
1	-1.0	0.0	0.50	1.0
2	-0.50	0.0	0.25	0.5
3	-0.25	0.0	0.1875	0.25
4	-0.0625	0.0	0.0625	0.125
5	0.0	0.0	0.031	0.0
6	0.0625	0.0	0.0625	-0.125
7	0.25	0.0	0.1875	-0.25
8	0.50	0.0	0.25	-0.5
9	1.0	0.0	0.50	-1.0

Table 12-1 Fuzzy Membership Values

Note how the rules near zero are finer-grained and closer together than those at the extremes (**Figure 12-5**). Note also how we have removed the plateau entirely, making the membership functions triangles.

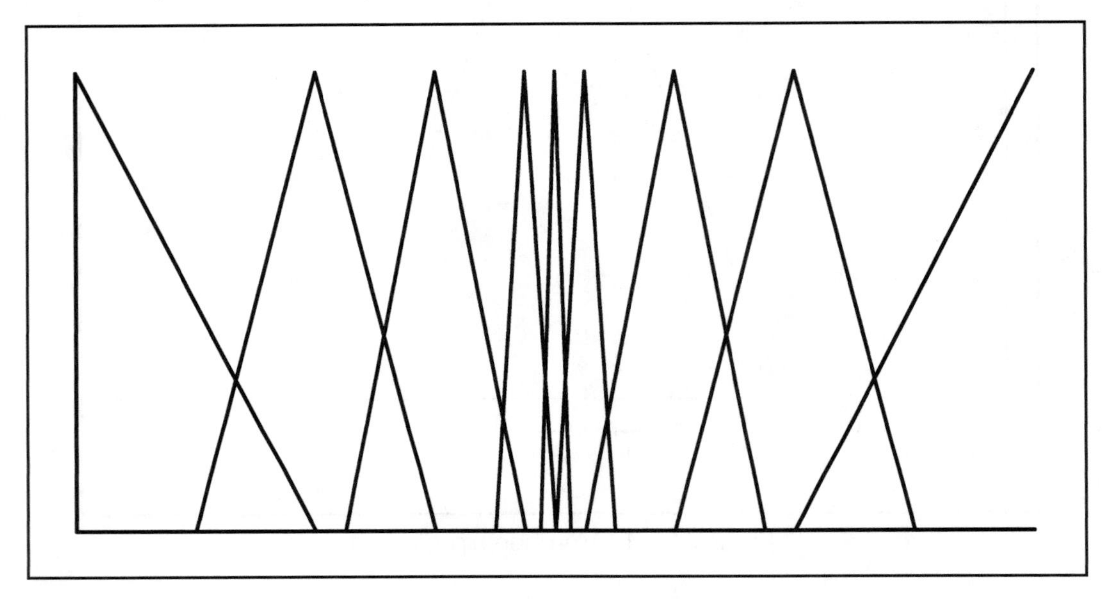

Figure 12-5 Fuzzy rules

DETERMINING MEMBERSHIP

Now, our input value of -0.32 only lies in the range of rules 2 (which covers -.25 through -.75) and 3 (which covers -.0625 through -0.4375). Since the input lies outside of the range of the other rules, it has zero membership in those rules and they can be ignored.

Now, to calculate the membership value *M* in rules 2 and 3, we need to do a little simple math:

$$\Delta = \left\| P_R \right| - \left| C \right\|$$

$$M = 1 - \frac{\Delta - P}{W - P} \ : \ \Delta < P, \ else \ M = 1$$

The value Δ (delta) is the distance of the input value from the center; this value is unsigned, we don't care which side of center it falls on. The notation is a bit clumsy though, taking the absolute value of both P_R and C and then the absolute value of their difference.

Since the plateau P is zero for our test, the formula reduces to a simple ratio of the distance from center Δ over the width of the triangle, actually, half the width of the triangle:

$$M = 1 - \frac{\Delta}{W}$$

So, for example, with Rule 3 our membership M is:

$$\Delta = \left||-0.32| - |-0.25|\right|$$

$$\Delta = |0.32 - 0.25|$$

$$\Delta = 0.07$$

$$M = 1 - \frac{\Delta}{W}$$

$$M = 1 - \frac{0.07}{0.1875}$$

$$M = 0.63$$

Continuing in this vein, we can determine the membership of the input for all nine rules (of course, seven default to zero since the input is out of range). **Table 12-2** includes the result value for the rule, which comes into play next.

#	M	R
1	0.0	1.0
2	0.28	0.5
3	0.63	0.25
4	0.0	0.125
5	0.0	0.0
6	0.0	-0.125
7	0.0	-0.25
8	0.0	-0.5
9	0.0	-1.0

Table 12-2 Fuzzy Membership Results

COMBINING THE RESULTS

Each row of the rule table represents one *fuzzy value* in this simple fuzzy system. The membership value M is a *weight* that is assigned the rule. The result R is still the result. Combining any number of results to one final combined result, let's call it R_C, is simple:

$$R_C = \frac{\sum M * R}{\sum M}$$

Simply add up the results of all M*R and divide by the sum of all M. In our case, there are only two rules where M does not equal zero, so we get:

$$R_C = \frac{0.14 + 0.1575}{0.28 + 0.63}$$

$$R_C = 0.33$$

The combined result is 0.33—the motor is turned on at about 1/3 power, forward.

AND BEYOND

It is possible to draw a graph that shows what the output would be for every possible input. After all, this type of fuzzy rule set is just one way of defining a transfer function from input to output, but then, most systems are.

To change the shape of this function, you need to adjust the R values, or perhaps the shapes or quantity of the rules.

This is a very simple fuzzy system—suitable for a basic reflex, perhaps. There are additional fuzzy logic operations and ways of combining rules or results. *Applied Robotics* talks a little bit about these fuzzy operators, but your best source of information comes from any of the many books devoted to the topic of fuzzy logic. Any book by Bart Kosko would be worth reading, since he's an evangelist for this technology.

ARBITRATION

In the fuzzy-logic example above, a number of individual rules were blended together to create a single result. Each of the rules overlapped a bit to provide smooth transitions between areas in the control surface. This, however, was a very simple example—a toy. In the "real world" of your robot, there may be any number of different rule sets, or other control systems, vying for control, each rule-set using different inputs and providing a different, and probably conflicting, control option as output. In these cases, there must be some kind of arbitration so the robot knows what to do out of this range of options.

The problem of arbitration and coordination is not unique to results of fuzzy-logic systems. The problem is universal—how does a robot, in the real world, with multiple conflicting choices and behaviors, decide what to do next? What is important and what is not?

META-FUZZY

In the fuzzy logic reflex example, there could be, for example, an additional layer of fuzzy logic rules that blends the results of the primary rule sets. Fuzzy meta-rules, perhaps in the form of "To the extent that this *context* is in effect, behave according to this *ruleset*" blending the results of different subsystems into a final command. For more information, check out either of the papers by Alessandro Saffiotti that are listed in the Bibliography section.

SUBSUMPTION

The example that most roboticists have seen for arbitration technology is the subsumption architecture of Rodney Brooks. Brooks pioneered the concept of *reactive* control of robots; layers of reflexes with no central director. The subsumption model arbitrated the different reflexive behaviors.

Brooks later refined the subsumption model, adding some global state information to help moderate the control, into his Behavior-Based Programming model.

In subsumption, each reflexive sub-behavior can be represented as a black box with inputs and outputs. The arbitration occurs when one reflex is allowed to *suppress* the inputs or *inhibit* the outputs of another reflex, effectively creating a hierarchy of reflexive behaviors.

Later enhancements to the model include *timers* in a reflex box so behaviors can be turned off after an appropriate amount of time, internal *registers* to hold local state information, and ultimately *hormones* which are registers that contain a global activation level that can be used by reflexes to moderate their behavior.

PANDEMONIUM

Another approach to coordinating multiple behaviors was pioneered by Oliver Selfridge of MIT, in his 1958 paper *Pandemonium*. In the late 1980s John Jackson expanded on the Pandemonium concept, moving it from a theory on perception and learning into the world of implementation.

In a pandemonium system there are any number of *demons*, each of which processes some of the system inputs and calculates both a control output and a measure of how much this demon wants the output to be realized (how loud it is *shouting*).

There are two areas where these demons reside—a large *stadium* of all demons in the system, and a smaller *arena* of active demons.

There are three types of demons: input, intermediary, and action. Some of each type of demon are in the arena at all times. There are also *links* between the demons in the arena, where these demons can reinforce or inhibit each other, based on the success of the actions being taken.

This structure of linked demons is, in fact, much like a neural network where the demons are nodes in the network and the links are the weights between the nodes. In practice, however, there are differences.

At any given step of time some of the loudest, or most active, sensory, intermediary, and action demons may be brought into the arena from the stadium, replacing some of the least active demons of the same types, which return to the stadium. Then, the loudest demons in the arena are allowed to control the action of the system for one step.

Depending on the success of this action, the links between the winning demons can be increased or decreased, allowing negative, or inhibitory, links. Then the process is repeated.

As demons are moved into and out of the arena, the action of the system adapts to meet the changing environment, based on the rewards or punishments received. For more information, see "Pandemonium as a Situated Cognitive Model," by Lee McCauley, and "Pandamat: Controlling an Animat with Pandemonium," by Jeff Whitledge.

PRODUCT OF EXPERTS

A related concept to pandemonium is the *Product of Experts* (PoE). The PoE model was not invented as an arbitration system but was designed to handle the problem of dimensionality in reinforcement learning. In spite of this, it has a definite resemblance to the pandemonium model.

Reinforcement learning is a control technique with a grid of all possible inputs along one or more axes and all possible outputs on one or more axes. In between, in the body of the grid, lie the *reward* values, also called the *policy,* that determine the system's behavior. The reward function holds the anticipated reward or punishment associated with entering that state.

As the state of the system changes, it leaves behind a trail of *activation* that decays over time. When some reward or punishment is applied to the system, it is spread out along this activation trail.

As the input changes during operation, the output that has the best history of success given this current state of the system is chosen. Reinforcement learning is a way of achieving the maximum reward given a history of previous efforts and the rewards received during them.

It is really more complex than this, and there are different ways to approach reinforcement learning. A detailed introduction to the subject can be found in Richard S. Sutton and Andrew G. Barto's *Reinforcement Learning: An Introduction.*

The key point here is that as the number of input states and the number of output states grows, the number of possible maps between them, the policy, explodes. This is the problem of dimensionality, and the problem is by no means limited to reinforcement learning. Any system that attempts to map from a complex set of inputs to another complex set of outputs will have similar issues.

The product of experts model makes the assumption that not all possible states will actually be significant. Each "expert" in this model takes some subset of input states and provides output answers based on these. Any number of experts may be created for the system, manually or automatically, and then the results of these experts are then combined. For more information, see "Using Free Energies to Represent Q-values in a Multiagent Reinforcement Learning Task," by Brian Sallans and Geoffrey E. Hinton, and *Training Products of Experts by Minimizing Contrastive Divergence*, by Geoffrey E. Hinton.

This tour of technologies continues, in more detail, in the following chapter.

CHAPTER 13

NEURONS AND NEURAL NETWORKS

This chapter is a continuation of the previous chapter on Simulated Intelligence. Here, we begin to delve deeper into the mysteries of the electronic brain. In this chapter we explore some technologies that truly do attempt to simulate the brain—technologies that have been brought to the fore by the field of connectionism. Connectionism is a branch of cognitive science that explores intelligence through the use of networks of artificial neurons—neural nets for short.

I want to reinforce here the important detail that *most of the work in these next chapters is speculative.* I am presenting work that I have found interesting and relevant to the search for robotic intelligence. It will take years to explore it all in any detail.

Before we start stringing together vast networks of neural simulations, we need to step back and ask a simple question. What is a neuron, and what does it do?

NEURONS

Neurons are long and complicated nerve cells that come in a variety of flavors. Neurons are the cells in our brain that do the bulk of our thinking.

At the input end of a neuron lies a bushy jumble of *dendrites.* Each of these dendrites is in contact with the output of other neurons—they accumulate the excitatory and inhibitory signals from these neurons. Throughout the dendrite tree and the cell body, the balance of certain ions changes in response to these signals; potassium, sodium, and other chemicals adjust and shift.

If things go the right way, the neuron will create an *action potential,* an all-or-nothing spike that travels along the *axon,* a specially coated highway for these pulses, and out along the various *axon terminals* to the dendrites of other neurons.

The connections between the inputs and outputs are called *synapses*—and these comprise small gaps between neurons that the signal jumps across.

The ways of the neuron are complex and still not entirely understood. However, a simple model of a neuron from the great squid was assembled through tedious trial and error by Hodgkin and Huxley in the 1950s. This model is presented here, but without any of the important theoretical background, as a toy to explore.

HODGKIN-HUXLEY NEURAL MODEL

This particular model is a variation of the Hodgkin-Huxley model that simulates the sodium and potassium balance, as well as the electrical state, of the neuron. This code comes from the depth of my notes, so I don't warrant its validity. Regardless, it is an entertaining simulation (**Figure 13-1**).

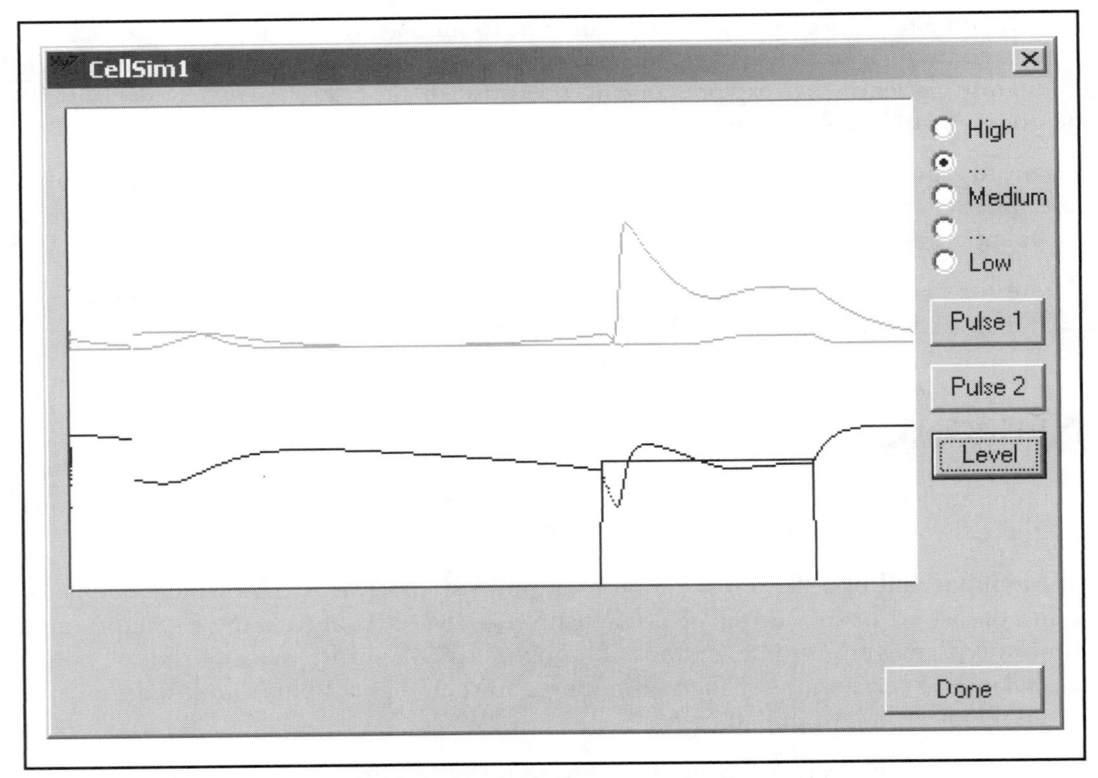

Figure 13-1 Hodgkin-Huxley simulation

First, you need to build a program framework that will call `Simulate()` and `Draw()` at regular intervals. This framework must also provide a way to stimulate the neuron with values ranging from 1 through 4 (give or take) applied via the `getPulse()` interface. My test harness creates pulse trains, each two steps wide, at different amplitudes and frequencies, as well as a voltage clamp lasting for 100 steps at different levels.

The variables in the simulation are defined in the header as follows:

```
double      m_K_act;    // N parameters
double      m_K_offon;
double      m_K_onoff;
double      m_Na_act;   // M parameters
double      m_Na_offon;
double      m_Na_onoff;
double      m_Na_inact; // H parameters
double      m_Na2_offon;
double      m_Na2_onoff;
double      m_KG;       // Conductance G
double      m_NaG;
double      m_KI;       // Current I
double      m_NaI;
double      m_leakI;
double      m_stimI;
double      m_totalI;
double      m_voltage;  // mV
double      m_time;     // mS
double      m_step;     // mS
```

Initialization is very simple, setting a few values and calling the `evaluate()` method to get the cell warmed up:

```
m_KG        = 0.367;
m_NaG       = 0.010;
m_stimI     = 0.0;
m_totalI    = 0.0;
m_voltage   = -70.0;
m_time      = -0.08;
m_step      = 0.04;
evaluate();
```

The `Simulate()` method is nothing more than this shell of code, plus anything you might need that is specific to your framework:

```
evaluate();
update();
m_time += m_step;
```

The `Draw()` method is up to you; however, mine tracks these four parameters:

```
middle = (m_extent.bottom - m_extent.top) / 2;
trace1 = m_extent.bottom + (int)m_voltage;
trace2 = m_extent.bottom - (int)m_stimI;
trace3 = middle - (int)(m_KG*10);
trace4 = middle - (int)(m_NaG*100);
```

The two methods that do the calculations are given in their entirety here:

```
void
CCell::evaluate( void )
{
   // Protection
   if (fabs(m_voltage + 45) < 0.001)
         m_voltage += 0.1;
   if (fabs(m_voltage + 60) < 0.001)
         m_voltage += 0.1;
   // Now evaluate
   m_K_onoff = 0.125 * exp((-m_voltage - 70) / 80);
   m_Na_onoff = 4 * exp((-m_voltage - 70) / 18);
   m_Na2_onoff = 1 / (exp((-m_voltage - 40) / 10) + 1);
   m_K_offon = 0.01 * ((-m_voltage - 60)
   / exp((-m_voltage-60) / 10) - 1);
   m_Na_offon = 0.1 * ((-m_voltage - 45)
   / exp((-m_voltage - 45) / 10) - 1);
   m_Na2_offon = 0.07 * exp((-m_voltage - 70) / 20);
   if (m_time <= 0.0)
   {
         // Setup for first step only
         m_K_act = m_K_offon / (m_K_offon + m_K_onoff);
         m_Na_act = m_Na_offon / (m_Na_offon +
m_Na_onoff);
         m_Na_inact = m_Na2_offon
/ (m_Na2_offon + m_Na2_onoff);
   }
   else
   {
         // All other steps
         m_K_act += (m_K_offon*(1.0-m_K_act)
- m_K_onoff*m_K_act) * m_step;
         m_Na_act += (m_Na_offon*(1.0-m_Na_act)
- m_Na_onoff*m_Na_act) * m_step;
         m_Na_inact += (m_Na2_offon*(1.0-m_Na_inact)
- m_Na2_onoff*m_Na_inact) * m_step;
         m_KG = 36 * m_K_act * m_K_act * m_K_act *
m_K_act;
         m_NaG = 120 * m_Na_act * m_Na_act * m_Na_act
```

```
* m_Na_inact;
   }
}
void
CCell::update( void )
{
   m_KI = m_KG * (-m_voltage - 82);
   m_NaI = m_NaG * (-m_voltage + 45);
   m_leakI = 0.3 * (-m_voltage - 59.4);
   m_stimI = m_dlg->getPulse() * 20;
   m_totalI = m_NaI + m_KI + m_leakI;
   m_voltage += (m_totalI + m_stimI) * m_step;
}
```

COMPUTATIONAL NEURAL MODEL

As you might guess from the math in the Hodgkin-Huxley model, it would take a lot of computer power to simulate any sizeable network of biological neurons. The connectionist systems don't use neurons resembling the Hodgkin-Huxley model, or any other biologically based model, but instead use a vastly simplified *computational model.*

There are two major areas of thought when it comes to neural networks. On the one side there are researchers who want to preserve the pulse-coded form of the biological neuron (see *Pulsed Neural Networks*, edited by Wolfgang Maass and Christopher M. Bishop). On the other side are the researchers who simplify the model even further and use a steady-state threshold model. While the pulsed neural models are fascinating, the bulk of the work in the field has been done with the threshold model so that is what we explore here.

The standard computational neuron (**Figure 13-2**) is a simple threshold device. There are any number of inputs I that can come from external signals or other neurons, a set of connection weights W, the threshold θ itself, and the output O, which may be sent to an output system or one or more other neurons. For sake of example, assume that the inputs to, and hence, output from, the neuron may be zero or one and the connection weights may be any value between -1 and 1.

A weight determines the effect of the input(s) on the state of the neuron. Positive weights are excitatory, and negative weights inhibitory.

The process of calculating the output of a neuron is quite simple:

$$X = \sum_{n}^{i=1} I_i(t) W_i$$

$$O(t+1) = \begin{cases} 1 : X \geq \theta \\ 0 : X < \theta \end{cases}$$

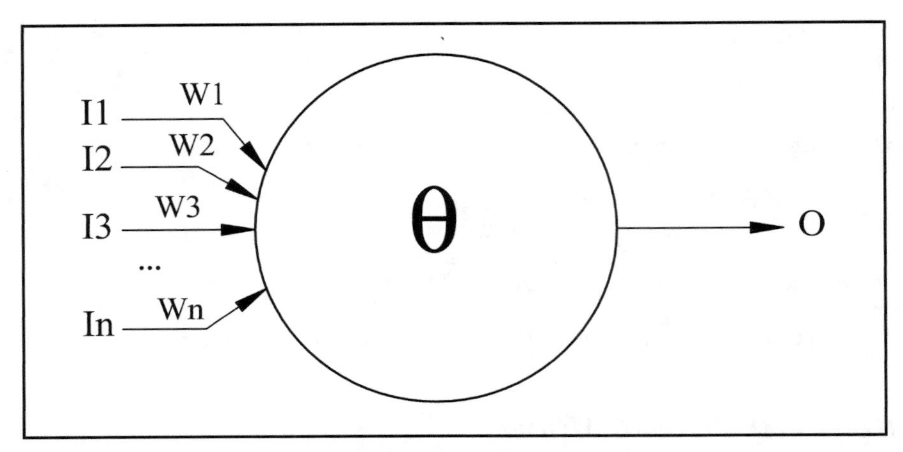

Figure 13-2 Computational neuron

This neuron operates in discrete time steps, with the output of the next time step calculated from the inputs of the current time step. The entire system of outputs is calculated internally and then updated simultaneously.

This simple device can perform the standard set of logical operations, and, by derivation, it is able to compute *anything* that can be computed by a Turing machine; they are computationally universal. By adding a simple binary input unit, whose value may only be zero or one, we can model these logical operations.

The OR neuron is shown in **Figure 13-3**. If either input goes to 1, the sum of the inputs reaches the threshold and the output becomes true.

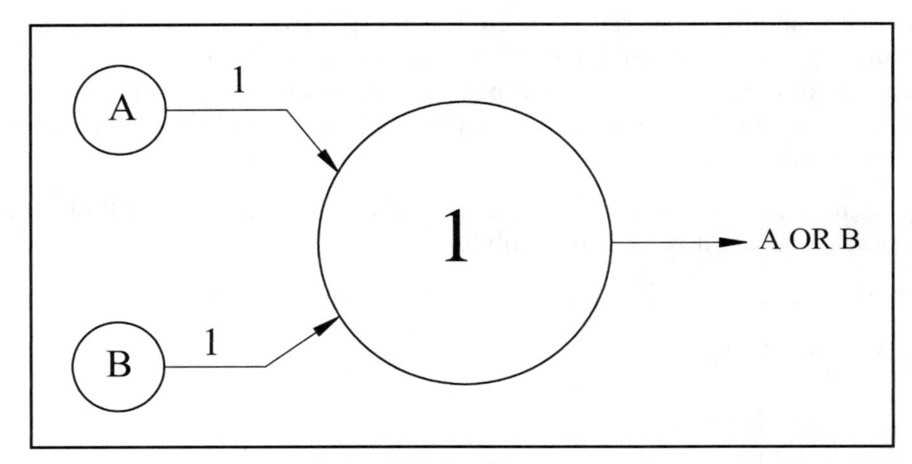

Figure 13-3 OR neuron

The AND neuron in **Figure 13-4** is very similar. However, both inputs must be true before the threshold reaches 2, turning on the output.

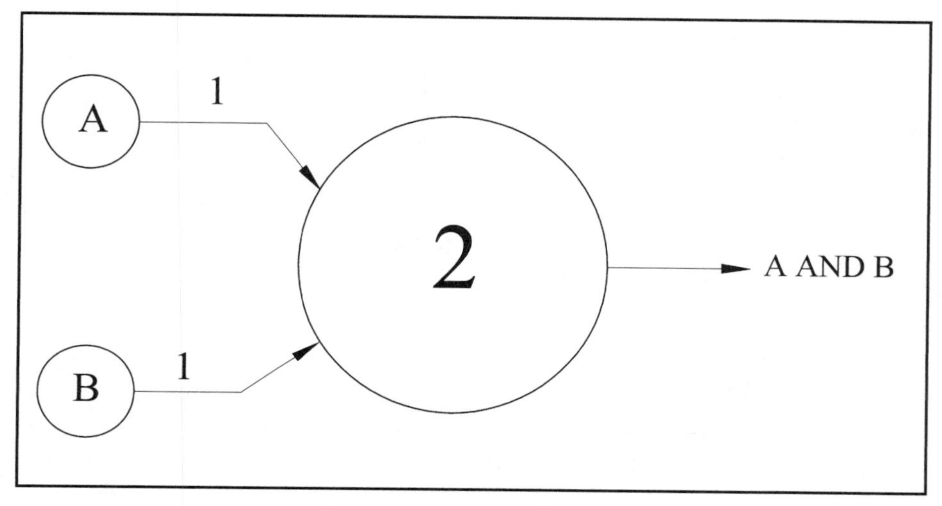

Figure 13-4 AND neuron

The NOT neuron in **Figure 13-5** is a little more interesting. This neuron is by default active, since the threshold is set to zero (no stimulation needed). The input, however, is inhibitory with a weight of -1. When the input goes true, it inhibits the neuron, turning it off.

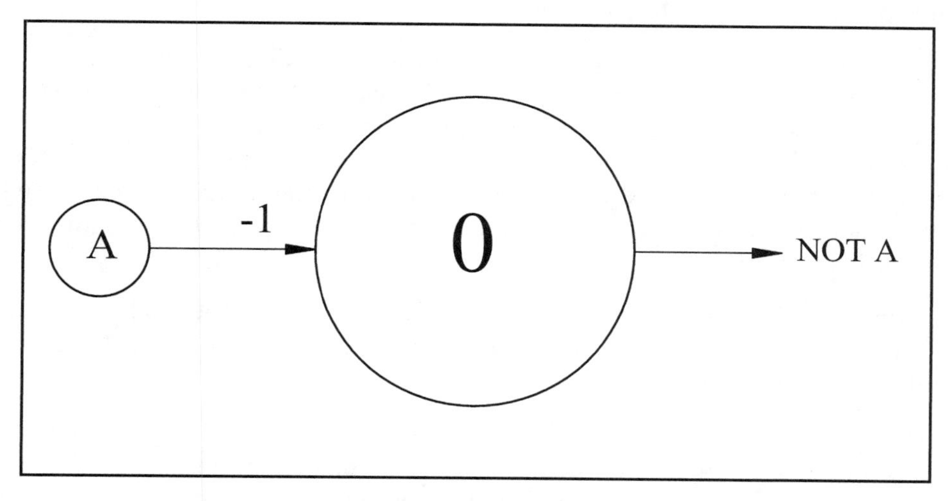

Figure 13-5 NOT neuron

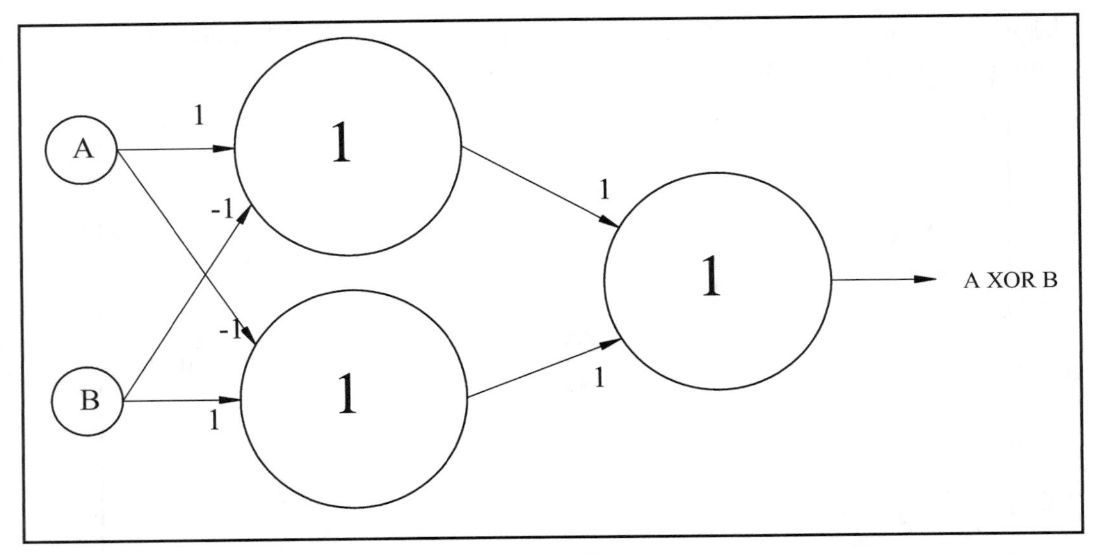

Figure 13-6 XOR neuron

A more complex neural circuit is shown in **Figure 13-6**, emulating the XOR logic function. Trace through the activation levels to see how it operates.

If you are interested in exploring this type of neural circuitry more, you will want to go to the Australian Wildlife Web site and download the Bug Brain educational program at www.australianwildlife.com.au\bugbrain. This is the most interesting, and affordable, program for this type of exploration that I have seen.

Learning in the computational neuron is performed by adjusting the weights of the neuron. There are many different schemes for this weight adjustment, some of which are explored in later sections. The simplest learning rule was described by D.O. Hebb in 1949, and bears his name. In Hebbian learning, a weight is increased if the output of the neuron is active at the same time as an input through the weight is active, reinforcing this synchronization. Of course, such a simple learning rule will grow the weights without bound, but it's a start.

Many neural models work with more complex neurons than the threshold pattern shown so far. The inputs to the neuron may be real values, and the output may be, and often is, a nonlinear transfer function.

In the analog configuration, each neuron is a vector comparator—the input values are one vector, and the weights are the vector the neuron is comparing against. The "fit" between these two vectors is like the dot product—and the output of the neuron is an analog signal that indicates how close the two vectors are.

ADAPTRODE NEURAL MODEL

While the standard computational neuron has only an immediate state and set of weights, the adaptrode neural model is in some ways closer to the biological model. The adaptrode has several levels of *memory* that affect its operation, corresponding to short-term and longer-term changes in the neuron in response to stimulus. While this model is an interesting diversion, it wanders too far from our central theme to be able to explore here. If you'd like to learn more, check out George Mobus's *Adaptrode-Based Neurons* and other papers.

NEURAL NETWORKS

No discussion on neural networks is complete without the picture of the standard three-layer net, such as shown in **Figure 13-7**. There it is. And that's the last of it for this book. Sorry.

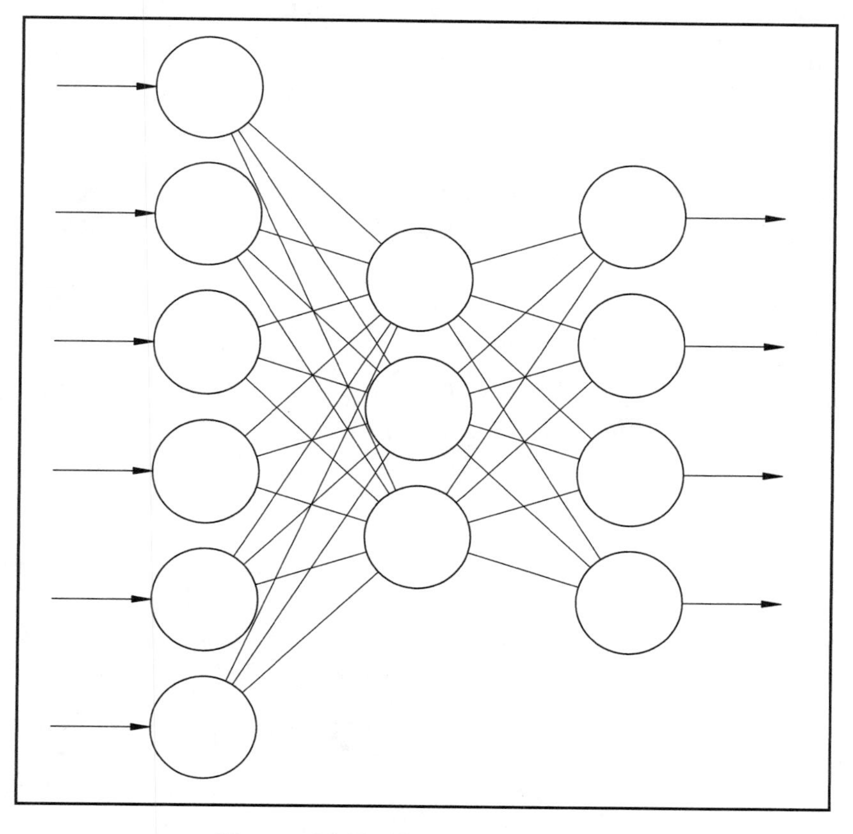

Figure 13-7 Three-layer network

Neural networks are, essentially, groups of connected neural models. These neural models have inputs, synapse weights, and activation behaviors as we have seen.

As we enter the world of the network we also become concerned with *teaching* the network's neurons to respond in a useful manner. There is an entire constellation of techniques relating to the architecture, that is, the shape and connections, of the network and the methods of teaching them.

There are two broad families of networks, those that are taught using *supervised* methods, and those that are taught *unsupervised.*

SUPERVISED VS. UNSUPERVISED LEARNING

In unsupervised learning, you turn on the network and start feeding it data and the network adjusts and adapts. One type of unsupervised network, which we will be paying more attention to later, is the Kohonen Self-Organizing Map (SOM).

In supervised learning, you first need to develop a comprehensive set of example data. This consists of both the inputs to the network and the desired outputs. The inputs are presented to the net and it is allowed to calculate some result. This result is compared to the desired result and, where it is in error, this error information is sent upstream through the network so that the weights can adjust themselves to reduce this error (back propagation learning). Many examples across the entire input space are presented, often thousands of times, until each input presented to the net generates an appropriate response.

With this background in mind, you may sympathize with my belief that robot brains are best served using unsupervised learning models wherever possible.

Regardless of whether a neural net was self-organized or laboriously trained by hand, the network will act as an efficient and fault-tolerant categorizer of data. It can usually take inputs that do not match any known example from the training set and create a reasonable response to them.

SELF-ORGANIZING MAPS

The starting point for our self-organizing neural nets is Teuvo Kohonen's Self-Organizing Map (SOM).

Having made a split between supervised and self-organizing nets, we can now make some further distinctions. There are three kinds of tasks we may put our networks to: (1) filtering, categorizing, and recognizing input; (2) thinking, planning, and deciding; and (3) generating and coordinating motor-control output.

SOM networks are a natural fit for the first task of categorizing inputs. The information that feeds into a robot has two problems that we need to address. The first is quantity—if the robot has very many sensors there will be a flood of data streaming from them into the brain. The second problem is quality—the information from these sensors, especially analog sensors, is not going to be consistent or reliable from moment to moment. If you put the robot in the same location twice, the input from the sensors is unlikely to be exactly the same both times.

The SOM is well suited to the task of mapping a large array of inputs down to a much smaller set of categories. In a sense, the SOM is recognizing patterns of input and flagging an output that matches that pattern. This is the *mapping* part of the name. Technically speaking, the SOM maps an input of a given dimensionality, that is, a given quantity of input values, down to an output with a much-reduced dimensionality, typically 2D, for easier visualization. This mapping eases the curse of dimensionality mentioned earlier and also creates an *abstraction* of the input data.

The *self-organization* aspect of the SOM comes into play when the network is trained. Your brain, and systems such as the SOM, makes use of the fact that information in our environment is not random. There are clusters or patterns of information presented to us, and related patterns tend to map together both in time and in our sensory state space. The SOM takes advantage of this natural clustering in state space.

ORGANIZATION

SOM networks are typically organized as a two-dimensional array of nodes (**Figure 13-8**), for simplicity in visualization. The network may be 1D, 2D, or any other organization that makes sense to the problem at hand.

The network may be arranged, in the case of a 2D grid, as a square grid or a hexagonal grid. The grid may be bounded at all four edges or with wraparounds at the edges, connected into a tube or fully wrapped up as a torus.

The nodes in the network are trained so that nodes that are near each other in grid coordinates have similar contents, or model vector weights, and hence respond to similar inputs. This is a topological mapping of inputs to nodes.

There are two ways to think about the SOM structure. On the one hand, there is the connectivity graph, which is the structural layout of the network nodes (Figure 13-8). With this view of the network you can think about which nodes are connected to other nodes and how far apart they are in terms of their grid neighbors. In this view there is no structural change through the life of the SOM network, though the values within the nodes will change as it learns.

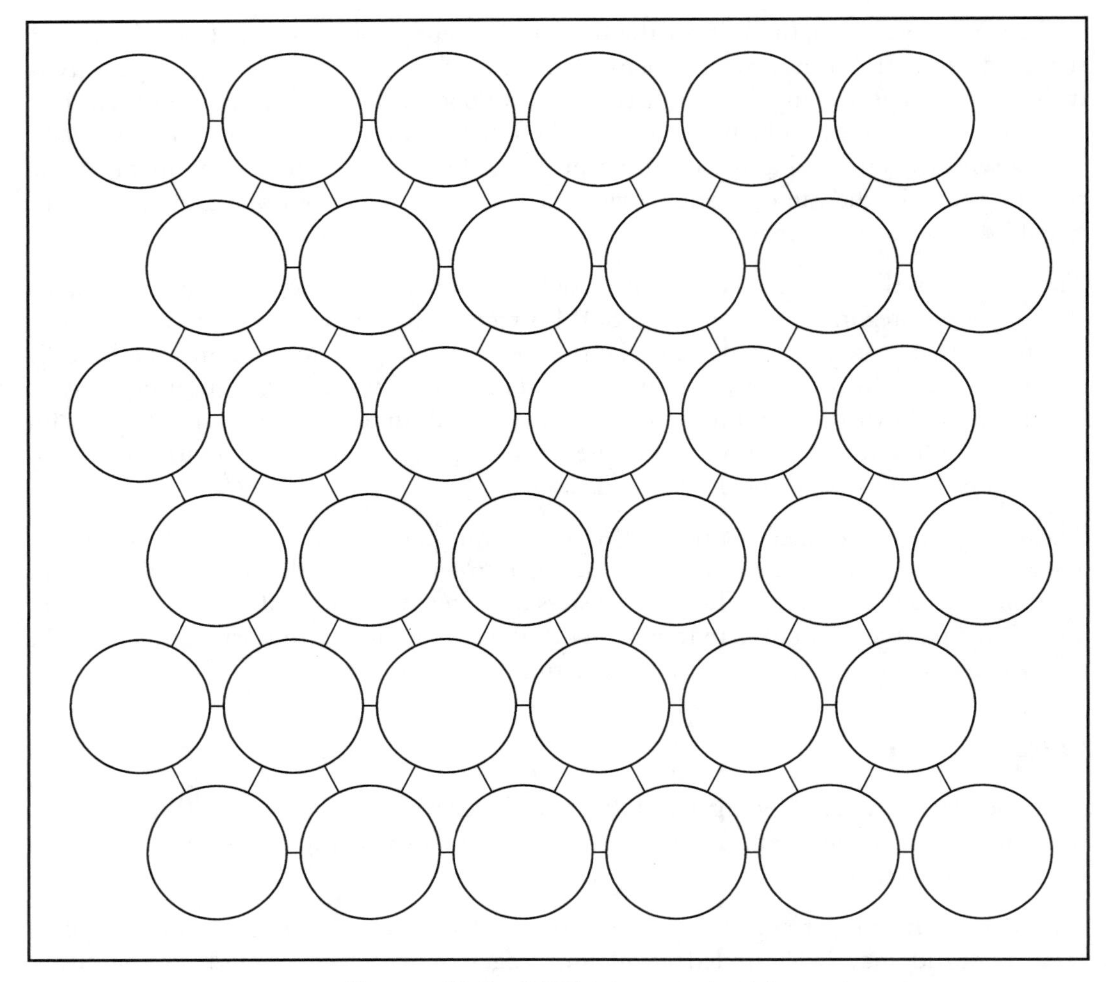

Figure 13-8 SOM hexagonal grid

On the other hand, there are the model vectors, the node weights, that, for appropriately small input dimensions, can be visualized as seen in shown in **Figure 13-9**. Each node in the network is, like in other neural models, an input classifier with a weight (model) vector that it compares to input data to get a nearness value.

In Figure 13-9, the diamonds spread around the display represent the 2D input data points that are used to train the network. The plus signs arrayed around the circle represent the 2D model vectors of the SOM nodes, initialized to this arbitrary shape.

The lines between the SOM nodes are the 1D connectivity of this circular SOM "grid."

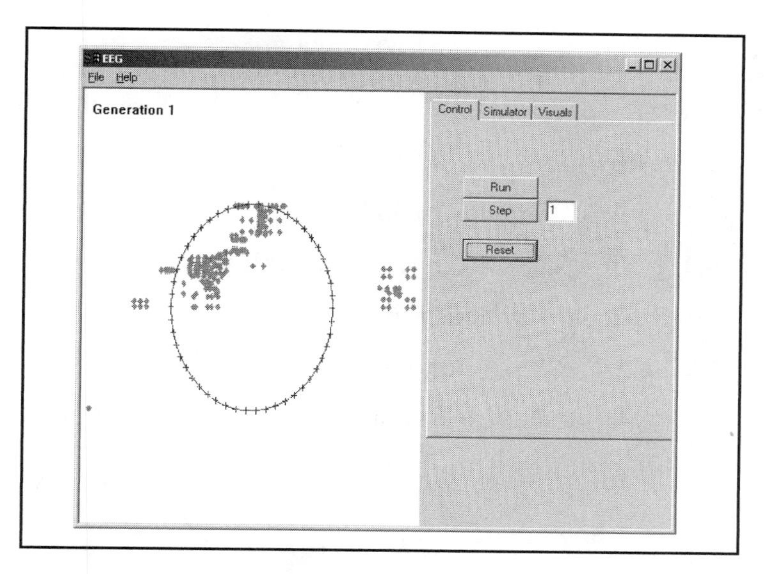

Figure 13-9 Model vector view of a 1D SOM loop

As **Figure 13-10** shows, the model vectors associated with each node change with training. From this view, the network appears to warp and shift to match the data samples, while the neighborhood connectivity, as represented by the lines between nodes in this linear network, does not change at all.

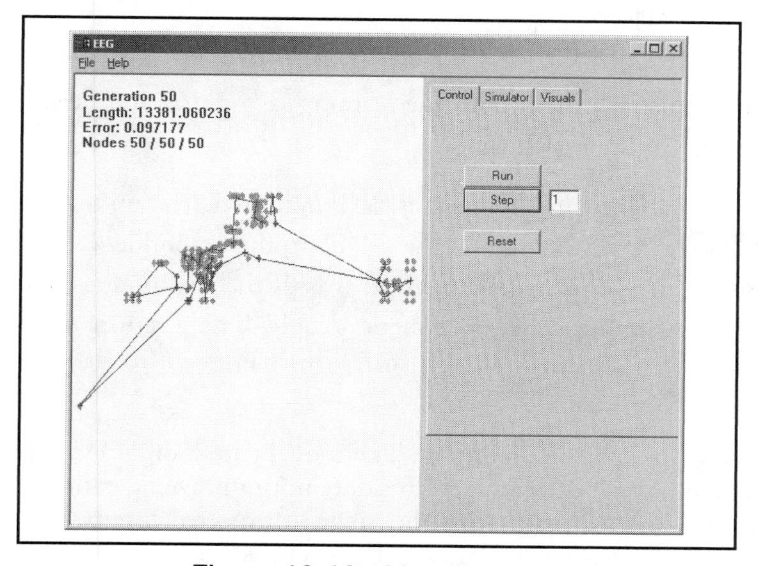

Figure 13-10 After fifty steps

OPERATION

Before we can explore the operation of the SOM, we need to formally define our terms.

t	Time step
i	SOM cell id (instance, or index) number
$m_i(t)$	SOM cell weight vector (contents) m instance i at time step t
$x(t)$	Input sample x at time step t
c	"Winning" cell instance; see $c(x)$.
$c(x)$	SOM cell instance that most closely matches sample x such that:

$$\forall i \left(\left| x(t) - m_c(t) \right| \leq \left| x(t) - m_i(t) \right| \right)$$

$\alpha(t)$	Learning factor at time t. $0.0 \leq \alpha(t) \leq 1.0$
$\sigma(t)$	Width of the neighborhood function h
r_i	Grid coordinate of SOM node instance i
$h_{c,i}$	Neighborhood activation function:

$$h_{c,i} = \alpha(t) \exp\left(-\frac{\left| r_i - r_c \right|^2}{2\sigma^2(t)} \right)$$

In the standard implementation, both the learning factor $\alpha(t)$ and neighborhood width factor $\sigma(t)$ decrease linearly with time.

The neighborhood function $h_{c(x),i}$ calculates the nonlinear activation that each neighbor of the winning SOM receives. The farther the neighbor, the less influence the winning SOM has on it. In fact, if $|r_i - r_c|$ is large enough, that is, if the trial node is outside the radius of influence of the winning node, the function could simply return zero.

Initialization

The first step of operation is initialization. Traditional papers on SOM networks initialize the nodes to random weights. However, this does not improve operation. A better choice is to spread your weights evenly across the range of expected inputs, or otherwise place your node weights in positions that reflect the expected inputs.

The SOM model provides for a smooth change from an initial *learning* stage of operation, to a later *categorizing* stage of operation. At the beginning, the learning factor $\alpha(t)$ is set to some fairly large value, for example, between 0.5 and 1.0. Each step of operation decreases $\alpha(t)$ by some factor so that it slowly approaches 0.

Likewise, the neighborhood size $\sigma(t)$ starts large and is slowly reduced with time to some small lower limit.

These changes bring the network from a place of high variability, where each change affects a large range of neighborhood nodes, to a place of small changes and limited overlap of effects.

Operation

Once everything is in place, you can start the process of presenting samples to the network and running the network update.

For each input sample $x(t)$ you must first check every SOM node to see which one is the closest match. Every node in the grid then has its weights updated using the formula:

$$m_i(t+1) = m_i(t) + (x(t) - m_i(t))h_{c,i}$$

This has the effect of moving the winning node's weights toward the input vector by an amount specified by the learning rate, and all neighboring node's weights by some lesser value depending on how far away they are topologically from the winning node.

After all of the nodes have been updated, the learning and neighborhood factors are reduced, and the next input is presented.

The first block of inputs serves to train the network and, once the learning factor has been completely reduced, the network proceeds to categorize the remaining inputs according to its structure.

ADJUSTMENTS

There are several issues involved with this SOM network implementation. One is the progression from learning to categorizing; biological networks manage lifelong learning, so some way of modeling this would be useful.

At each step, the winning node must be found using a global, that is, external, scan of all the network nodes. An enhancement to the algorithm would get rid of this explicit scan, perhaps using a local function of the distance between the input vector and each node's weight as another factor in the update equation plus using some form of lateral inhibition between active nodes. This would, in effect, create multiple "winning" nodes, providing more complicated network states. The chief benefit would be that the network could be calculated using only local, or at worst, nearby, information.

The winning node, of which there is only one in this model, serves, in a way, as the point of attention of the network. This focus of attention is reset from scratch at each input step. In a biological model, however, attention tends to have a certain sense of momentum instead of flitting about randomly.

The SOM is also a pure classifier—it maps inputs to outputs with no reference to previous) states or any other form of history. This limits the SOM's usefulness.

The number of nodes in the network is fixed from "birth," while in a biological model neurons grow and die depending on their context.

Finally, there may be other models of node weight update that can be added into the picture, including surface tension between nodes, to smooth the network's growth; repulsion between nodes, to prevent overlearning; and the conflicting attraction between nodes, to try and minimize the distance between nodes, another smoothing method.

Some of these concerns are addressed throughout this and other chapters. Others I leave as an exercise for the diligent reader.

ATTENTION

One of the issues of the plain-vanilla SOM model above is that of *network attention* (for more information, see "Attentional Networks" by Michael I. Posner and Stanislas Dehaene). A model of attention applied to the SOM gives a sense of continuity across time of the winning nodes *c(x)* of the network (see "Dynamic Extentions of Self-Organizing Maps" by Josef Göppert and Wolfgang Rosenstiel). This attention modification makes use of the continuity across time of inputs in a physical environment.

In the plain SOM model, the input vector at a given step in time *x(t)* is used to select a winning node from the entire set of nodes in the network. The winning node *c* is taken to be the one that is closest to the input vector *x,* stated here somewhat differently as:

$$c(x) = \min_{i \in n}\left(\left|x(t) - m_i(t)\right|\right)$$

This winning formula is based entirely on the instantaneous Euclidian distance of the input value to the trial node weights $\left|x(t) - m_i(t)\right|$.

The modified SOM, to account for smooth transitions in attention, does not weight each node in the network equally in its search for a new winning node. Instead, it adds a distance factor D_c that adjusts the probability of a neuron winning the new input based on that neuron's topological grid distance from the current center of attention A.

First, we need to define the center of attention. It is simply the node that won the input in the previous time step:

$$A(t) = c(x(t-1))$$

Given these two new factors, attention, and distance from attention, the winning node is now determined by:

$$c(x) = \min_{i \in n}\left(D_c(r_A - r_i) * |x(t) - m_i(t)|\right)$$

Where D_c is a function of distance, such as one of these two functions:

$$D_c(d) = 1 + \frac{d^2}{D\sigma}$$

$$D_c(d) = \begin{cases} 1 : d < D_\sigma \\ \infty : d \geq D_\sigma \end{cases}$$

In short, the likelihood that a node m_i will become the new winner c, and hence next center of attention A, depends on how far that node is from the current center of attention. This forces the attention A to move smoothly from its current position, pulled across the network by new inputs over time.

GROWTH

Another issue with the SOM network is deciding how many nodes it should contain. Preferably, you can create a skeletal initial network that will then grow as needed to represent the inputs presented to it. For more information, see *Growing Cell Structures: A Self-organizing Network for Unsupervised and Supervised Learning* by Bernd Fritzke, and "A Resource-Allocating Network for Function Interpolation" by John Platt.

There are four issues at hand with network growth. First, you need to determine when a new node needs to be added to the network. This leads to the second issue: where that node is placed and how it is hooked into neighboring nodes. To prevent overgrowth of the network, you may need to remove nodes as well. The third issue is:

when are nodes killed? Finally, four: how do you reconnect the network once you've removed a node from it?

Given an SOM network, we ultimately want to grow new nodes where the input vectors are the most likely to occur, that is, where the input probability density is high, and remove old nodes where inputs rarely occur, that is, where the input probability density is low.

TOPOLOGY

To address network growth, we begin by looking at the network a little bit differently.

As before, the SOM connection grid can be of different dimensionalities k. Our two-dimensional grid from before is shown in **Figure 13-11**, with an emphasis on the triangular cell structure. The topology of the network is thought of as a k-dimensional simplex, such as lines for k=1, triangles for k=2, tetrahedrons for k=3, and hyper-tetrahedrons for greater values of k.

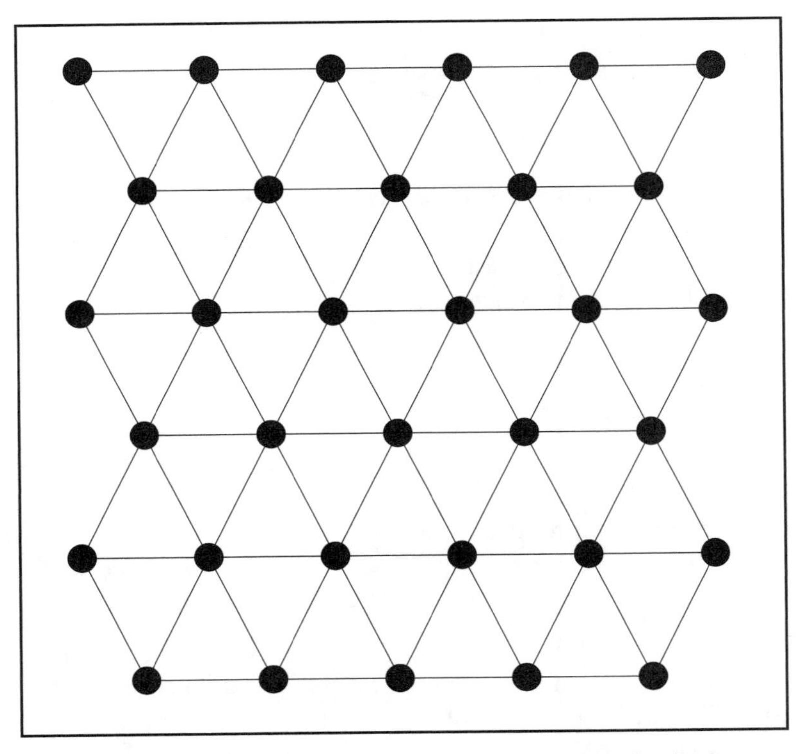

Figure 13-11 Cell structure for dimensionality k=2

For each network topology the $(k+1)$ vertices are the cell nodes (neurons) of the network. The $k(k+1) \div 2$ edges are topological neighborhood relations, that is, the connections between the nodes.

OPERATION

Every node i, of course, has an n-dimensional weight vector m_i associated with it. This is seen as the position of the node in the input state space.

We add an activation counter t to the node that keeps track of how often that node is activated.

During the evaluation of the network, a new step is added after the winning node c is determined. In this step, we increase the activation level of the winning node τ_c and decrease the activation levels for all other nodes τ_i.

$$\tau_c(t+1) = \tau_c(t) + 1$$
$$\tau_i(t+1) = \tau_i(t) - \delta\tau_i(t)$$

The activation decrease is controlled by δ, the forgetting factor. This should be fairly large so that activation levels decay slowly.

ADDING A NODE

Over time, assuming a relatively constant statistical distribution of inputs x, those nodes that are visited regularly by the input will stabilize at some level of activation. Nodes that are visited more often than most will develop a spike in their activation level, while nodes that are rarely visited will tend toward zero. These spikes and dips can be used to determine when to grow or kill a node in the network. Another way to look at this is to calculate the relative activation h_i of each node:

$$h_i = \frac{\tau_i}{\sum_j \tau_j}$$

Over time, there may develop a node m_q, with an unnaturally high relative frequency h_q, that should be divided into two nodes to better handle the activation load.

The newly added node will be placed between m_q and its furthest neighbor in input space m_n. Node m_n is selected from the immediate topological neighbors of m_q where:

$$m_n = \max_i(|m_q - m_i|)$$

The new node m_r is inserted on the edge between m_q and m_n, with new topological connections added between it and the mutual neighbors of m_q and m_n, preserving the topological form.

The weight vectors of the new node m_r are calculated to be midway between its parent nodes:

$$m_r = \frac{(m_q + m_n)}{2}$$

The new node's activation counter t_r can be calculated to be some accurate representation of that node's expected activation level for its newly demarcated area of input space—see Bernd Fritzke's book, mentioned earlier, for more details. The new node can also simply steal activation points from either its original parent τ_q, the same nodes it got its weights from (τ_q and τ_n), or even all of the nodes it is connected to. Ultimately, the input variations themselves will smooth out the actual counters to valid levels.

Deleting a Node

While inserting a new node is fairly intuitive, deleting a node is a bit more complex since we need to make sure the network topology is intact after the deletion. The node to be deleted is first identified by its unnaturally low activation level.

Before the node is removed, what is left of its activation counter is distributed to its immediate neighbors, based on their distance in input space, if desired, or just spread around.

Then, all hypertetrahedrons, triangles, in our case, that this node was a member of are deleted. If you just deleted the node, and the edges that connect to it, the network can be left with an unstable structure (as shown in **Figure 13-12**).

The somewhere more involved, but accurate, process of deleting a node is illustrated in **Figure 13-13**. All of the triangles that are attached to node A are evaluated and deleted. Edges that are shared by marked triangles (of A) and unmarked triangles (not of A) are, of course, retained.

State and Sequences

The final enhancement to the SOM is also the one with the most impact on its capabilities. A pure SOM network is a simple classifier, mapping high-dimension inputs to a lower-dimension output. With the addition of a feedback loop the SOM develops a sense of *state*; a marker indicating its recent history that can influence its future directions.

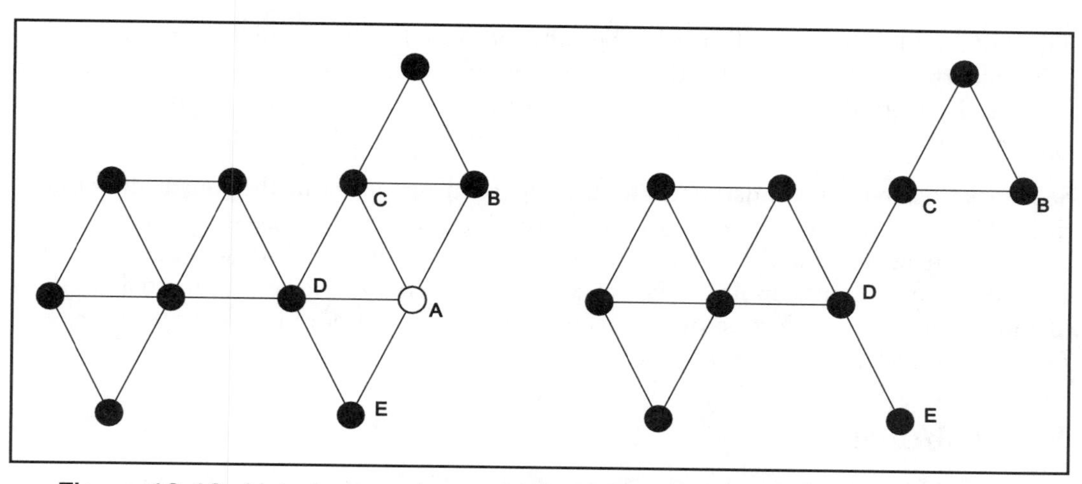

Figure 13-12 Naively removing node A. Edges DC and DE are not part of a triangle so the structure is inconsistent.

This section is based on, and heavily simplified from the work of Gary Briscoe (currently at the University of Wisconsin-Oshkosh—quite a move from his graduate university in Australia) and his Adaptive Behavioral Cognition (ABC) model. This is a simplified view of the model that won't jump right off the page into a successful implementation. For details, see Briscoe's Ph.D. thesis, *Adaptive Behavioral Cognition*, and *Machine Learning and Image Interpretation*, edited by Terry Caelli and Walter F. Bischof.

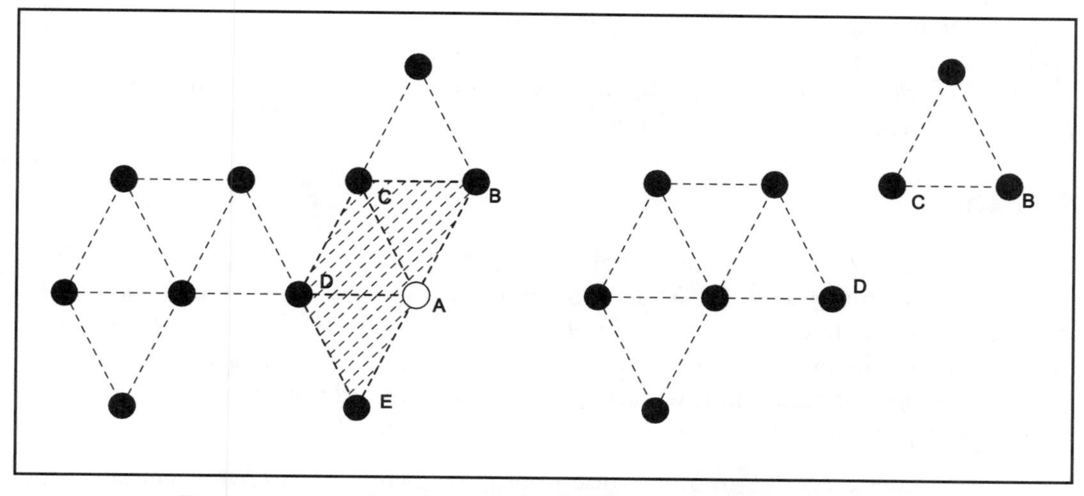

Figure 13-13 Removing the triangles that contain node A. Though the network is split into two parts, it is topologically correct.

The ABC model began its life as a biologically motivated architecture for computer vision using recurrent SOM networks, though it has wider applicability. The key to the system is its ability to find, learn, and ultimately reproduce *sequences*. The basic module of the system is, thus, called a LAPS (Learning And Production of Sequences) module.

Sequential behavior is fundamental to any organism operating in the fourth dimension, time. There are several aspects to the problem of sequences. A successful organism must be able to manage sequence prediction; sequence generation, which is essentially the same as preduction, but in a more active role; sequence recognition; and sequential decision making, which is the selection of actions in a sequence. For more information, check out *Introduction to Sequence Learning* by Ron Sun.

VISUAL MODEL

Since ABC began as a model for vision, we begin our exploration by looking at the problem it was designed to solve.

Though we have a sense of our vision as being all-encompassing, absorbing an entire scene as a single unit, that isn't how it actually operates. The fovea in our eye is the only part that can see in any great detail—for example, when you read a book, the only part that can really read the text is the fovea. The fovea has the highest density of your 130 million or so photoreceptors, including most of the color-receptive cones. These photoreceptors converge into a mere one million ganglion cells, with the foveal receptors getting the most attention.

Though the fovea does most of our "seeing," it only covers about 2° of your 210° visual field—about twice the area covered by your thumbnail held out at arm's length. This biological efficiency is actually very handy since it limits the amount of visual information you have to process in a simulated intelligence as well.

SACCADES

You see your environment, not as a single picture, but as a series of tiny foveal snapshots, in a coarse and blurry context. Your eyes are normally in constant motion, known as *saccades*. Between these saccades are brief pauses, called fixations, where the image is captured before the eye moves on to the next fixation. The rough visual context around the foveal image contains color, texture, and more importantly motion cues, and can be used to help plan the next saccade.

Your eye saccades every 200 to 300 milliseconds, giving you three to five image snapshots per second, with the motion itself taking about ten percent of the time. Your eye moves very quickly in a ballistic motion lasting only 20 to 30 mS.

The sequence of events during one saccade/fixation sequence is fairly straightforward. During the fixation, the visual data is not only being processed by your recognition system(s), it is being used to plan the next saccade. Just prior to the saccade your visual sensitivity is suppressed. The saccade is a brief firing of the eye muscles, driving the eye to its new position. Once in place, visual processing is unsuppressed and the cycle repeats.

FIXATIONS

The sequence of fixation, or the *scan path*, is a learned response to visual input. Each scan path across an image is characteristic for a given person viewing a given picture. Conversely, each person will have a different scan path for a different picture, and different people have different scan paths for a common picture. This suggests that scan paths are learned by each person uniquely.

We seem to recognize complex images not as a whole but as a fixed series of snapshots. As such, the spatial relationships within the image are converted, by way of the scan path, into a temporal sequence of smaller images. Of course, this avoids the issue of our exceptional ability to recognize very familiar scenes and faces essentially instantaneously, in a single fixation.

Fixations tend to rest on areas of an image where there is a high gradient of change in luminance, such as edges or other bold changes in the image. From this general rule, we go on to learn that skilled professionals are able to fixate on particular important details of a relevant image, while laypersons will scan it in a more uniform way. Experts, then, have well-trained fixation points when it comes to images relevant to their experience...again hinting that vision and visual recognition is a trained activity.

Not only do we see the world as a series of small snapshots and not as a comprehensive tableau, our memory of a scene is not kept as a massive visual 3D representation of the space around us. Instead, the world itself serves as a sort of external memory that can be revisited at will by our roving eye. Internally, we seem to store the snapshots, or at least, the higher-level recognition associated with the snapshots, plus the peripheral cues and saccade sequence associated with them. Our short-term semantic memory of the scene helps give us the illusion of "seeing" it all at once.

FEEDBACK LOOP

The ability to feed the previous state of the network into current processing is the key to being able to learn and reproduce sequences of events. In turn, learning and being able to reproduce learned sequences are valuable skills in a brain. This section describes the Learning and Production of Sequences (LAPS) module that is used as a later building block, much like the SOM is being used here.

SOM VARIATIONS

Bear in mind that the LAPS model is using a generic fixed-grid SOM without any of the growth or attention mechanisms explored above. These refinements would be interesting extensions to the standard LAPS architecture.

The SOM network used by the LAPS does allow for one extension, though the LAPS model doesn't explore it very far. The winning node for an input can be determined not by a scan of all nodes by a global overseer, the "winner take all" method, but through a local suppression algorithm. Using local neighbor suppression, the SOM will have multiple active nodes for any given input, instead of a single active node.

When an input is presented to the SOM, each node in the network "turns on" to the degree that its input weights match the input vector. Normally, the global managing code scans the network and picks out one node to be the "winner" and turns off all other nodes. In a local system, each node is given an opportunity to suppress the activation of a handful of its nearby topological neighbors.

The amount of suppression depends on both the level of activation of the suppressing node and the distance the neighbor is from the suppressing node. Immediate neighbors are heavily suppressed, while more distant neighbors, four to ten steps away depending on the network, are suppressed very little.

The standard winner-take-all strategy of the SOM network provides slim pickings for downstream networks to work with. A local SOM suppression method provides a richer array of active nodes to act as inputs for downstream networks.

Another variation looks at moderating the winning node(s) of the SOM. Depending on your training situation, you may not want the same nodes to win for two inputs in a row. If two input vectors are similar to each other, they could capture the same SOM node, and not be properly distinguished from each other downstream. To prevent this, you can add a *refractory* period for winning nodes, either the globally selected winner, or those locally active nodes that rise above a threshold. This refractory period would begin by absolutely disallowing a node from becoming active for one cycle or more. After this there is a relative refractory period where the node is simply suppressed to lessening degrees.

One last look at SOM variations approaches the problem of the decreasing learning and neighborhood factor used by the SOM. You continue to learn all of your life. However, the emotional content of a situation (among other things) does affect your ability to learn. Emotion, mood, motivation, and instinct have a powerful role in perception and cognition. Highly emotionally charged events are learned faster. There is also a strong link between vision, memory, and the emotional centers, so an emotional context is linked with the other sensory inputs.

LEARNING SEQUENCES

An architecture for learning sequences of inputs is shown in **Figure 13-14**. There are three processes at work in this model: the classification of vectors by SOM networks, the conversion of SOM networks into linear vectors, and the Hebbian training of the output vector.

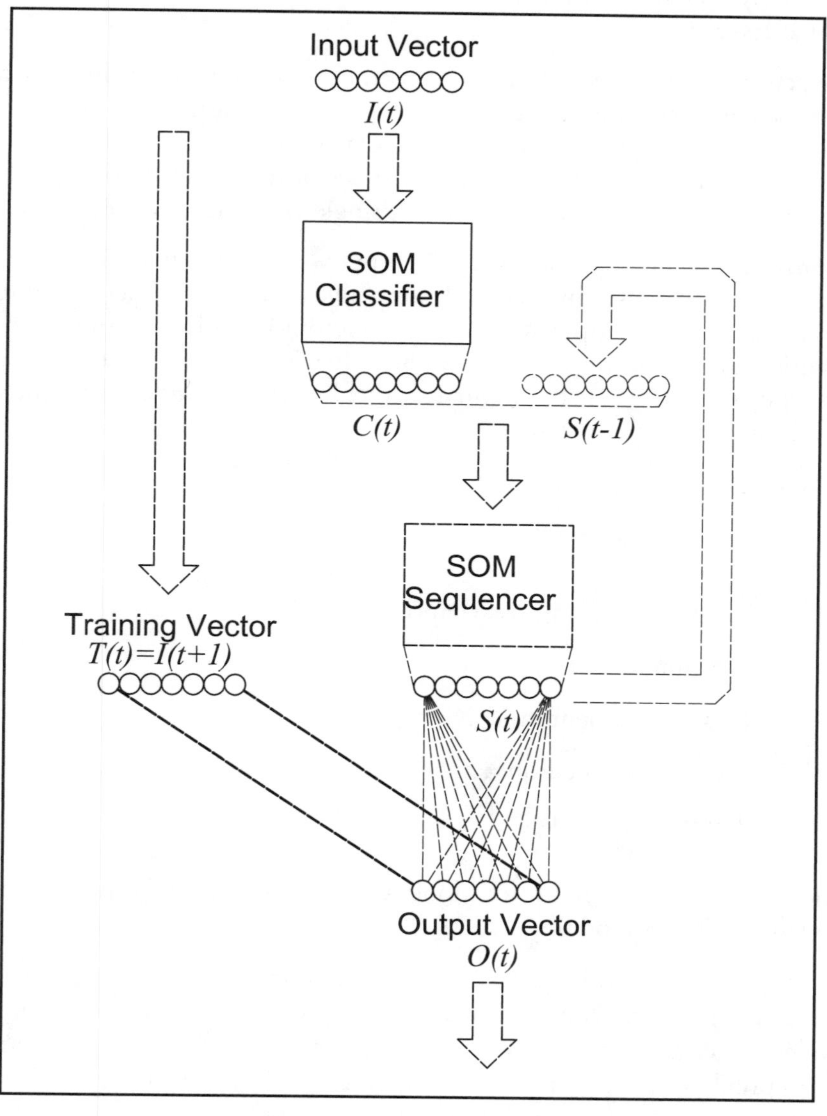

Figure 13-14 Learning sequences

As we saw earlier, an SOM network takes as an input one or more values organized as an *input vector*. Each node in the SOM network has a series of input weights, one weight for each entry in the input vector. The nodes in the SOM network are activated based on how closely they match the input, and during training the weights of highly activated nodes, and their closest topological neighbors, are adjusted so that they more closely match the input vector. Once the SOM is done being trained, it acts as an input classifier.

If the node activations in an SOM are to be used as input to another downstream SOM, they need to be converted to a suitable vector form. This is done by the simple expedient of reorganizing the N-dimensional topological SOM grid into a long 1D activation vector. For example, given a 2D SOM grid that has i rows of nodes arranged in j columns, the columns (or rows) can be concatenated into a single ($i * j$)-long vector.

Hebbian learning is a training rule where the input weights of a node are increased when both the inputs and outputs from that node are active at the same time. However, this can create a situation where the input weights grow without bound, so in this application the rule is modified to also decrease input weights when the inputs and outputs are out of synchronization, such as when the input is active but the output is not, and vice versa. This generalized Hebb rule is stated as:

$$w_i = w_i - \eta(x_i - x_0)(t - t_0)$$

Where:

w_i	Weight pre- and postsynaptic neurons
η	Learning factor
x_i	Presynaptic neuron activity
t	Postsynaptic neuron activity
x_0, t_0	Tuning constants

Bearing these operations in mind, let's look at Figure 13-14 again and watch an input travel through the LAPS module.

1. Input data *I(t)* for this current time step t is presented to the classifier SOM in the usual manner. Once the classifier has processed the input vector, the i by j nodes are concatenated into a classification vector *C(t)*. This concatenation is conceptual—there is no need for an additional data structure to be created to store it.

2. The vector *C(t)* is combined with the previous sequence state *S(t-1)* through the simple expedient of appending *S(t-1)* onto *C(t)*. Again, this is a conceptual step.

3. The sequencer takes the classified input vector plus the previous state vector as one long vector to be reclassified as a position in a sequence. The nodes in the sequencer are concatenated to make the output sequence/state vector *S(t)*.

4. The presynaptic activation levels from *S(t)* are trained against this step's training vector *T(t)* using the Hebbian rule described above. This training vector holds the next input *I(t+1)* so that the input *I(t)* prompts as a result the next input *I(t+1)*. The activation levels of the nodes caught in this crossfire are the output of the LAPS module, and should eventually be equal to the training vector *T(t)* for any given input *I(t)*.

PRODUCING (LEARNED) SEQUENCES

Once the LAPS module has been trained, it can be used to play back the sequence. The architecture for this playback is shown in **Figure 13-15**. Let's watch it in operation.

1. To start things off, we seed the input vector *I(0)* with the first vector in the sequence.

2. The input is classified and the results vectorized, giving the output vector *C(t)*. This has the previous state vector *S(t-1)* (which will be zero for the first step) appended to it and these are sent as a whole to the sequencer.

3. The sequencer categorizes its input *(C(t) + S(t-1))* to get the current state *S(t)*.

4. The well-trained output vector takes *S(t)* and generates *O(t)* from it...which should be *I(t+1)* from the original sequence.

5. The output is fed back to the input and the process repeats at step 2.

Figure 13-16 shows the LAPS module in schematic form, showing the two operating modes from Figure 13-14 and Figure 13-15. Reducing the complex LAPS diagram into a simple icon helps organize things when it comes time to incorporate it into a larger system. The inputs can be the raw input vectors, which works for systems shown so far, or they can be vectorized SOM classifier results, which will work better for systems described later.

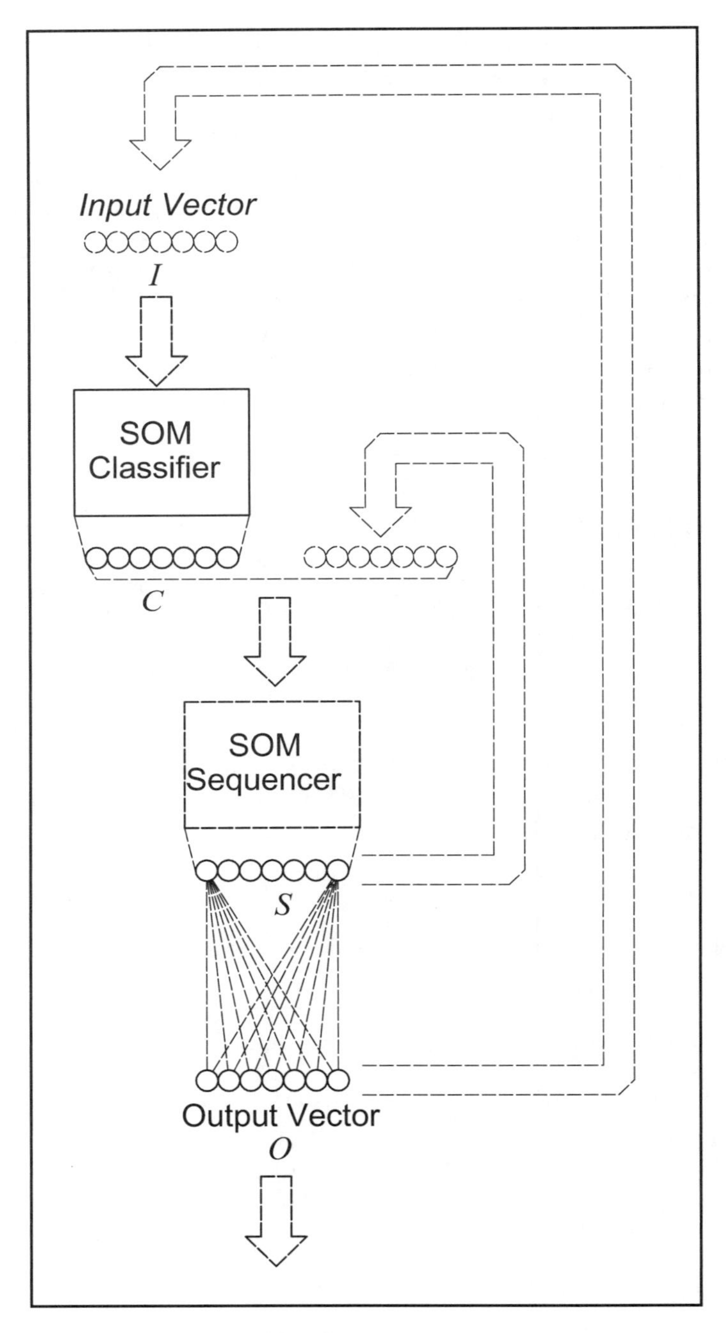

Figure 13-15 Producing a learned sequence

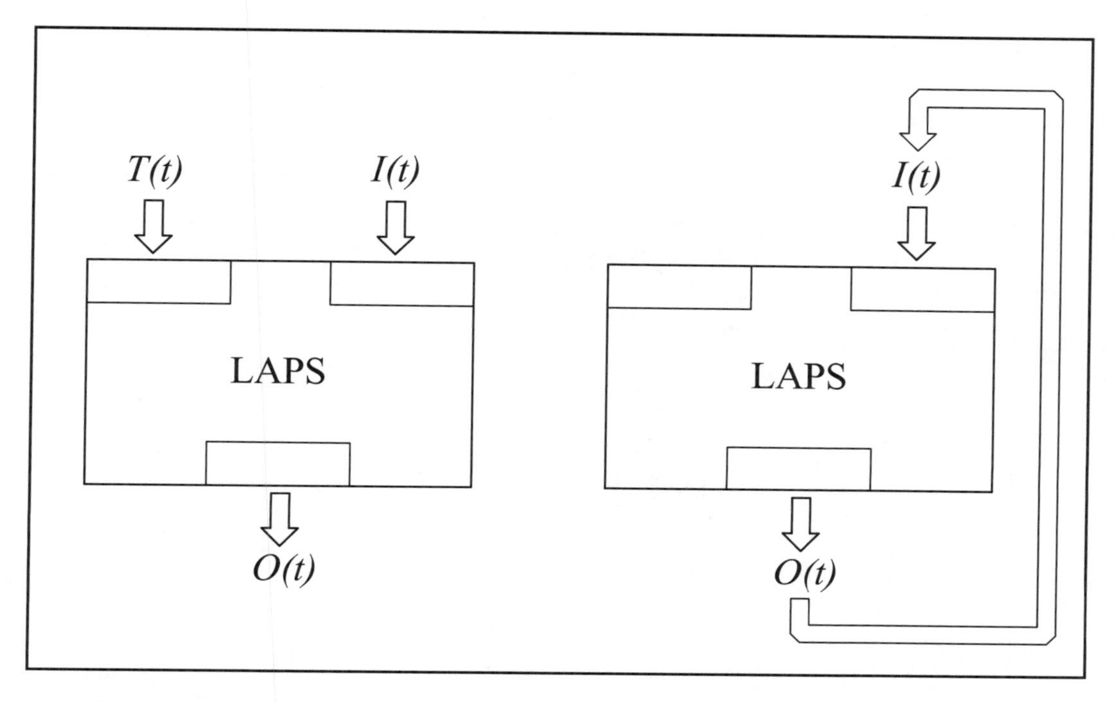

Figure 13-16 Schematic view of LAPS module in Learning and Producing modes

LAPS

A slightly different look at the LAPS operation allows us to tie the input and training vectors together, as shown in **Figure 13-17**. This is *essentially* the same system as described in Figure 13-14. Lets watch it do two loops of operation. Note, however, that the output to input loop is not discussed in this processing...it's simply a foreshadowing of things to come. In some cases, however, you may want to *merge* the output back to the input, so the actual input is some combination or average of the external input and the internal feedback. This is like a self-talk loop, where the net is mumbling to itself while it is also listening to outside input.

Note that in these diagrams, the icons for the vectors are not drawn to scale. Here is a sense of how big things can become. Say the input vector I has i entries. The classifier then has x by y nodes in a grid, each with i weights in their input. The output vector C has (x^*y) elements in it plus the (q^*r) elements from the sequencer S. Of course, each q by r node in the sequencer has $((x^*y)+(q^*r))$ input weights. The output vector O should have the same number of nodes as the input vector I, giving a significant reduction in size from the (q^*r) nodes in the sequencer.

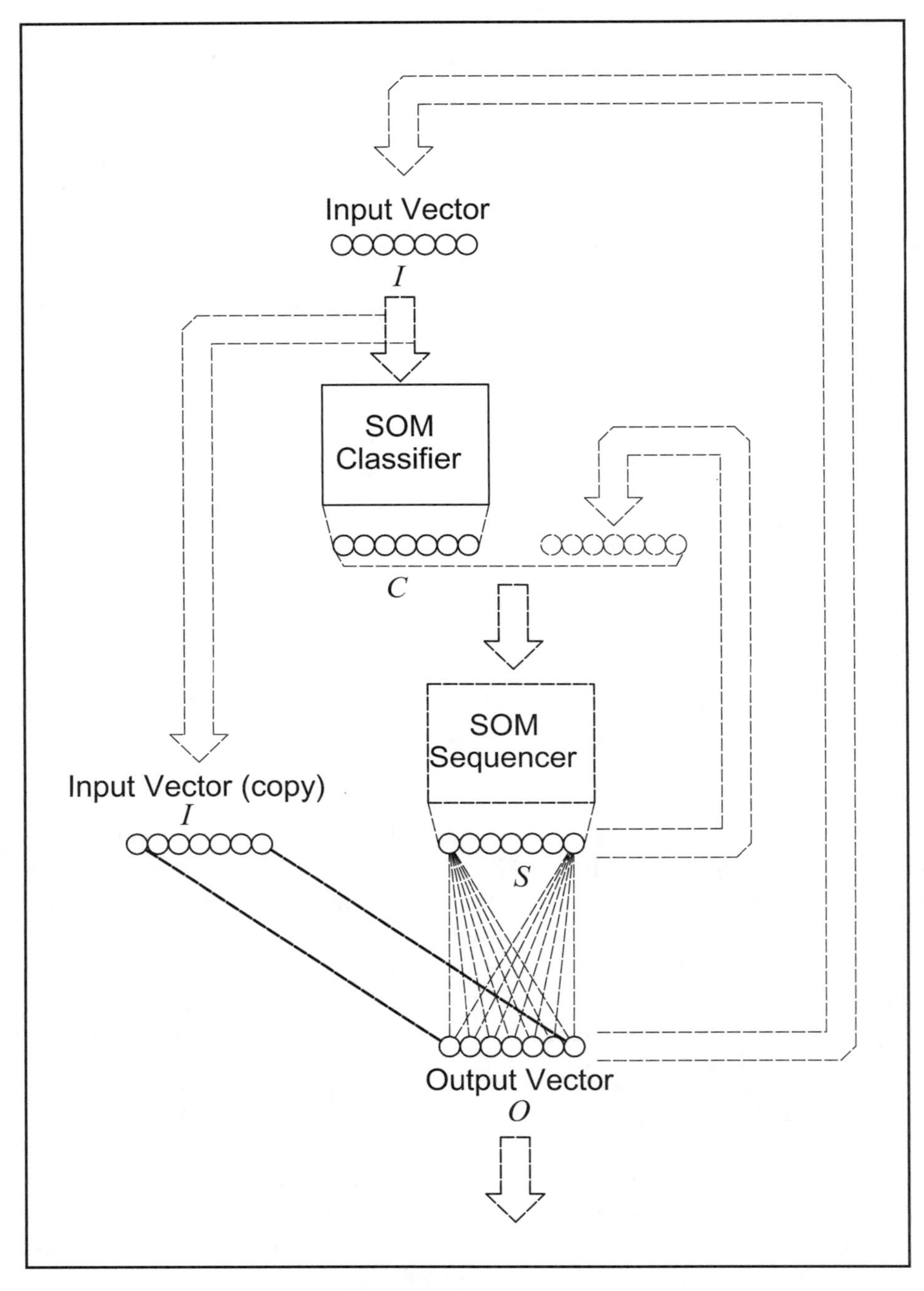

Figure 13-17 LAPS

1. Time t=0. The first input *I0* is presented to the classifier and is copied down to the training vector (see **Figure 13-18**). The classifier creates its vector *C0* based on the input.

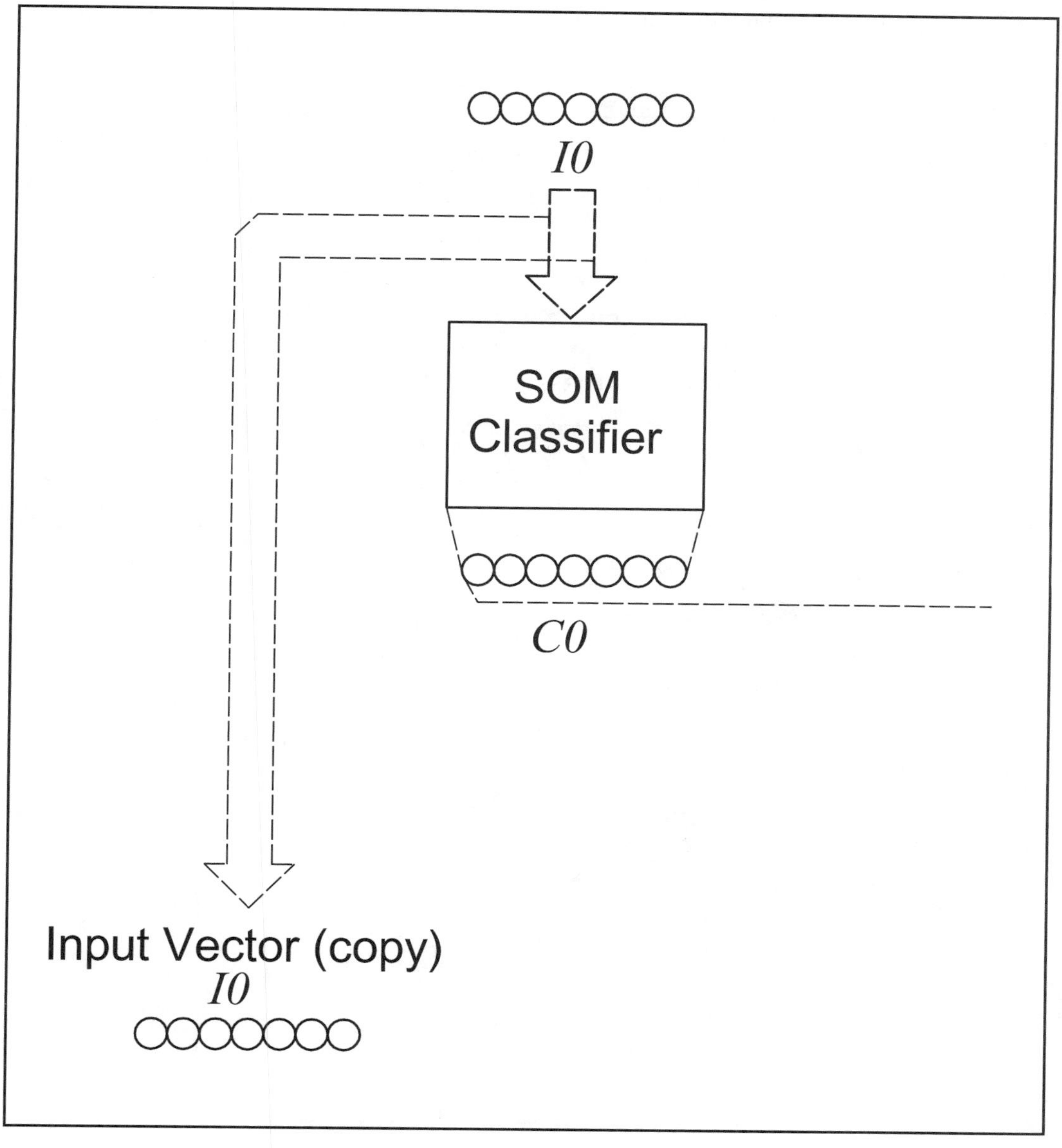

Figure 13-18 Time t=0, Input I0

2. Time t=1. Input *I1* is presented to the classifier and is copied down to the training vector (see **Figure 13-19**). The classifier creates vector *C1* based on input *I1*. Meanwhile, the sequencer creates its vector *S0* based on *C0*, which it received in the previous step. This *S0* is cloned and added to *C1* as the next step's input to the sequencer. The output vector *O0* is trained against *S0* with the current training vector, which is *I1*...the "next" step in the sequence.

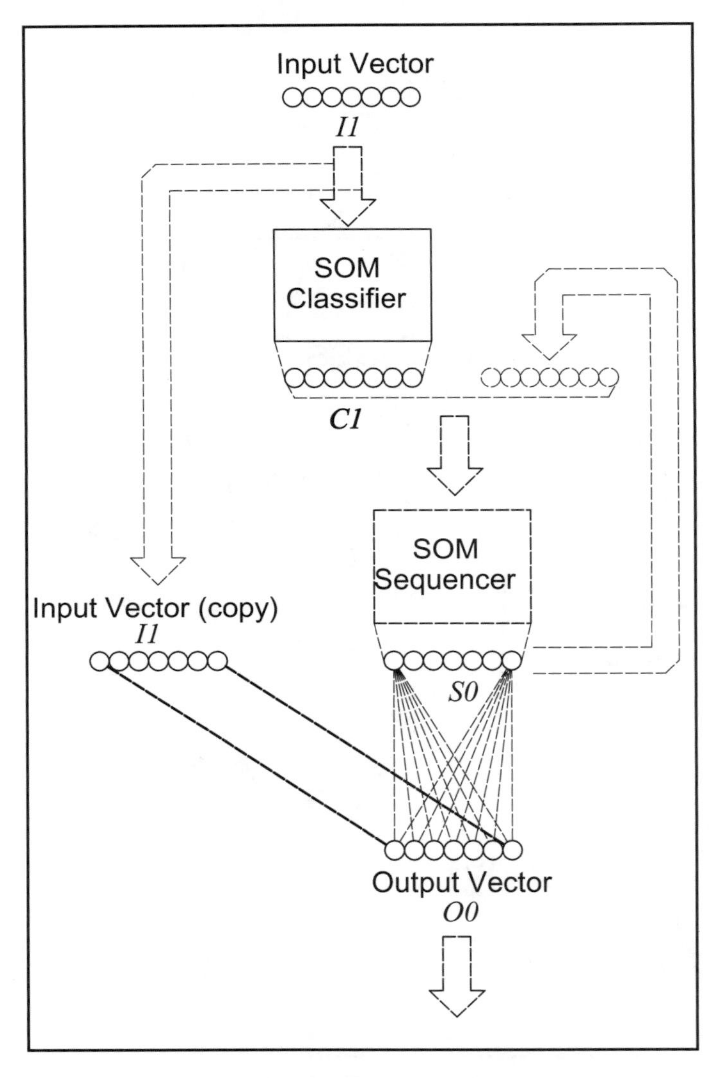

Figure 13-19 Time t=1, Input I1

3. Time t=2. Input *I2* is presented to the classifier, generating *C2* (see **Figure 13-20**). The sequencer generates *S1* based on its pending input *C1* and *S0*. *I2* is also at the training vector, where it is being combined with *S1* to create *O1*...the next step in the sequence.

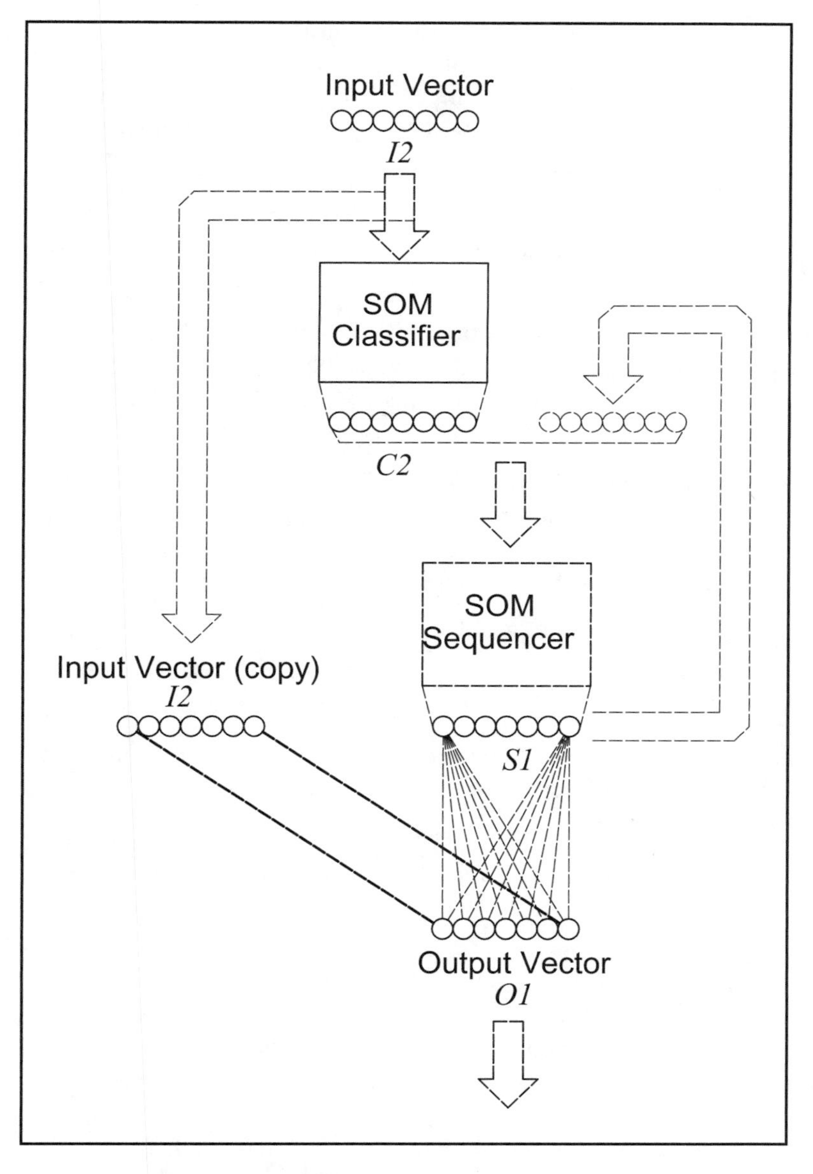

Figure 13-20 Time t=2, Input I2

Dynamic LAPS

Two extensions to the LAPS model give it additional power, but make it even harder to describe. Both of these extensions add to the context of the input being processed.

Where we reduced the LAPS to a single input in Figure 13-17, we now add another input back to the module. The original input value is the *direct* input, the input we are learning from. This new input is an *associated* input that provides context from other sensory modalities or subfilters. Its use is discussed later, in the ABC model.

One of the powers of the LAPS module is in predicting the future—and knowing the future can help in processing the now. The other input that will feed the module is an internal prediction vector, which is finally incorporated into the model in a more formal manner.

Both the association (context) information and the internal prediction loop are added to the model with biological justification...but you will need to go to the greater body of literature on the subject for an analysis of this. I am here to present the *how* more than the *why*.

The Dynamic LAPS module in **Figure 13-21** shows the last extension to the LAPS model. First, the output vector is formally introduced into the module, appended to the input vector to provide state. Note that in Briscoe's *Adaptive Behavioral Cognition* this vector is duplicated; appended once to the direct input and again to the association input.

The other extension is the association input vector. This vector allows you to tie the Dynamic LAPS module into a larger system and provides additional context for the processing of the direct input.

Only the direct and association inputs are propagated to the training vector, while the entire set of input vectors, direct, association, and feedback, are used in the classifier.

Variations on this module put additional SOM classifiers between the input vectors and the SOM shown here, to supplement the "prediction" vector.

Motor Control

So far we have only discussed learning and playing back sequences. How does this fit in to the whole scan-path visual model discussed earlier? Assuming that the input is appropriately filtered and preprocessed visual data, the output can be partitioned so that part of it drives the "eye" muscles. This way, an input stimulus will trigger some muscle response, which changes the stimulus, and so forth. Mind you, this is mostly just thinking out loud. **Figure 13-22** shows this possible arrangement. Note that motor control really

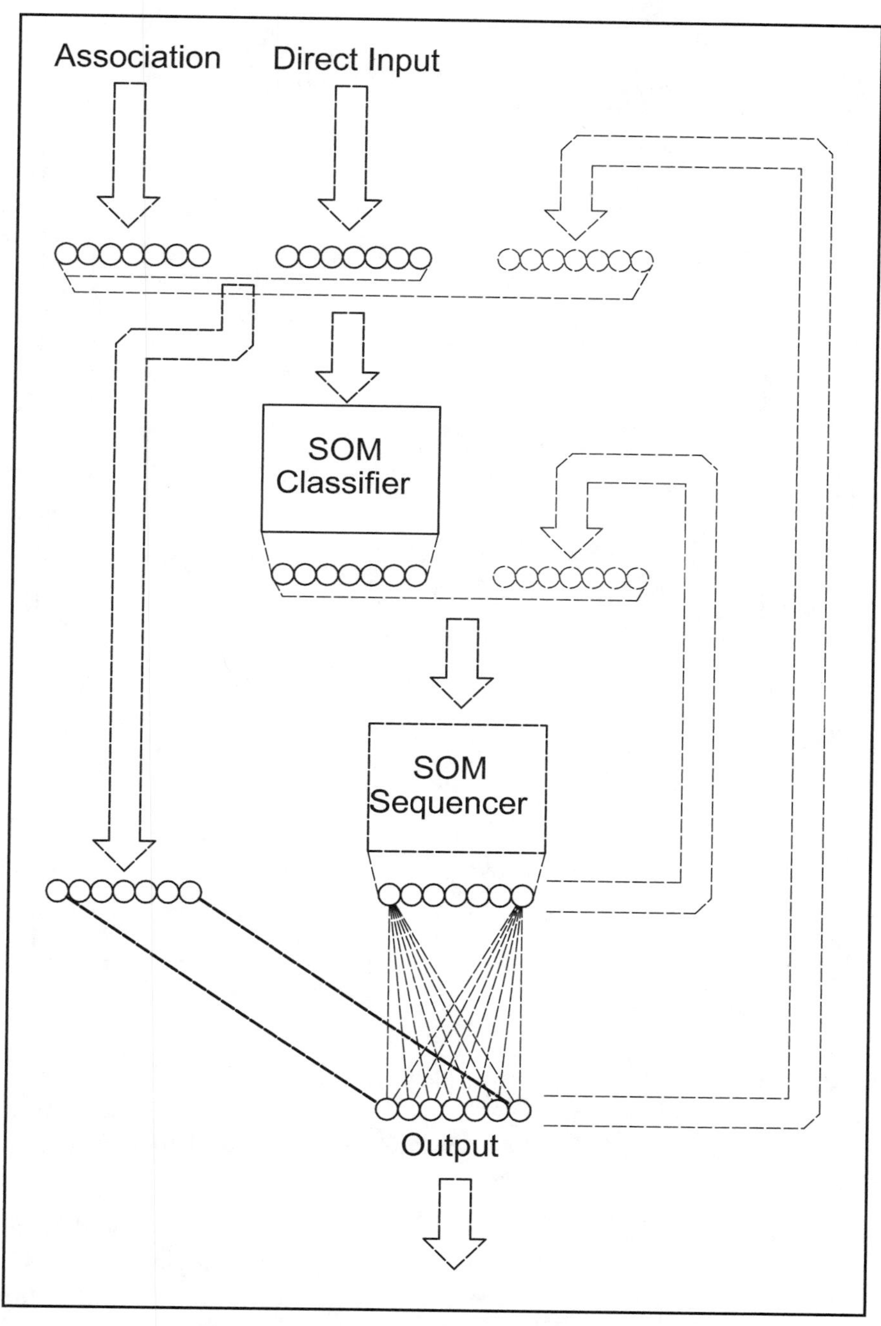

Figure 13-21 Dynamic LAPS module

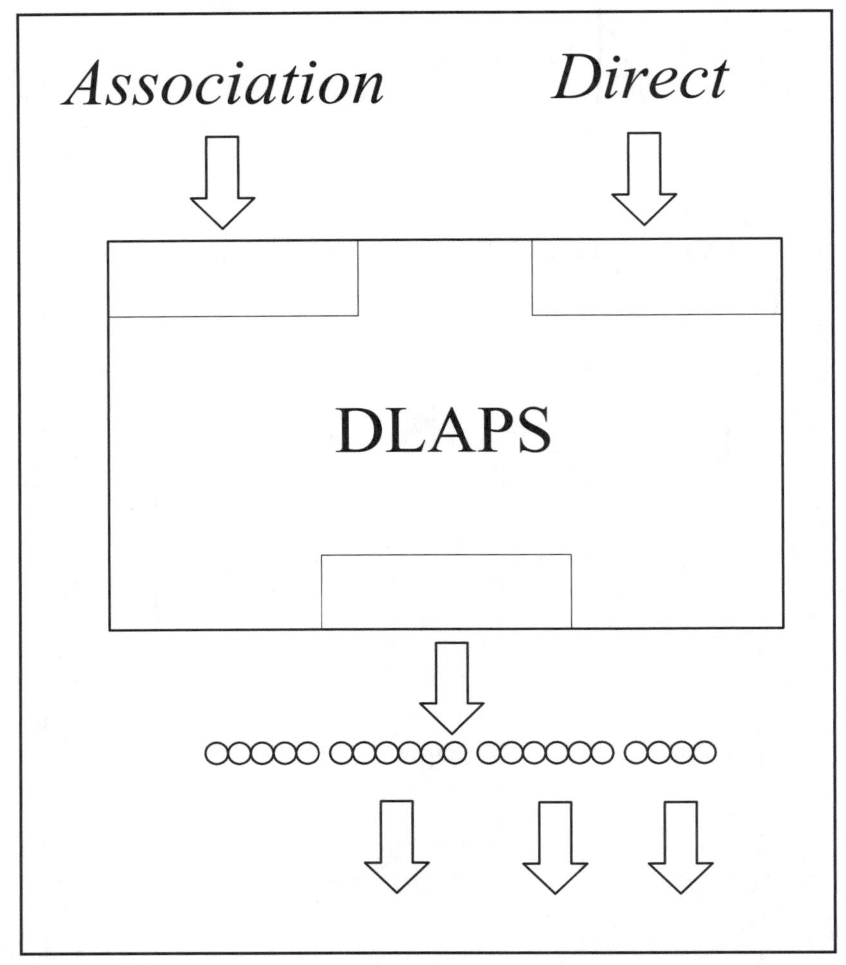

Figure 13-22 Possible scan-path network

only applies to our interpretation of the model's output. The direct input would be from the retina, giving the results of the eye's current status, and the output would be split up to drive various muscles of the eye. The association input could include information from various position sensors around the eye.

Internally the module is a flurry of activity and feedback...and there would be additional processing of the output vectors so they make "sense" to the muscle control. Of course, the eye-motor control could be a separate network, tied in through an association input. This would be an interesting model to explore in the robot.

ABC MODEL

The Adaptive Behavioral Cognition (ABC) model is a framework that ties together the DLAPS modules across multiple sensory domains and into a larger cognitive system.

Sensory input is, at least in the case of vision, processed heavily before it reaches our cognitive centers. Without getting into the specifics, visual signals are split up and spread all over the brain, where they are characterized in different ways. Some areas of the brain look for areas of color; others look for sharp edges or lines at various angles; yet others look for the ends of lines. Texture, motion, and tens, if not hundreds, of other relatively crude sensory filters are applied to the stream of visual data. Each filter, we assume, ultimately creates some kind of map or classification of its input, preferably in the form of a relatively small vector, to be used in downstream processing.

A more advanced model might use one or more image recognizers to recognize higher-level aspects of the visual stream—up to and including face recognition. This information would be carried into the visual stream along with the other, simpler, results.

The filtered information is fed into an SOM classification layer. This breaks down the information further, reducing its dimensionality a bit more and organizing it according to the statistical distribution of the world.

In the ABC model, two or more of these filtered and categorized data streams may associate with other filtered data, reinforcing or inhibiting information based on context.

Figure 13-23 shows this sensory association. The node values from the upstream classifier are impressed 1:1, that is, without any weights or other modifications, on its downstream association layer. The cross-stream, or associated, nodes, however, visit the association layer through adjustable weights.

Each node from an SOM classifier has a weighted link to *every* node in the cross-stream association layer. These weights are adjusted using Hebbian rules, as discussed earlier for output vectors. This filtered, categorized, and associatively reinforced data is then vectorized and passed downstream for even more processing.

This new concept of the association layer can be applied to the LAPS module, as well, as shown in **Figure 13-24**. Of course, things are beginning to get insanely complex here. The new *imprint* output (for lack of a better name) is essentially the same as the vectorized output. This new representation allows us to think about the module differently in the next step. We will stay with the DLAPS schematic symbol, since we are only using one of the outputs. Bear in mind, however, that it is using the imprint output and not the vectorized output.

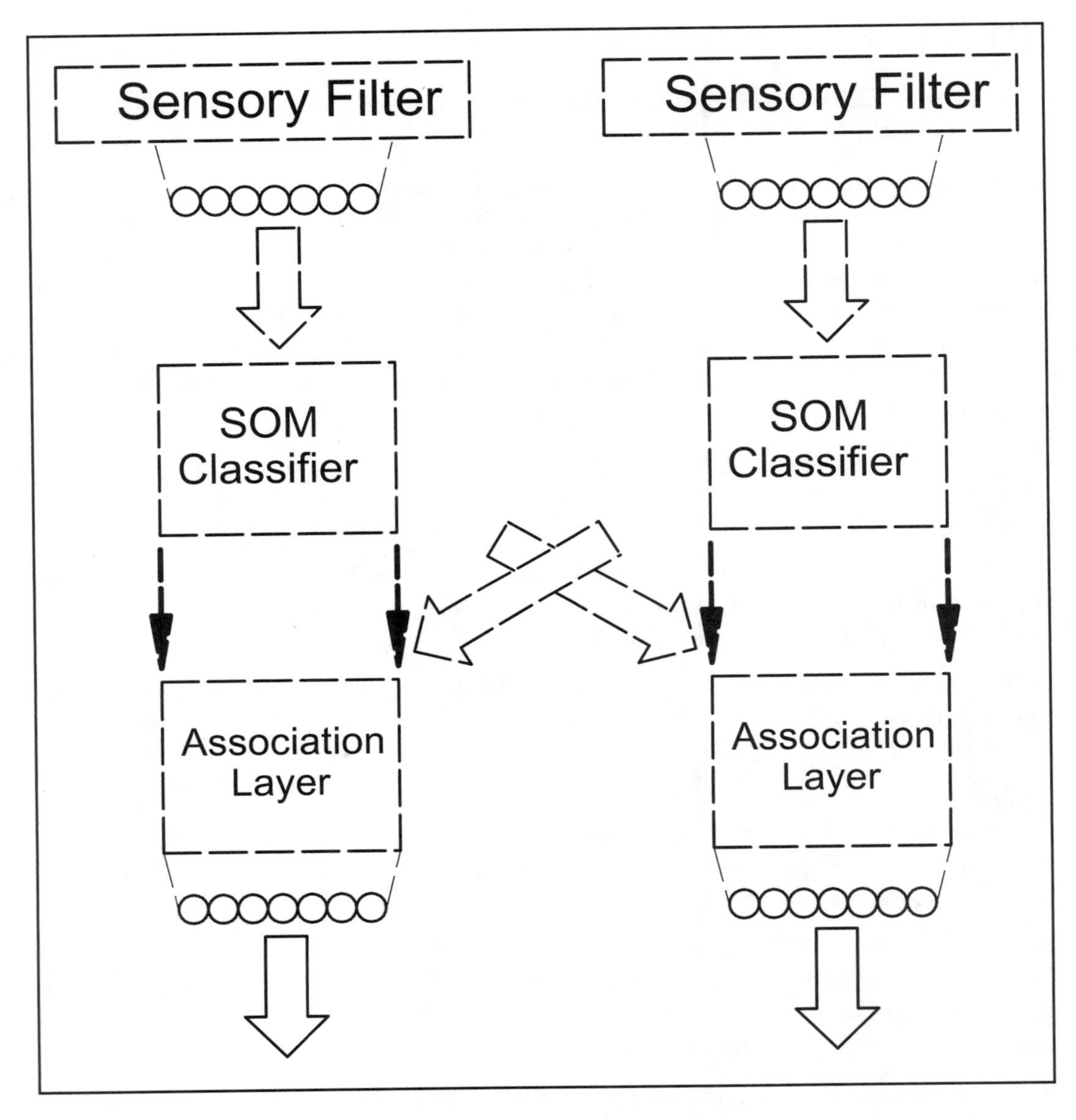

Figure 13-23 Sensory association

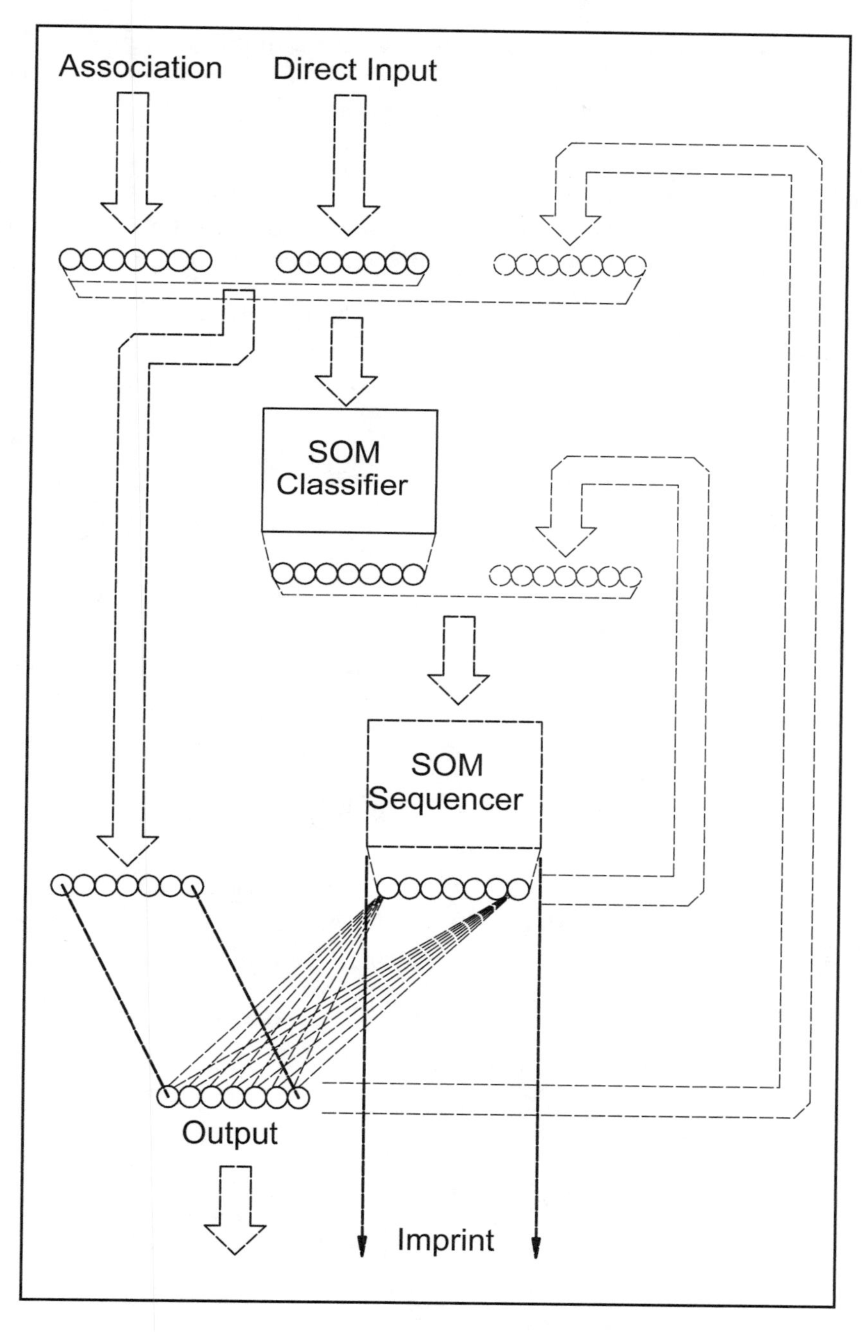

Figure 13-24 Dynamic LAPS, revisited

Just as there will be associations between filtered sensory inputs in a given sensory area (for instance, vision), there will be associations across sensory areas (for instance, vision + touch). **Figure 13-25** shows an example of this, including subsequent processing for motor control.

The filtered inputs for one sensory area are associated and processed in their respective areas. The results of this processing are then imprinted on a downstream association layer and then associated with the results of the cross-stream sensory result.

These associations are then further processed until, ultimately, they result in changes in the motor state.

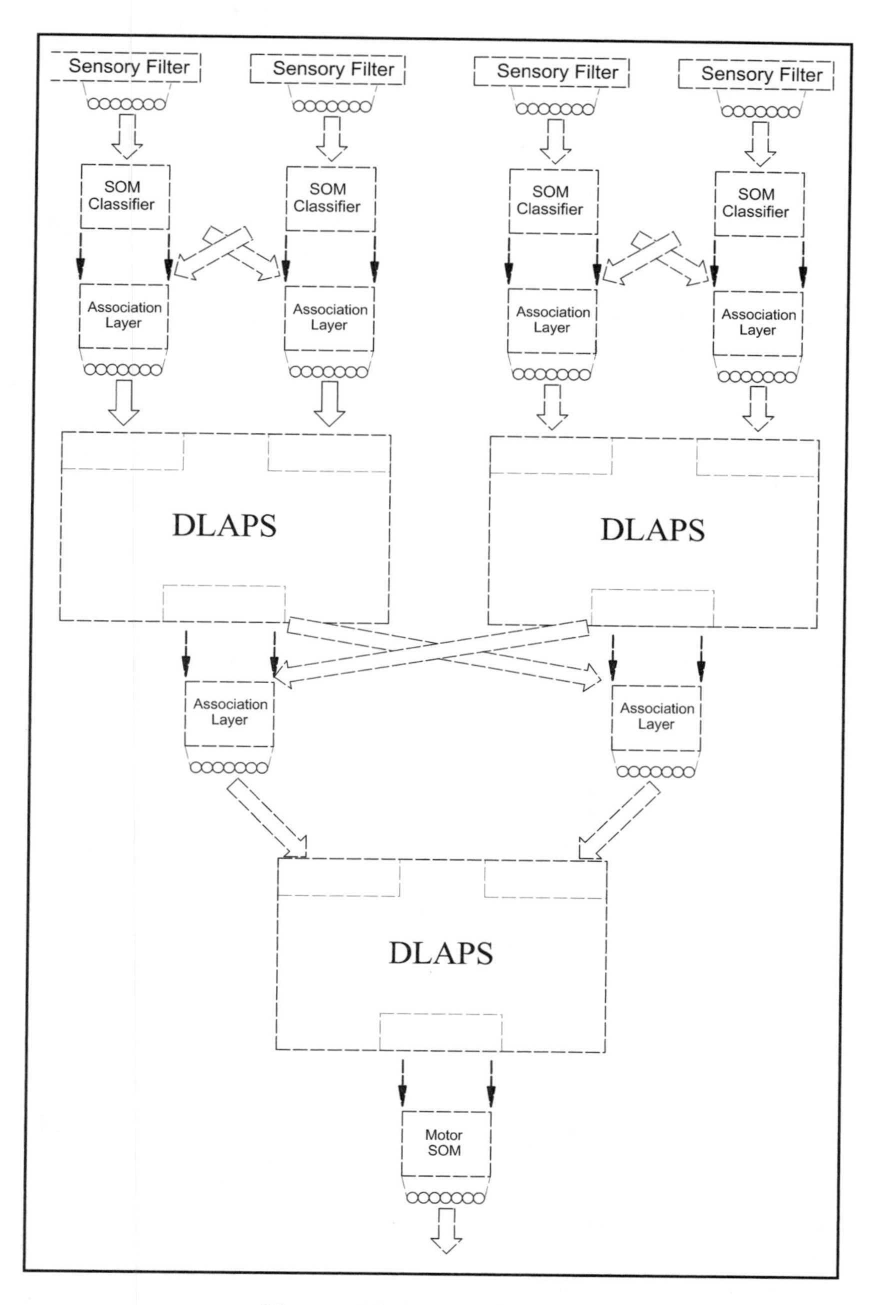

Figure 13-25 ABC model

CHAPTER 14

LEARNING AND NOTICING

The previous chapter looked at neural learning from a mechanical point of view—the adaptation of weights needed to implement a SOM network or a Hebbian neural node. We explored the ABC model, which uses these simple rules to create a sophisticated system that can learn and reproduce sequences. And we just touched the surface of the problem.

In this chapter we look further into learning, memory, and the act of *noticing things* in the context of neural systems. While the information presented in this chapter will not lead you directly to implementation, it provides food for thought and direction for research. Ultimately, a comprehensive model of cognition and memory will need to incorporate ideas from all of these areas. What we don't do is delve into anything that will give the robot personality, nor do we go very far into the dynamics of reward, punishment, or instinct.

In a certain sense, personality can be based in what the organism (or robot) likes and does not like, for example, its internal reward structure. In another sense, personality can be faked. But on the third hand, personality is going to be ignored.

LEARNING

Learning is not just about memory but also about the various factors that influence the storing of stimulus in that memory. How is it that some brief episodes in our life are burned indelibly in our minds, and others are just a blur, indistinguishable from other blurs in our lives? How does our pet robot brain learn what is good and what is bad? These and other thoughts are explored below.

MEMORY

Though we may think we understand memory, seeing as how we all have one, the mechanisms behind memory have proven to be elusive.

Biological memory has evolved, along with our other cognitive and perceptual mechanisms, to help its parent organism deal with a complex and changing (yet structured) world. The principle value provided by memory of past events is its ability to help predict the future.

Before learning can take place, the organism must be able to perceive the world around it. The raw signals taken in by these sensory nerves are combined, split, massaged, processed, and then (here's the learning bit) these compressed inputs are correlated with each other and ultimately filed for future reference. From a symbolic point of view, memory is recording associations between attributes, or between objects and attributes. To learn more about this theory, see "Why There Are Complementary Learning Systems in the Hippocampus and Neocortex: Insights from the Successes and Failures of Connectionist Models of Learning and Memory," by James L. McClelland, Bruce L. McNaughton, and Randall C. O'Reilly.

Using the notation explored for the ABC model in the previous chapter, **Figure 14-1** provides one look at this association of symbolic attributes (though a rather different look at it than seen in McClelland et al, who are using a back-propagation network). The *object* is a vector with one node per object, such as a living thing, plant, tree, daisy, robin, etc., and the *association* is a similar vector for abstract relationships, such as is-a, is, can, has. The *attribute* at the bottom is a Hebbian vector with a matching training vector (not shown). The attribute vector mirrors the object vector with an additional list of attributes, for instance, living, green, red, grows, moves, swims, flies, barks, or petals.

In training, objects, associations, and attributes are all specified, and the network learns (via Hebbian learning, back-propagation, or whatever) the associations between them all. In operation, the object and association are presented, with the attribute popping out the end.

In spite of ducking the question of how the various vectors are created and generated, this model lets us explore different aspects of memory. For example, how can we get the system to remember the largest number of associations?

It turns out that the secret to stuffing a lot of information into the memory is two-fold. First, the learning rate must be set fairly low, which means that a training example must be presented many times before it is fully recorded. That, of course, rules this model out for episodic learning, where memory stores the information in a single presentation.

Second, the different examples must be intermixed, so the knowledge from the training examples is built up gradually, in layers. If one example is presented repeatedly until it is learned, then the second example is presented until it is learned, and so on, the system will exhibit *catastrophic interference*—where early associations are completely lost because they are overwritten by later associations.

Both of these things, small learning factor and interleaving examples, make episodic learning a problem.

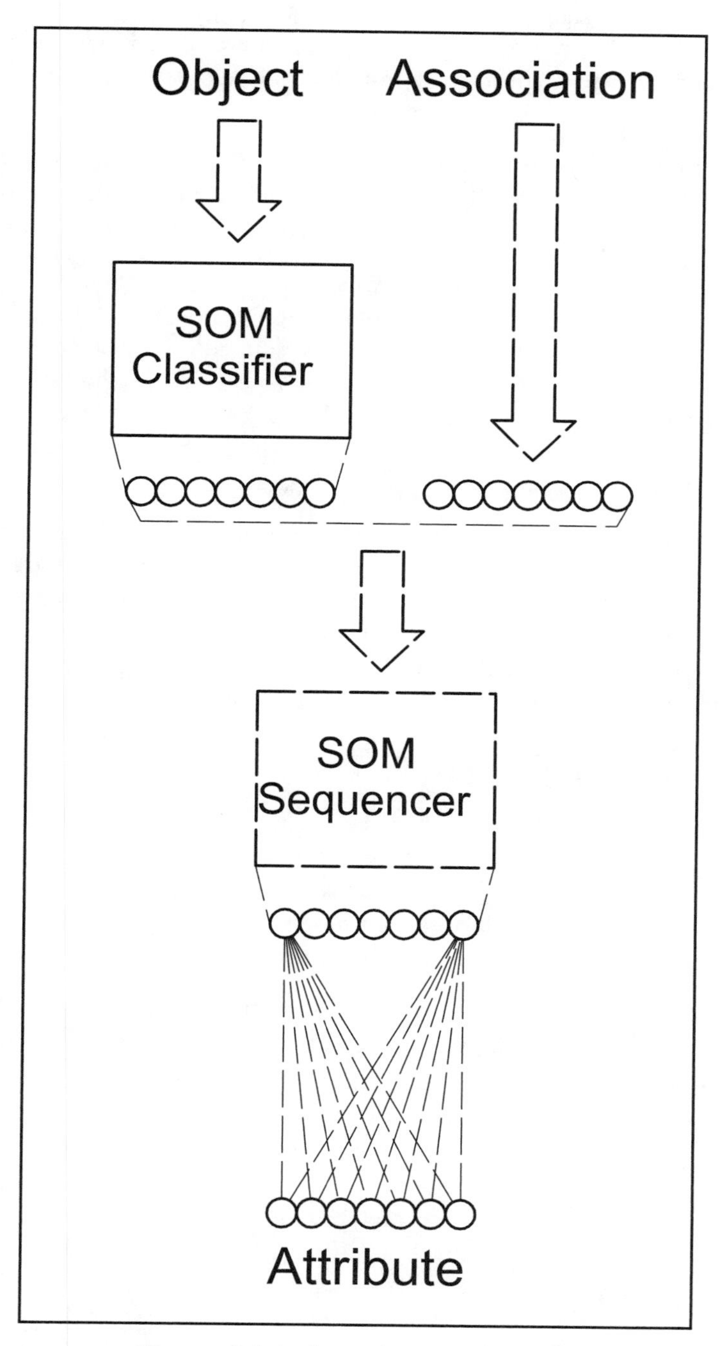

Figure 14-1 Learning associations

The solution to this problem is a two-layer system, where there is a *working memory* that learns very quickly but doesn't store very much, and a *long-term memory* that learns very slowly but retains a huge amount of information.

The working memory can record various inputs and correlate them across a short period of time. This memory will learn very quickly, but will retain the information for only a short period of time. Working memory would not necessarily benefit from experience or long-term learning changes.

The working memory, whose role mirrors that of the hippocampus, then feeds the long-term memory, playing back its memories (via self-talk loops, dreaming, or other mechanisms) to the long-term memory, whose role mirrors that of the neocortex. Since the working memory can provide a playback loop over time, the long-term memory has both of its criteria fulfilled: interleaved examples, plus learning over time with a small learning factor.

Long-term memory itself is not necessarily a single thing, but will be a number of related information stores. Some theories break long-term memory into two parts: *semantic* and *declarative* memory. Semantic memory stores implicit, or procedural, information—generic facts and data, such as the capital of France, the formula for salt, or how to repair an internal combustion engine. Declarative memory stores explicit, sometimes called episodic, information—personal and detailed events and scenes, such as a recollection of an evening by the river, or your first kiss. Declarative memory may also be involved in the conscious recollection of an event.

The actual process of using memory has been divided in some theories into three parts. The first part is the *central executive* that is the main controller that guides the overall process of attention and recall. This executive is aided by the *phonological loop* that maintains about two seconds worth of speech-based information in sort of a self-talk cycle, as well as the *visuospatial scratch pad* that holds visual and spatial information.

The two holding areas, the phonological loop and the visuospatial scratch pad, use the same neural space as the sensory input, so both the self-talk and the visual imagination are forms of self-induced hallucination.

Reward and Punishment

Artificial neural systems (and, for that mater, biological neural systems) learn through feedback. In crude terms, they learn through reward and punishment. Neural systems are designed to maximize the positive feedback—rewards—while minimizing the negative feedback—punishments, whether natural consequences of behavior such as the burn from touching a hot stove, or imposed punishments such as the spanking for breaking a vase.

When you design a robot, you need to put some thought into its policies of reward and punishment. Evolution has spent millions of years designing our feedback systems; we don't have that much free time. One area where reward and punishment are explored in great detail is the field of Reinforcement Learning, and it would be great if RLs capacity to learn through sparse feedback could be generalized to fit the ABC model. The definitive work on reinforcement learning has to be Richard S. Sutton and Andrew G. Barto's book *Reinforcement Learning: An Introduction*, which is also available online at www.anw.cs.umass.edu/~rich/book/the-book.html.

As an example, how do you set up the reward and punishment of a mobile robot, so that it does not run into walls? There are several ways this could be done as described in "Rapid Reinforcement Learning for Reactive Control Policy Design in Autonomous Robots," by Andrew H. Fagg, David Lotspeich, Joel Hoff, and George A. Bekey.

PUNISH HITTING WALLS

When the robot hits a wall, it receives a negative reward—punishment. Soon, the robot is unable to decide what to do and becomes a dithering wreck. Since it is only told what does not work, with no reward to reinforce correct behavior, the system spends its time trying out new actions, fruitlessly searching for something that works.

REWARD NOT HITTING A WALL

The network soon finds a happy place; the one-move minimum action that satisfies the basic requirement and returns a positive reward, but not doing much else. For example, it may just turn to the right endlessly, moving in an endless happy circle. The network gets locked into this action and cannot adapt to new situations, such as when a new obstacle is placed in its path.

PUNISH HITTING A WALL AND REWARD NOT HITTING A WALL

Okay, since neither of the simple rules work, let's try doing both. There are two subcases to this policy.

■ Punishment is Less Than the Reward

This policy tends to create short cycles of motion with an overall reward, not unlike those seen when there was a reward without punishment. When the robot is faced with an obstacle, a common solution is to just hit it, back up a bit, go forward, and hit it again. Since more cycles are devoted to contact-free rewarded motion than the punished collision, the robot is more than happy to stand there and bang its head against the wall.

■ Punishment is Greater Than the Reward

This turns out to be the best policy. Make it hurt to hit walls, more so than the feel-good for not hitting them. The reward makes it profitable to explore, since doing nothing generates no feedback at all, but the stiff penalty for collisions makes it worth the robot's time to not incur them.

INTERACTIVE TRAINING

Instead of just writing in a fixed feedback policy and turning a mobile mind loose in a possibly dangerous and uncertain environment, you can take an active role in training it, as described by Paul Martin and Ulrich Nehmzow in their paper "'Programming' by Teaching: Neural Network Control in the Manchester Mobile Robot".

One way to teach your robot is to have it learn associations between inputs and actions while you drive it under remote control; guiding its hands, as it were. This is a form of reinforcement learning in real time, with the training feedback being the human control signal. Your control could be absolute, where the robot is just an observer as it is toodled around the environment, or the control inputs could be just an advisory, blended with the robot's own impulses to create the final action.

Central to this training mechanism is an associative memory, where input sensory signals are correlated to output motor signals. This memory associates sensory inputs with motor actions while the human controls these actions with the remote control. Once unplugged from the teacher, the sensory inputs will then control the motor outputs directly, based on what it learned during training.

Using this system, it is possible to quickly train the robot to perform a number of actions, such as avoiding walls, pushing boxes, cleaning an area, or learning a route.

EPISODIC LEARNING

Episodic learning is an important feature of our memory, where we can record in detail an important scene or fact from a single glimpse or presentation. In the backpropagation world, the SOM world, and, in fact, most of the neural network technologies, episodic learning is something of a holy grail. Most networks have a definite tradeoff between capacity and speed of learning, as discussed briefly in the section on memory above.

A cognitive model of associative memory, which, essentially, is what we are studying here, should have a variety of features if it is going to provide a decent copy of the biological process it is mimicking. First, all storage and retrieval should be based on local processing, that is, possible to implement from a single cell's point of view. Events and associations related to them should be able to be stored in a single presentation—the very definition of episodic learning. The representation used should be distributed, using continuous-value nodes (that is, not a binary, on/off system), not unlike what we see in the SOM model. Finally, any memory interference effects and degradation of memory should follow those patterns that have been discovered from the study of biological brains. See "Trace Feature Map: A Model of Episodic Associative Memory," Risto Miikkulainen, for more information.

CONVERGENCE-ZONE

The memory solution introduced earlier used a rapid-learning model based on the human hippocampus that then proceeded to spoon-feed a slow-learning model based on the neocortex. Another model based on the hippocampal component of this two-part system is the Convergence-Zone Model, discussed in detail in Michael Howe and Risto Miikkulainen's "Hebbian Learning and Temporary Storage in the Convergence-Zone Model of Episodic Memory".

In the convergence-zone model, as shown in **Figure 14-2**, the SOM classifiers cook all of the sensory inputs equally. The resulting feature maps are all projected onto a single binding layer, reminiscent of the association layer of the ABC model, and fulfilling essentially the same role. The binding layer uses Hebbian learning to find synchronicities between the various features projected onto it.

During recall, one or more feature maps project to the binding layer and reactivate the nodes that were trained previously. These binding nodes then project *back* up to the feature map and reactivate the original nodes used to train the binding originally. This, then, re-creates the experience in these feature maps, creating the hallucination that the event is occurring as learned—giving a vivid, multisensory memory of the event.

To stay true to the hippocampal episodic model, the learning weights for the Hebbian learning are at or near unity. The relationship between the sparse features in the maps is stored directly in one try. Oddly enough, this high learning rate also provides the greatest storage density in tests of the model.

Because of the rapid learning rate needed for episodic learning, the node weights inside the model can easily grow to excessive values. To compensate for this, we must also incorporate some form of "forgetting." Of course, because of this, older associations will fade from memory—but this is expected for this phase of memory.

There are a number of methods you can use to "forget" or otherwise normalize the connection weights in the model. Regardless of the method used, the results are similar, so it is best to stick to fast and simple techniques. The simplest method is to simply decay all node connection weights slowly over time.

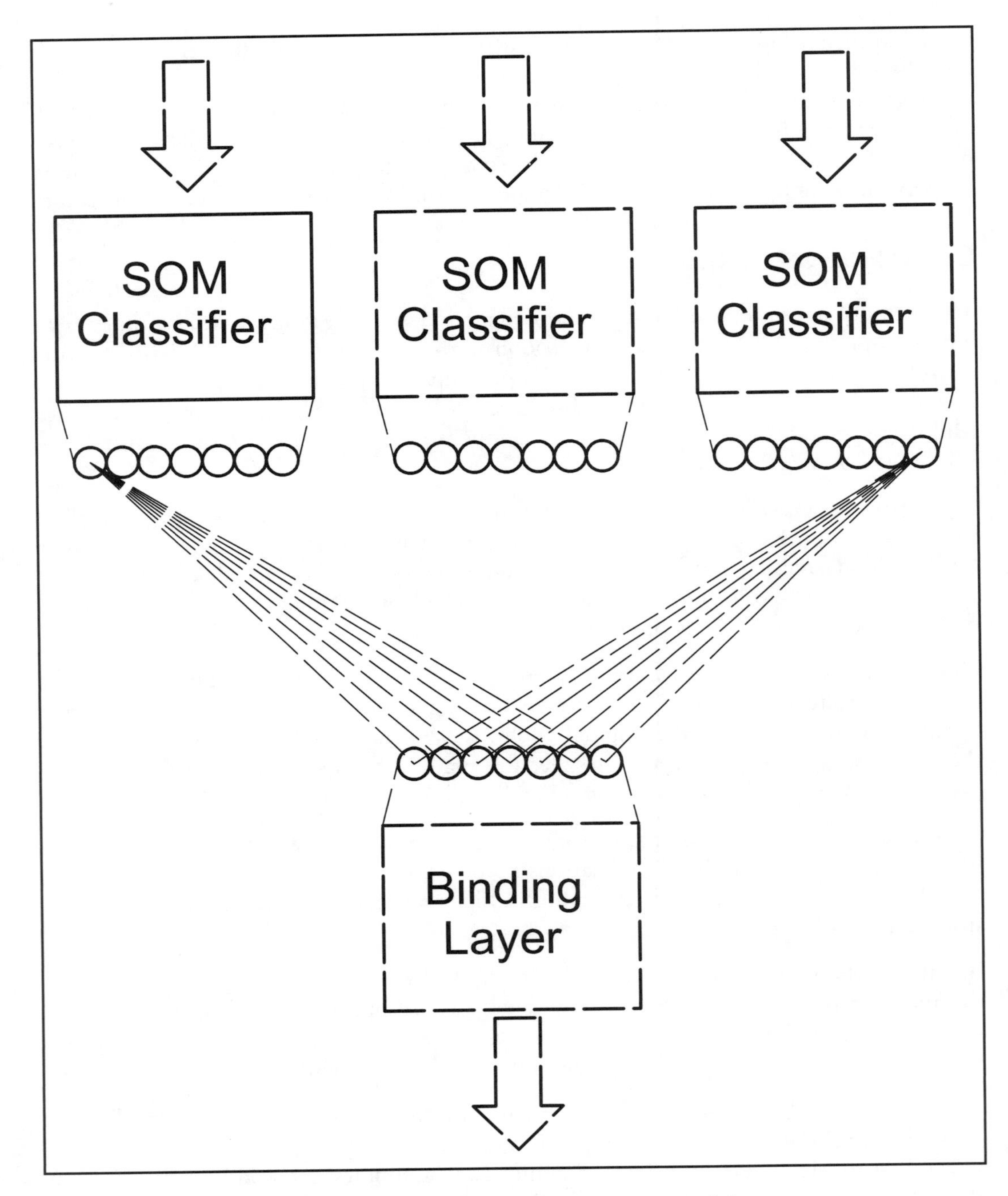

Figure 14-2 Convergence-zone model

TRACE FEATURE MAP

A different approach to episodic learning is the trace feature map. The trace feature map, as described by Risto Miikkulainen, is a variation of the vanilla SOM we have been using through the last couple of chapters.

The trace feature map model uses a variation of the SOM algorithm. The layout and meaning of the nodes is the same as the SOM model discussed in previous chapters; however, the activation and inhibition of neighbors is different.

The first variation allows us to eliminate the global check for a winning node by providing a sensible algorithm for activating the nodes in the network. Given these values, some of which will be familiar:

m_i	SOM cell weight vector *m* instance *i*
x	Input sample *x*
d_{max}	Maximum distance $\max\left(\left\|x_j - m_j\right\|\right)$ (discussed in text)
s_i	Similarity measure

The activation of a node is based on its similarity with the input, normalized by d_{max} that describes the maximum difference that can exist between a weight vector and an input. For example, for a 4-element input vector (a 4-dimensional input), with values from 0 to 1, d_{max} would be the distance from (0,0,0,0) to (1,1,1,1), or $\sqrt{4}$.

The similarity measure is calculated as:

$$s_i = 1 - \left(\frac{|x - m_i|}{d_{\max}} \right)$$

The activation of the node is then a sigmoid function based on s_i. Neural network scientists just *love* the sigmoid function; it provides a nonlinear response that works well in these systems, and limits the activation to the range of 0 through 1:

$$n_i = \frac{1}{1 + \left(e^{(\delta - s_i)\beta}\right)}$$

Where:

n_i	Activation level of SOM cell *i*
β	The slope of the sigmoid
δ	The displacement from the origin (if any)

In addition to the normalized activation of the SOM node, we add in a lateral inhibition to help focus the response to the input:

$$s_i = s_i + \sum_j \gamma_{i,j} n_j (t-1)$$

Where:

$\gamma_{i,j}$	The lateral connection weight between nodes *i* and *j* (see text)
$n_j(t\text{-}1)$	The activation weight of a neighboring node from the previous time step

Putting this all together, the activation n_i of a given node is given by the equation pair

$$s_i(t) = (1-\theta)\left(1 - \frac{|x(t) - m_i(t)|}{d_{\max}}\right) + \theta\left(\sum_j \gamma_{i,j} n_j (t-1)\right)$$

$$n_i(t) = \frac{1}{1 + \left(e^{(\delta - s_i)\beta}\right)}$$

where θ is the balancing factor that adjusts the ratio of local activation and neighborhood inhibition for this network.

The joker in this deck is the lateral connection weight $\gamma_{i,j}$. This defines a connection weight between nodes in the network, much like the connection weights between a node and the input vector. The lateral connection weight is the new aspect of the trace map.

The lateral connection weights, or *trace,* are used to funnel "near misses" to the appropriate node. The lateral weights are slowly adjusted with each training, approaching the constants of γ_E for positive excitatory weights when n_i is less than n_j, and γ_I for negative inhibitory weights when n_i exceeds n_j. The weight adjustment depends on both the learning factor and the level of activation.

During training, θ=0 and the activation of a node depends only on the inputs. Later, during playback, θ takes a positive value (such as 0.5) and the lateral connections come into play, pulling slightly off-target inputs over to the precise node they were trained for.

MAPPING

Most of our attention has been on the subject of locating the robot's mind in *time*, that is, learning, recognizing, and generating sequences of events or actions, and not enough attention has been paid to locating the robot's mind in *space*. Knowing where you are, what is nearby, and how to get somewhere else is important for any mobile organism. In a sense, though, the passage through space is a lot like the passage through time—at each step of the process, the sensory inputs will change in a more-or-less structured and well-defined way. Recognizing where we are in space is a task of matching the temporal pattern of sensory inputs. Heh. Written as if it were easy!

Random exploration of the environment requires the robot to mostly just avoid running into things or falling down stairs. More complex behaviors require the robot to locate itself in space, that is, determine its current position relative to a map or other system of landmarks, and its position relative to some goal position, plus some idea of how to get there from here. A side effect of this may be the creation and use of a topological map of some kind.

The only information available to a mobile robot, or any other mobile organism, for that matter, is the flow of redundant, noisy, and ambiguous sensory information. From this, though, we are able to determine where we are in space. How can our robot use this flow of data, though, to create a stable, meaningful, and useful representation of its position in space? One way is through episodic learning. This method is discussed in detail in Ulrich Nehmzow's "An Episodic Mapping Algorithm for Mobile Robot Self-Localization: 'Meaning' through Self-Organization."

EPISODIC LEARNING

In episodic learning, the system recognizes its position in space based on the current perception of the world, plus the last t_n previous perceptions. This algorithm is similar in many ways to hippocampal mappings found in rats.

There are, of course, some shortcomings to the method presented here. One is that successful recognition is dependent on the robot taking a particular trajectory through space. Another problem can be the effect of "freak" or anomalous perceptions on the stability of the episodic map.

Figure 14-3 shows the architecture for episodic learning. The SOM categorization units are the same as you have used before. (Pick a model, any model.)

Raw sensory input vectors are fed into the first classifier. The vectorized result of this classification is impressed onto the perception trace. The winning node(s), however, are merged with the perception trace in such a way that the trace keeps only the last t_n most recent perceptions. Old perceptions are unmerged from the trace over time, to keep things tidy. When run in this manner, there is no precedence or relationship between the perceptions on the trace.

The perception trace is then fed into another SOM categorizer, which will identify, to its best ability, the robot's location in space as determined from the t_n perceptions it works with.

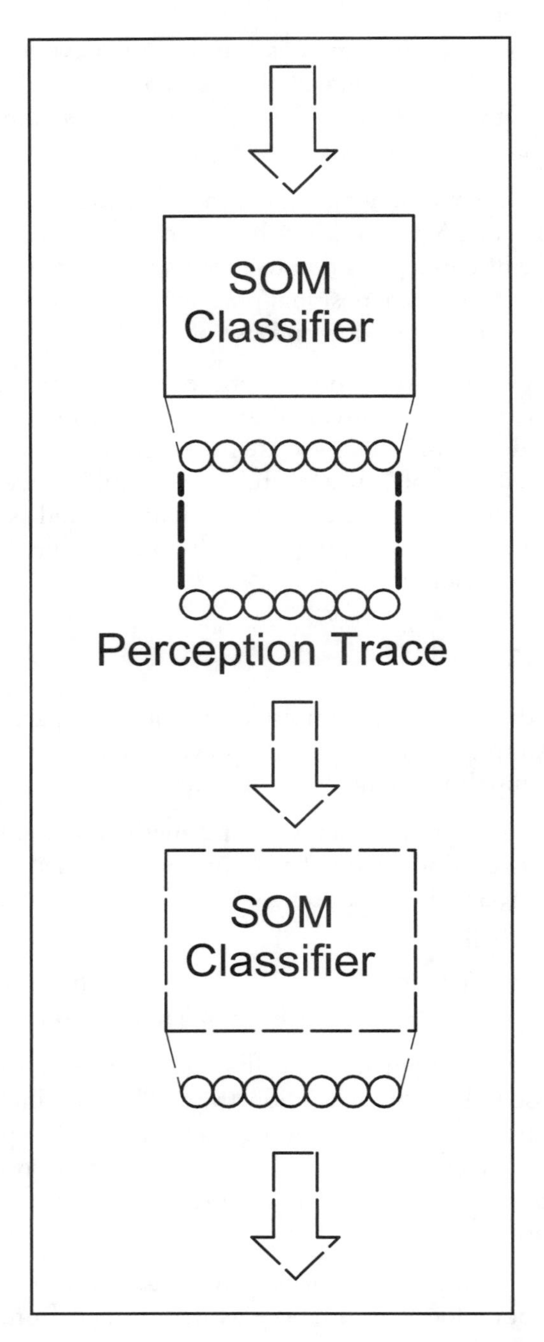

Figure 14-3 Episodic mapping architecture

SEMANTIC HIERARCHY

Benjamin Kuipers's spatial semantic hierarchy, discussed in his paper, "A Robot Exploration and Mapping Strategy Based on a Semantic Hierarchy of Spatial Representations," tries to impose a sense of greater order onto the topological map.

Many robotic mapping schemes attempt to create a geometric, or metric-based, map of the environment based on the robot's odometry-based path through the environment, coupled with sensory readings from such senses as sonar and optical distance sensors. However, both odometry and most sensor readings are fairly inaccurate, creating problems during long journeys.

The semantic hierarchy starts with the same approach as the episodic learning algorithm above, building a topological, or landmark-based, map of the environment. The robot stores the basic map in the form of perceptual landmarks and the paths used to travel between them.

There are four levels of information in the hierarchy, as discussed in Chapter 12's section on the Brain as Semantics.

The topological mapping systems are similar to the ABC saccade model explored in Chapter 13. The brain fixates on an "interesting" location or configuration of sensory inputs, categorizes it, and stores it away in memory. Then, it takes a path to a new interesting location (saccades for the eye, or a brief journey for the robot), categorizes it, and stores it away in memory, associated with the path, or motor impulses, used to travel between the two locations. Repeat until bored.

There are two types of movement strategy at play during the robot's travels. The first strategy is used to move the robot and its sensory apparatus closer to a recognized or otherwise interesting position in space; the *distinctive place* (DP). This is a *hill-climbing* strategy, working to optimize the interest value of the sensory input. Once the robot is in as interesting or distinctive a position as possible, it recalls the existing location, or marks the new location.

How you implement this hill-climbing behavior, and the metrics behind what is "interesting" and "distinctive," is a large problem to tackle on its own. However, there are some factors that contribute to making a position in space interesting. It could be due to symmetry of shape, such as in a corner, some type of discontinuity, such as when a wall ends or an opening is found, or some other symmetry or organization in the degrees of freedom or number of open places around the robot.

Once the robot has marked its distinctive and interesting location, it then gets to choose one of many *local control strategies* (LCS) to move it to the next possible location. The specific strategy to use is chosen based on its relevance, via internal algorithm or perhaps memory of what it did before, to the current situation. These LCS algorithms can include such low-level behaviors as wall following, center-of-hall traversal, random wandering, and so forth. The LCS is used to escape the current DP and, hopefully, will find a new DP. The LCS choice is recorded as the path between DP_t and DP_{t+1}.

The LCS is the mobile equivalent of a postural primitive, as explored in the Cog project at MIT—a single internal "button" that the control system can push in order to achieve a coherent, complex posture or behavior. There is evidence in nature for both the postural primitive and the internal pattern generator.

Unfortunately, I must leave you with just these few hints about mapping. It is a complex subject that requires more space than I can commit to in this book.

NOTICING

In order to escape the tyranny of the forced learning curve imposed by almost all SOM implementations, where the learning factor and learning neighborhood decrease with time during training, until they are fixed at small values during operation, it is necessary for the robot to be able to *notice* that something is interesting or significant. The learning factors can then be keyed to these interesting events.

NOVELTY

One simple measure of interest is in the *novelty* of an input (see "A Real-Time Novelty Detector for a Mobil Robot," by Stephen Marsland, Ulrich Nehmzow, and Jonathan Shapiro). Novelty is the measure of how well an input matches, or, more accurately, *doesn't* match, the prediction for that input.

Recall in Figure 13-21 the prediction loop in the LAPS module. A novelty detector would add a step to compare this prediction to what actually occurred and flag the input as interesting only to the degree that it was different from the prediction. The detection of novelty is not unlike the comparison used in the Hebbian output vector against the training vector (again, from Figure 13-21).

The flip side of novelty is habituation. When the sensory inputs do not change significantly over time, they lose interest and fade into the backdrop of attention—ignored.

SYNCHRONY AND RHYTHM

Another measure of interest can be found in the synchrony of events—when disparate sensory events occur at more or less the same time. Usually such synchronizations are significant, so it pays to attend to them.

One possible real-time system is described in "What is a Moment? Transient Synchrony as a Collective Mechanism for Spatiotemporal Integration," by J.J. Hopfield and Carlos D. Brody. In that model, the transient synchrony of the action potentials of a group of neurons is used to signal "recognition" of a space-time pattern across the inputs of those neurons.

Regardless of how you achieve it, it is important to look for correlations of events across time—not just on the order of a few milliseconds, the natural decay rate of a neural transmission, but across times up to a half-second long if not longer.

Now, to find a way to implement this in the context of the SOM models explored so far. Good luck!

RHYTHM

Related to synchrony is rhythm, the recognition and structuring of repeating sensory sequences (see "Learning to Perceive and Produce Rhythmic Patterns in an Artificial Neural Network," by J. Devin McAuley). Left to their own devices, human subjects tend to group repeating stimuli in patterns of twos, threes, and fours, indicating a need to impose rhythmic order where none may otherwise exist.

Of course, any subtle variation of tonal quality is used to lock onto a rhythm grouping.

Once a rhythm is recognized, there is also a strong predictive element to it—people sense a beat in the pattern even if it is not explicitly played, a feature used by drummers to create interest in their drum patterns.

Sounds, and possibly, by extension, other repeated stimuli, need to occur a bit farther apart than 50mS to be distinguished as separate beats and not part of the tone itself. Likewise, gaps between notes that are longer than 1.5 to 2.0 seconds are heard as disjoint events and not part of a connected sequence. These experimental results set the time boundaries on our experience of rhythm.

Spontaneous tapping in subjects occurs at a pace between 1.1Hz and 5.0Hz. When asked to *reproduce* a tapping rate, subjects did best at 1.7Hz, which is 600mS spacing, so this is a good, representative *natural* rhythm; a center point that we deviate from to reproduce and recognize other rhythms.

One approach to rhythm is by way of a *pacemaker model*, where there is spontaneous internal rhythm, such as the 1.7Hz natural rhythm, that interacts with external signals. An accurate pacemaker model should demonstrate the three attributes of biological pacemaker systems, as discovered in animal experiments. These attributes are entrainment, assimilation, and generalization.

In entrainment, external stimuli become evident in internal electrical activity; the brain "pulses" in time with the stimulus. Assimilation is demonstrated when the external stimulus stops or skips a beat and the neurons trigger at the entrained frequency anyway. Generalization is where the presentation of a somewhat different stimulus, at a different frequency, still triggers an earlier conditioned response—one learned pattern matches not only the stimuli it was trained with, but variations on it as well.

One such pacemaker model is explored here. This integrate-and-fire model consists of four equations: input activation level, spontaneous activation level, total activation, and the trigger threshold level.

All of the equations operate in discrete time steps of duration Δt. Input to the system is accumulated through the input activation level equation:

$$x(t) = x(t + \Delta t)e^{-\tau_x \Delta t} + i(t)$$

where:

$x(t)$	Input activation level at time t
τ_x	Input decay rate
$i(t)$	Input at time t

The spontaneous activation level $s(t)$ of the pacemaker, independent of any input, is given by the equation:

$$s(t) = s_l + (s_0 - s_l)e^{-\tau_s t}$$

where:

s_0	Minimum (origin) level of spontaneous activation
s_l	Maximum (limit) level of spontaneous activation
τ_s	Spontaneous activation decay rate

The spontaneous activation level is initially s_0 and, over time, approaches the asymptotic limit s_l.

The total activation $y(t)$ is simply the sum of the input and spontaneous activations:

$$y(t) = x(t) + s(t)$$

Finally, the trigger threshold *h(t)* is given by the following equation over time:

$$h(t) = h_l + (h_0 - h_l)e^{-\tau_h t}$$

with the variables and values following the same pattern as the spontaneous activation function.

In operation, time starts at t=0, and each equation is calculated each Δt timestep. When total activation *y(t)* exceeds the threshold *h(t)* the pacemaker triggers and all variables reset back to their initial values for the next cycle.

Though McAuley goes into great detail, only the gist of the operation is presented here. The key is in the way the system is tuned relative to an input pulse. When an input pulse is received by the system and it occurs before the system was able to spontaneously trigger, it may force the system to activate prematurely and, at the same time, speeds up the spontaneous activation rate of the pacemaker.

When an input pulse is received soon *after* a spontaneous pulse, the spontaneous activation rate is reduced. The spontaneous firing rate is adjusted by increasing or decreasing s_l.

Of course, if a pulse occurs far away in time from a spontaneous pulse, as defined by a window of opportunity, or *entrainment window*, it is not considered to be relevant to the pacemaker and is ignored. The effect of an input pulse on the pacemaker is stronger the closer that pulse is to the spontaneous pulse. The amount of change applied to s_l, the entrainment window width, and the magnitude of the node's activation are variable and/ or subject to training.

Once you have built a trainable pacemaker node, you can organize many of them into a pacemaker SOM (PSOM) arrangement. Sensory inputs, then, will activate and train the PSOM much like a traditional SOM network. The patterns of activation, possibly integrated across the maximum time frame of sensible rhythms, can then be fed into recognition networks in the same way as the categorized sensory vectors in the ABC model.

Alas, this discussion on dynamic neural systems is all too short—a search through the literature (and here, the Internet is your friend) can expand your horizons on this subject. Also, check out the book by Wolfgang Maass on pulsed neural networks, you won't regret it.

CHAPTER 15

VISION AND LANGUAGE

Though we hooked up a camera "eye" in Chapter 11 (using software on the www.simreal.com Web site), the mere presence of such a device on the robot does not give it *vision*. Unfortunately, reading this chapter won't give your robot vision, either—but it should give you some insight into the issues, problems, and directions of study if you want to use visual information to guide your machine.

Visual information is both a blessing and a curse. The blessing of vision is that the camera or eye collects an enormous amount of information about the world without having to go out and touch it. That, unfortunately, is also the curse—the flood of data coming in from the visual sensors can easily overwhelm even the most advanced computer system. The human brain uses something like 30% of its resources in processing vision (though in blind people, that part of the brain is used somewhat differently), so clearly it's a daunting task.

Similarly daunting is the subject of language on computers. This is a huge field, involving speech synthesis, speech recognition, parsing and part-of-speech tagging, and dialogue management. This chapter provides a few measly references and directions for research.

VISION

Figure 15-1, adapted from *Memory: From Mind to Molecules*, by Larry R. Squire and Eric R. Kendal, is the cryptic brain area and connection diagram that is required in any discussion on vision. I apologize for not defining TEO and TF in the image, but my sources neglected to name them.

Visual information takes two basic pathways through the brain. Entering at brain area V1, it splits up into a ventral (bottom) and dorsal (top) route. There are different theories as to how the two streams of visual information are then processed, some of which are discussed in the book by Squire and Kendal mentioned above and "The Visual Pathways Mediating Perception and Prehension," by Melvyn A. Goodale, Lorna S. Jakobson, and Philip Servos.

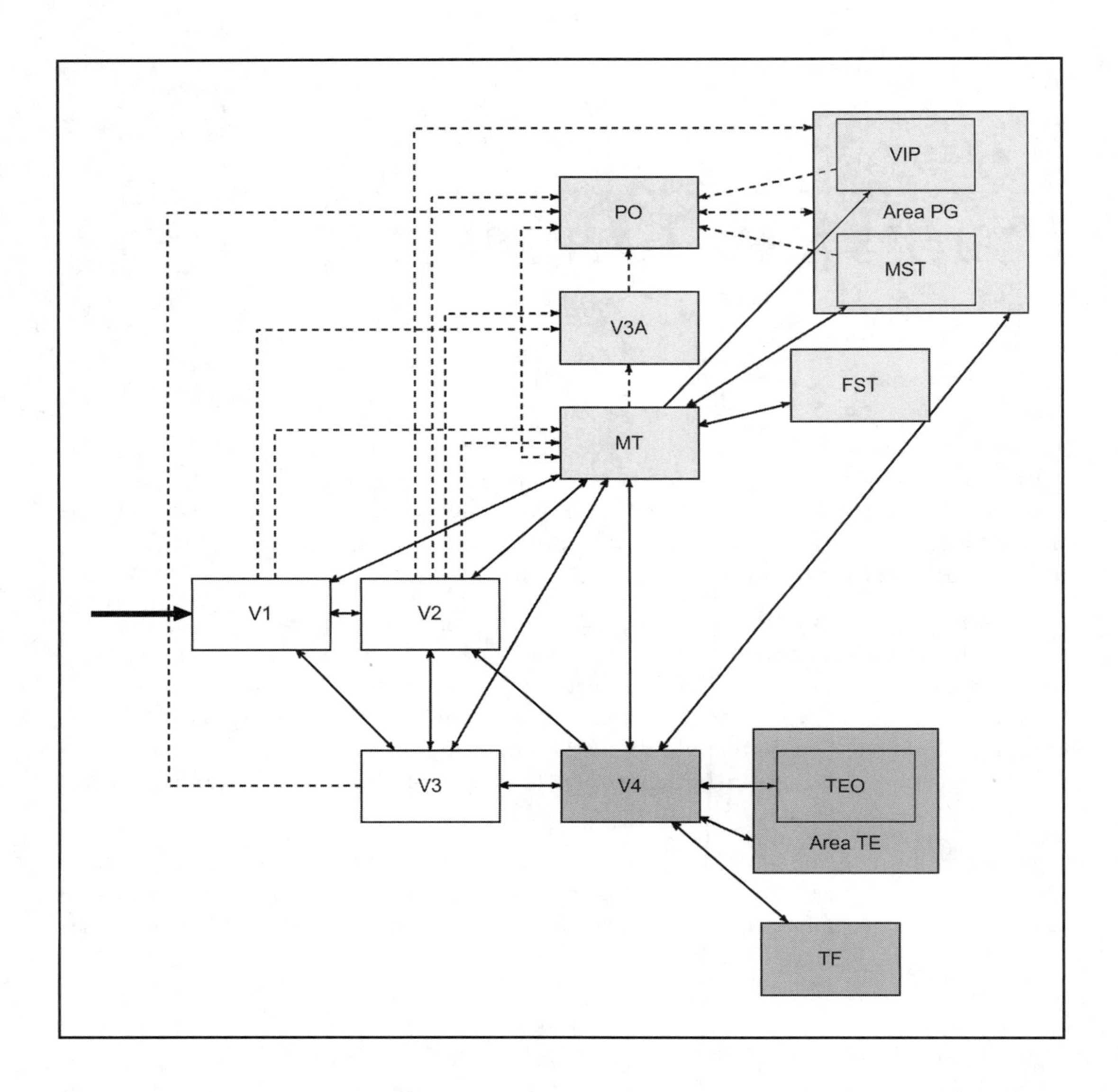

Figure 15-1 Some visual areas and connections. V1-V4, visual areas; FST, fundus of superior temporal area; MST, medial superior temporal area; MT, middle temporal area; area PG, parietal cortex; PO, parieto-occipital area; area TE, inferior temporal cortex; VIP, ventral intraparietal area.

The ventral stream leads to the temporary areas of the brain and is processed with respect to the question, "What is this?" The ventral stream is analyzed for the form and quality of objects; the recognition of and identification of objects or, perhaps, the recognition of various enduring *characteristics* of objects, as well as their spatial relationships to each other.

The dorsal stream leads to the parietal areas of the brain and is processed with respect to the question, "Where is this?" or, alternately, "How am I located with respect to this?" The dorsal stream is analyzed for the spatial locations and relationships of objects. This visual stream is used to mediate our interaction with the environment, as feedback to our physical, goal-directed actions.

Depth Perception

One route to depth perception is via the use of two cameras, giving a stereo view of the world. Correlations between the data from two cameras can provide depth information.

Of course, that's not the only way to do it. For very small creatures, the spacing between their eyes is too small to give much stereo separation. Or perhaps your eyes don't have much overlap in their visual fields. Ultimately, one camera is cheaper than two.

Anyway, as your eyes move, objects at a distance cross the visual field at a different rate than objects that are close to your head. This visual-field information flow, or parallax, can provide the depth information you need to navigate. For more information about how locusts (and robots) use parallax, see "Look Before you Leap: Peering Behavior for Depth Perception," by M. Anthony Lewis and Mark E. Nelson.

The first step is to determine, on a pixel-by-pixel or zone-by-zone level, how information is changing in the visual field. There are two ways to go with this. On the one hand, a *directional* vector can determine which way something is moving in the visual field. On the other (and simpler) hand, a simple *nondirectional* scalar value can indicate just that something is changing and to what extent, though not in what direction.

Given the mostly coherent nature of our environment, a simple "temporal differentiation of brightness," that is, the change in light intensity from one frame to the next, can provide an estimation of motion, given a sufficiently textured environment. This is easily done by taking the difference in value between the black and white pixels of one "frame" of input and the next.

For a directional field flow analysis, we can turn to Reichardt correlators, a fairly simple model for determining the direction of visual flow (there's a good discussion in "Accuracy of Velocity Estimation by Reichardt Correlators," by Ron O. Dror, David C. O'Carroll, and Simon B. Laughlin). Though it is imperfect, sometimes simplicity is preferable to perfection. This visual flow mechanism was discovered through many years of research on invertebrates, and has also been applied to the problem of motion detection in humans, birds, and cats.

A simple form of this correlator is shown in **Figure 15-2**. The visual information impinges on the receptors *A* and *B*. These receptors (e.g. pixels) are separated by some angular

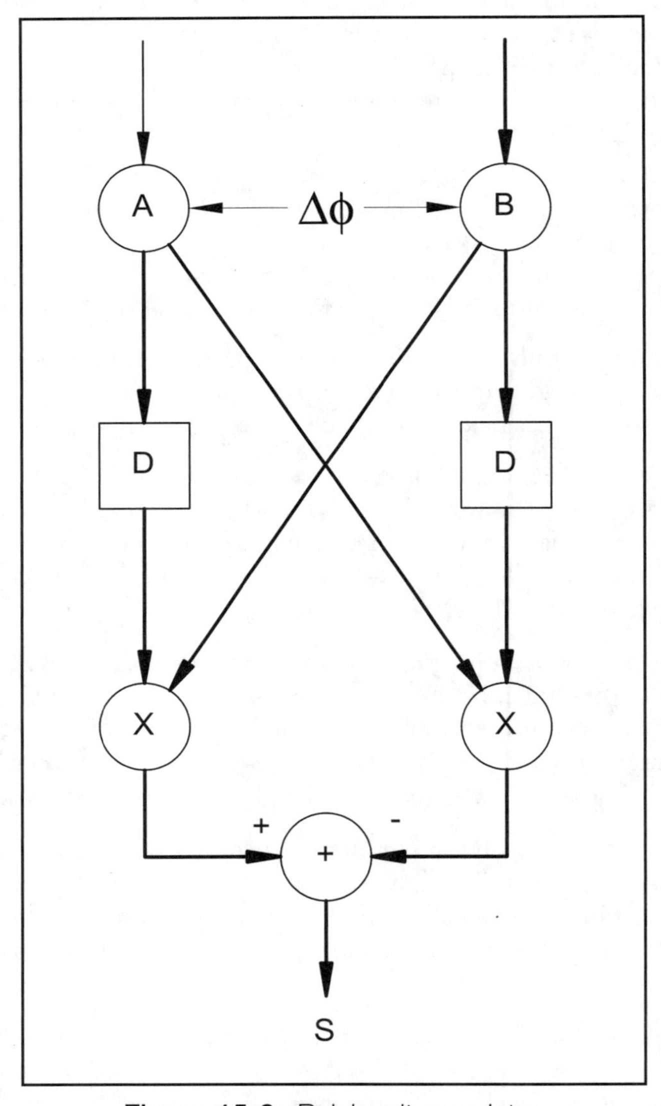

Figure 15-2 Reichardt correlator

distance $\Delta\phi$ either horizontally, vertically, or at some angle, depending on what axis of motion you are trying to detect. The signals from the receptors are each split, with one path leading to a temporal delay D and the other crossing over to a multiplier.

The delayed signal from one receptor is multiplied against the direct signal from the other receptor. The results of the two multiplications are then combined to create a signed value indicating the magnitude (speed) and direction (along the *A-B* axis) of the visual flow.

Needless to say, there are many ways to tweak this model—adjusting the time delay D, fiddling with the spatial or temporal filtering of the receptors, not to mention changing their response curve, and so forth. For our current purposes, the nondirectional model should be sufficient.

Given some measure of visual speed s, it is possible to reconstruct a depth estimate z:

$$|z| = \frac{1}{|s|}$$

Using the spacing value $\Delta\phi$ and, if it is known, the rate of the head's motion, this estimate can be used to calculate real units of distance. This all assumes that the viewer, not the object, is moving, and that there is not a significant rotational component to the motion.

Visual Coordination With the Body

Using eyes to guide the body's motion is one of the two basic uses for vision, the other being visual recognition of objects, faces, and so forth. We are going to look at visual coordination of motion in three steps: visual navigation, correlating visual field changes with head motion, and finally using visual information to guide reaching and pointing behavior.

Visual Navigation

For a mobile robot, anything that keeps it from banging against the walls, running over the dog, or falling down the stairs is a welcome addition. Since most of us use our eyes to navigate around the world (and I must exclude those people reading this in braille from this list, alas), a camera mounted on the robot to help it get around the world seems a natural addition. One approach to visual navigation is explored in "'Programming' by Teaching: Neural Network Control in the Manchester Mobile Robot," by Paul Martin and Ulrich Nehmzow.

The basic process is fairly simple. A camera image is processed (heavily) to create sixteen receptive fields, each of which is used as input to an associative memory, such as the LAPS system described in Figure 13-14. The output of this memory consists of two output nodes—forward speed and rate of rotation. A second input, the training vector, forces the rotation and forward motion, training the memory on how to respond with regard to the sensory input. The user teaches the robot by driving it around while the robot's memory learns how it should react to its sensory inputs.

The video processing described by Marin and Nehmzow is clever, and can be performed fairly simply (see **Figure 15-3**).

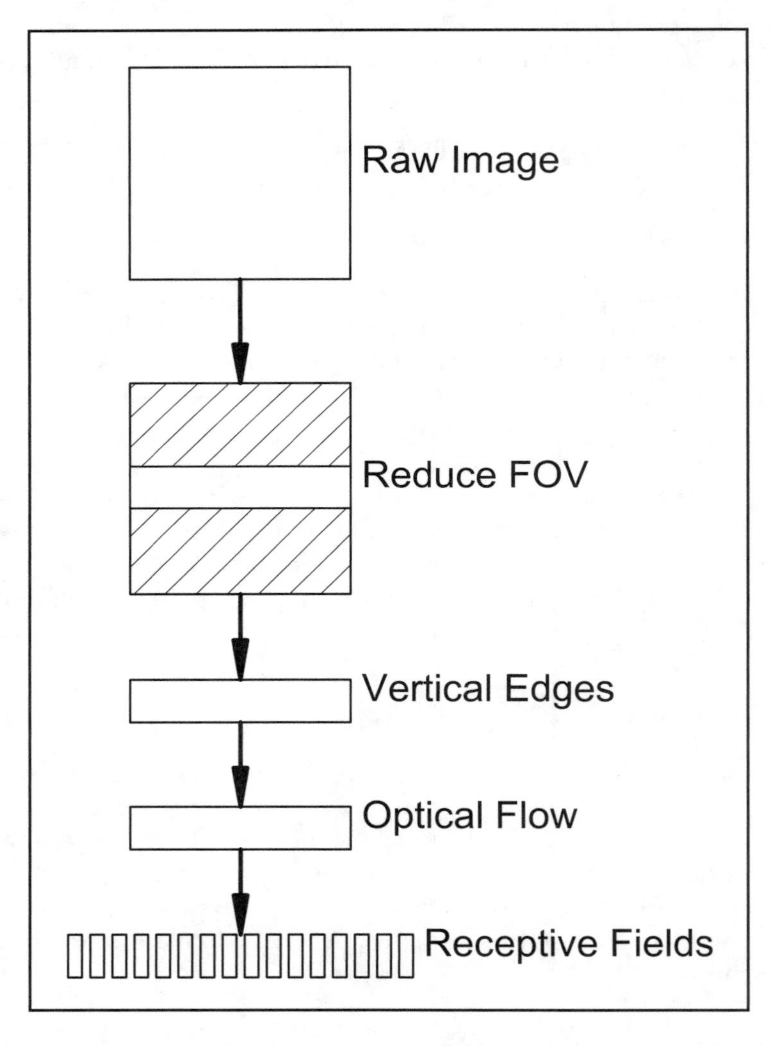

Figure 15-3 Vision processing, camera image to receptive field signals

The camera's output, for this example, a 320x200 array of 8-bit gray-value pixels, has its field of view (FOV) reduced to a short horizontal strip at about the horizon.

This 320x50 strip is then processed to enhance the vertical edges, since these edges represent the major obstacles to be avoided—walls, doors, table legs, and so forth. To do this, the image is convolved against a 3x3 matrix. In short, a convolution is where one pixel's value is based on the values of its neighbors. In this case, the convolution, or neighborhood value, table is:

$$\begin{bmatrix} 1 & 0 & -1 \\ 1 & (0) & -1 \\ 1 & 0 & -1 \end{bmatrix}$$

Next, the optical flow is determined. Of the two methods discussed in the previous section for performing this operation, a simple differentiation method, with threshold, is sufficient. The pixel value from the current frame is compared to the pixel value from the previous frame. If the difference is greater than the threshold value, eight in our example, the flow value is set to 1; otherwise, it is set to 0. The threshold helps provide some immunity to background noise, and the reduction in scale, 0 or 1, versus 0 through 255, prepares us for the next step.

The visual field is now broken into sixteen separate receptive fields, each covering 20x50 pixels. The value of each field is calculated by adding up the value of all pixels within that field's area of influence. This array of sixteen values is then used as the sensory input vector.

This test system was used in an environment filled with bold vertical stripes—so a more refined version may be needed in the real world. One of the primary benefits of this visual input system is its simplicity and speed of computation. As such, it is a good starting place for more advanced experiments in vision. For example, a similar navigation technique is described in the paper "Visual Navigation in a Robot using Zig-Zag Behavior," by M. Anthony Lewis.

Correlating Vision against Head Motion

Where the previous experiment used a fixed camera for navigation, the camera platform on our robot provides a pan-tilt environment. If the camera is allowed to move around, it becomes important to correlate the camera's motion with the perceived motion in its visual field. With this correlation, it becomes possible to compensate for the head (well, eye) motion and get world motion information. For details, see "Learning Maps Between Sensorimotor Systems on a Humanoid Robot," by Matthew J. Marjanovic.

There are three components to the problem: head motion generation, visual motion detection, and correlation between head motion and visual motion.

The basic architecture of this system is shown in **Figure 15-4**. Most of the boxes in this system use technology that we have already explored—except for the central compensation network. For more information on the technologies used in this system, check out the many papers written about the Cog Project at the Humanoid Robotics Group at MIT. On the one hand MIT likes to do everything the hard way, but on the other hand they have *really interesting* projects.

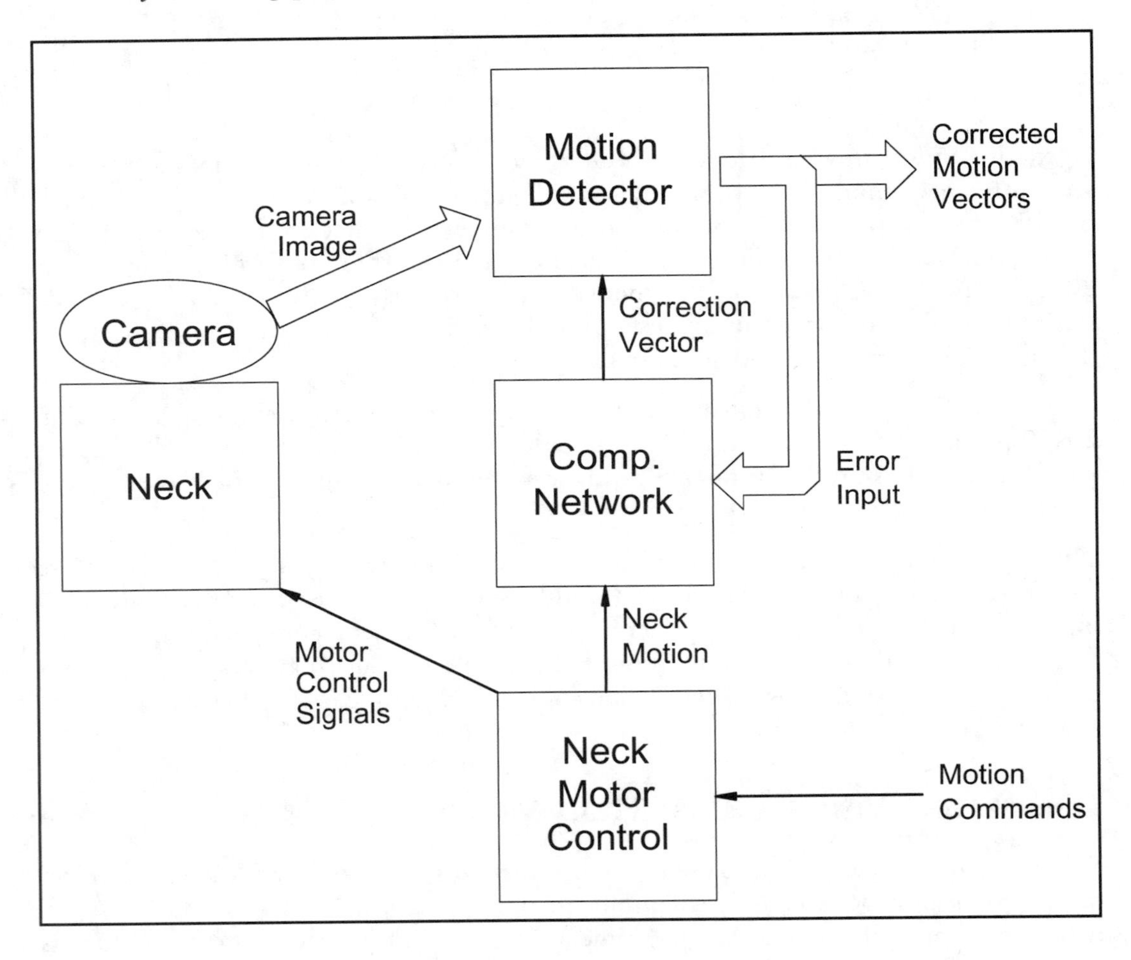

Figure 15-4 Vision motion correlation system

The motion of the head is created by the neck's motor-control reflex system, which responds to motion commands. These motion commands will typically come from a higher-level brain system, guiding the head's motion with purpose. For the purpose of experiment and training the network, however, these motion commands can come from a simple random head-motion generator.

The neck motor controller sends motor signals to the pan-and-tilt servos in the neck. It also sends the neck's pan and tilt position and/or motion information to the compensation network, as inputs.

The neck servos, of course, change position in response to the control signals. This moves the head and, in response, the visual images captured by the head's camera change. The motion in the camera's image consists of a composite of object motion within the scene and scene motion induced by the neck's action.

The camera image is sent to (among other places) a motion detection system. The MIT motion detector is significantly fancier than the motion detectors described so far. It takes many 16x16 pixel patches from the visual field and moves them around the data from the previous frame trying to find a correlation. Using the best correlation between frames, it creates a motion vector centered on the patch. Repeat until happy. The end result is one or more (or many more) motion vectors within the visual field—a motion field. A similar result would be to have several Reichardt Correlators, each tuned to a different direction, with their results combined into a single motion vector detector.

The motion vectors are then corrected, using a single correction vector from the compensation network. In effect, the correction vector is subtracted from the motion field. This corrected motion field is passed on to the brain for further processing, as needed.

An interesting trait of the visual field is that, in general, if the head and body are not moving, neither is the visual field. Sure, there may be motion here and there as people and animals move around, but in general, and over time, there won't be much motion. Because of this, the sum total of motion across the visual field can be used as an error signal for the compensation net. Given the neck motion and an error signal, the compensation network attempts to create an output, the correction vector, which reduces the error signal to zero. Since most visual motion is a direct result of neck motion, this network can eventually learn how to compensate for the results of neck motion in the visual field.

The core of this system, though, is a network that can learn to create the correct outputs with nothing more than the two neck motion inputs plus an error signal. One such network might be found in the technology of reinforcement learning (RL).

In the simplistic view, a reinforcement learning system takes any rewards or punishments and distributes them *back in time* to any nodes, based on how much they participated in the decisions that led to the reward or feedback. Each node/decision pair has an activation trace that decays with time. This is used to parcel out feedback. These nodes, in turn, try to make decisions based on the ultimate likely reward that will result from them. The nodes themselves are arrayed in a state grid, where each axis of the grid is defined by the relative value of one dimension of the input. I can't do justice to the field of RL here, especially since it is a detour from the main technology line explored in this book. Fortunately you can find the classic treatise of the subject (*Reinforcement Learning: An Introduction*, by Richard S. Sutton and Andrew G. Barto, MIT Press, 1998) online; the last time I checked, you could read it at www-anw.cs.umass.edu/~rich/book/the-book.html. I've spent the last several chapters avoiding trying to condense this book into a few concise paragraphs plus diagrams—so you'll make me feel a lot better (and assuage my guilt at skipping over RL here and there) if you'll promise to go there now and read the Sutton and Barto book. Okay, great. Good, wasn't it?

VISUALLY GUIDED REACHING AND POINTING

The third step in our visually guided behavior trilogy is the most complex—hand-eye coordination. Some humans practice for *years* and never master this with any grace, so I hope you can forgive me for being brief here and simply pointing in a direction for further research for your robot studies.

In theory, coordinating the action of an arm using feedback from the eye is similar to the problem addressed in the previous section on correlating visual motion with head motion. In practice, of course, it is rather more complex.

MIT's Cog project, of course, has something to say on the subject of visually guided reaching (see "Self-Taught Visually Guided Pointing for a Humanoid Robot," by Matthew Marjanovic, Brian Scassellati, and Matthew Williamson for more detailed information). A saccade map is created first, as a variation of the head-motion map described above. From there, the model branches out to create a hand-eye motion map that is trained to put the tip of the arm into the center of the visual field using similar techniques.

The saccade map tries to learn the relationship between head motion and visual position—that is, what head motion will move a given off-center spot in the visual field to dead center. To train the saccade map, some training algorithm will pick a spot in the camera's visual field at random, so long as "random" means points distributed evenly over the entire visual field. It takes a 16x16 "snapshot" of the pixels around this point and then the neck motor control system is allowed to center the eye on this point, using a preinitialized "best-guess" saccade map. Once the motion is complete, the saccade-map trainer does a reality check using the captured snapshot to see how close the saccade came. This error feedback is used to adjust the saccade map until it works smoothly.

With a decent saccade map in place, we can exercise the reaching motion. An important aspect of watching one's arm do work is to be able to identify it as *your* arm and under *your control.* To do this, we start with a variant of the motion-detection schemes described throughout this chapter. In this case, the head is held still and the arm is moved. The various motion vectors from the visual system are grown blob-like until there are one or more semicoherent areas of motion. The blob with the fastest motion is considered to be the tip of the arm, which is the part of the arm we are trying to position.

The arm motion itself is controlled with a system of *postural primitives.* Each primitive, of which there are only a few, four in the example case, is, essentially, a knob that can be turned up or down. At the "full-on" position, each knob encodes the joint positions needed to attain a unique, and extreme, position of the arm. One knob encodes for a relaxed "rest" position, and the other three knobs encode three extremes in the arm's workspace. As the relative activation level of these four primitives is adjusted, the arm interpolates throughout its entire range of motion. The use of postural primitives allows for very simple control values to drive a fairly complex piece of equipment.

Now, like learning the saccade map, the robot learns the arm map. Given a head position, the robot attempts to put the tip of its arm into the center of the visual field. If the point of interest isn't in the center of this visual field, it saccades first so it is, and then attempts to reach it.

A related system of reaching is described in Patent 4,884,216, "Neural Network System for Adaptive Sensory-Motor Coordination of Multijoint Robots for Single Postures," by Michael Kuperstein, but I'm not going to go into that here, either. If you have survived this far, I fully expect you to trudge online, to www.uspto.gov, and print out the patent for detailed review on your own time.

And on this surly note, we bid adieu to vision processing.

LANGUAGE

Language is bound up intimately with the subject of knowledge representation. Language, after all, is a way of sharing knowledge between two entities. There are different schools of thought with regards to language and knowledge. On the one hand is the traditional symbolic school of thought, where language and knowledge are *merely* abstract tokens that are transformed by specific rules...the "Chinese Room" analogy. On the other hand, there is the approach that insists that language be grounded in physical, sensor-based reality before it can make any "real" sense to the robot.

An excellent book on symbolic language processing is *Natural Language Understanding* by James Allen. On the lexical-grounding front, the ABC visual model also applies itself nicely to the subject of language. A quick browse through a good technical bookstore (online or offline) will yield a number of resources in the venerable and heavily researched field of natural language processing.

Avoiding, for the moment, the question of knowledge representation and language *meaning*, we can look to Microsoft for a suite of free (yes, free, from Microsoft) tools to aid in the mechanical generation and identification of spoken words. Its Speech SDK (as of this writing, version 5.1, found at www.microsoft.com/speech/) includes both speech-recognition and text-to-speech (TTS) capabilities. Giving spoken commands to your laptop-powered robot, and having it talk back to you with witty yet humble quips, is now a simple task, assuming you can get around the bane of mobile robot's microphone noise. Noise from both the robot's motors and the environment around it can handily kill any attempts at verbal direction. To counteract this, you can hook up a wireless microphone to the computer in order to wow the neighbors—just be careful about interference with the other wireless systems on the robot!

Once you get past the hurdle of spoken language, things are back on steadier ground. Of course, the optimum way to implement language processing in your robot is through neural techniques, as expounded through the various sections of this book. But if you want impressive results with less computing horsepower, you can look towards the world of chatter-bots. Chatter-bots can add both personality and a certain level of expertise to your machine.

Based on that ancient wheeze "Eliza," modern chatter-bots have moved beyond their role as simple toys and curiosities, and are applied to everything from games to online technical support (www.verbot.com). Simon Laven is an ardent supporter of the technology, and he provides a good list of systems and technologies on his Web page at www.simonlaven.com. He also provides a good definition of what a chatterbot is: "A Chatterbot is a program that attempts to simulate typed conversation, with the aim of at least temporarily fooling a human into thinking they were talking to another person."

Type "chatterbot" into a Web search engine and you will have enough reading material to pass many a quiet evening. If you want to examine the innards of one of these, I recommend you visit the Alicebot open-source project (alicebot.org). Combined with the Microsoft Speech SDK, you might be able to create a clever-seeming robot without breaking a sweat!

APPENDIX A

BATTERIES

PRIMARY BATTERIES

This section adds to the discussion in Chapter 2 on carbon/zinc and alkaline batteries.

CARBON/ZINC

George Leclanché invented the carbon/zinc battery in 1866. By 1868 it was adopted by the Belgium telegraph service and ultimately went on to be the standard for portable batteries around the world. The original Leclanché cell was a wet cell, with the electrodes immersed in a liquid electrolyte. Later developments moved the electrolyte to a wet paste, giving us the carbon/zinc "dry" cell. A "heavy-duty" version uses a zinc-carbon-zinc chloride chemistry, for a higher capacity.

This battery has a limited shelf life and is susceptible to leaking its corrosive electrolyte.

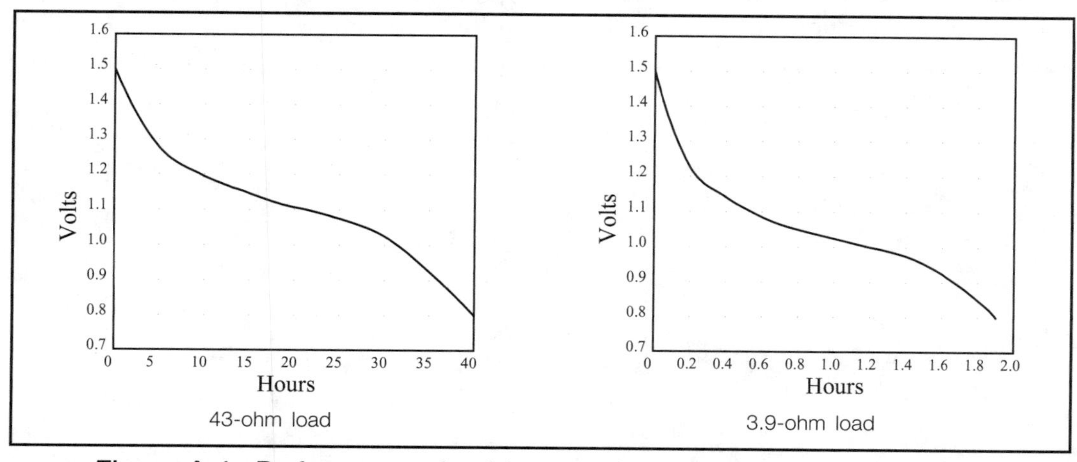

Figure A-1 Performance of a AA carbon/zinc battery at different loads

Cell Voltage (nominal)	1.5
Cell Voltage (discharged)	0.8
AA-cell Capacity (mAh)	950
C-cell Capacity	3,000
D-cell Capacity	5,900
Internal Resistance (Ri in Ohms)	0.4 – 0.8
Energy by Weight (Wh/Kg)	9
Recommended Discharge Load	n/a
Shelf Life (to 80% capacity)	2-3 years

Table A-1 Carbon/zinc battery characteristics

ALKALINE

Alkaline batteries, as a class, were developed between 1895 and 1905 and were finally commercialized in the mid 1950s. This coincided with the rising popularity of electronic flash units in small portable cameras, which required the high power output the alkaline chemistry provided.

Though the alkaline cell has a similar theoretical energy density to the Leclanché cell, it achieves much higher values in practice. Alkaline cells are more expensive, but they are also more leak resistant, have a longer shelf life, and have better low-temperature performance than equivalent Leclanché cells.

While the alkaline battery has traditionally been a primary battery, new developments have given us a rechargeable alkaline. One drawback of the rechargeable alkaline is its capacity fade. After each discharge, the battery will lose some of its capacity. After about 25 cycles, it is at 50% capacity; 50 cycles sees it at 20% capacity, where it appears to stay until the 100-cycle point at the end of its rated life.

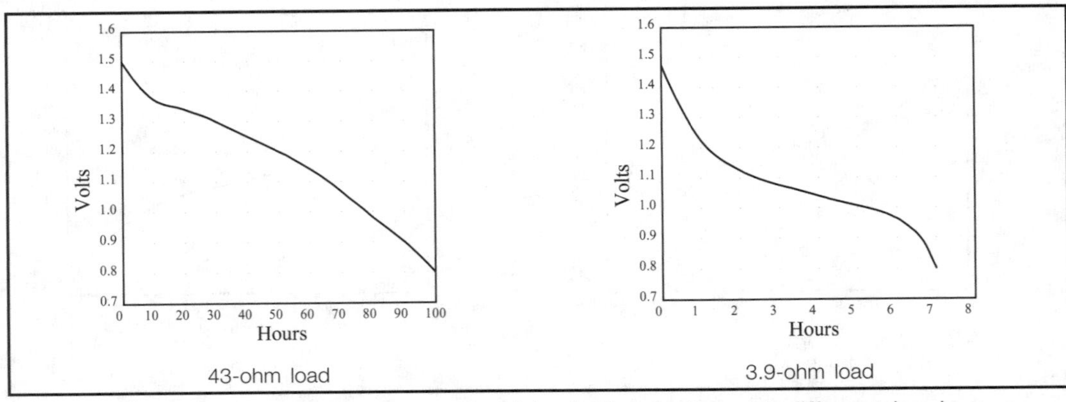

Figure A-2 Performance of a AA alkaline battery at different loads

Cell Voltage (nominal)	1.5
Cell Voltage (discharged)	0.8
AA-cell Capacity (mAh)	2,800
C-cell Capacity	8,300
D-cell Capacity	18,000
Internal Resistance (Ri in Ohms)	0.2 – 0.6
Energy by Weight (Wh/Kg)	75
Recommended Discharge Load	620W..39W
Shelf Life (to 80% capacity)	5 years

Table A-2 Alkaline battery characteristics

SECONDARY BATTERIES

This section adds to the discussion in Chapter 2 on lead/acid and nickel/cadmium batteries, and adds information about nickel metal hydride and lithium ion batteries, and introduces the concepts of fuel cells.

LEAD/ACID

Rechargeable (secondary) batteries became practical in 1860 with the invention of the lead/acid battery by Raymond Gaston Planté. In 1881, Fauré (and others) improved the yield of the lead/acid cell by substituting a lead oxide paste for the pure lead of the Planté cell.

The largest problem associated with this battery is the damage caused by leaking acid. German researchers addressed this problem in the early 1960s by developing a gelled electrolyte. Working from another direction, other researchers developed a way to completely seal the battery, preventing leaks. Either way, the sealed lead/acid (SLA) battery needs little or no maintenance, which, while costing more, can be an advantage in some situations.

A completely sealed battery, whether it is a gel-cell or not, also prevents hydrogen gas from escaping when you recharge the battery, which is an improvement in safety when the battery is to be used indoors, such as on a robot or wheelchair. A gelled battery won't leak even if it is punctured, but it can also have a slightly lower energy density than its liquid counterpart, at about 80% or so.

Deep-cycle batteries are a special variety of lead/acid battery that can be discharged to low voltage levels without coming to harm. Deep-cycle batteries are typically used in marine or wheelchair applications. Regular car batteries are designed for short bursts of high-amp use to start the vehicle, with no deep discharges allowed. The electrode plates in a deep-cycle battery are made thicker and less porous than the car battery, and will last two to four times longer than the car battery in deep-cycle applications. "Dual marine" batteries are a compromise of the two types.

Deep-cycle batteries can also be rated in terms of "Reserve Capacity" (RC) rather than amp hours (Ah). You can easily convert between the two:

$$Ah = RC * 0.6$$

$$RC = \frac{Ah}{0.6}$$

Cell Voltage (nominal)	2.0
Cell Voltage (discharged)	0.8
Internal Resistance (Ri in Ohms)	.02
Energy by Weight (Wh/Kg)	30
Cycle Life (recharge cycles)	200-500
Shelf Life (to 80% capacity)	6 months

Table A-3 SLA battery characteristics

NICKEL CADMIUM (NICD)

The technology behind the nickel-cadmium battery was invented in 1899 by Waldmar Jungner, but the battery didn't reach commercial use until the 1930s when new electrodes were developed. The original version of the NiCd battery used a vented, unsealed cell that required regular maintenance. In the 1940s they perfected the sealed NiCd cell, though the cells do retain a need to breathe a bit, which is maintenance free, and the battery came to the fore in the 1950s. In 2000, it accounted for more than 50% of the world's rechargeable batteries for portable applications. Today's NiCd batteries can take a lot of abuse, both mechanical and electrical, and are cheaper than other batteries in cost per hour of use.

The capacity of a NiCd isn't seriously affected by the discharge rate. If you extract current from the cell at a lower-than-specified rate, you get a little more life. Extracting current from the cell at a rate ten times the specified rate only lowers the capacity to about 70% of its rated level, so a 1000Ah battery would only give 700Ah.

This battery has a surprisingly high capacity for current delivery. The AA battery shown has a recommended maximum continuous current draw of 9 amps, with 18 amp pulses allowed.

There are two issues you face when you use an NiCd battery. One is the dreaded "memory effect" (which doesn't seem to plague other batteries), and the other is "cell reversal."

Though hotly disputed in hobbyist circles, the memory effect is very real in some, but not all, NiCd batteries. This effect appears because the battery retains the characteristics of previous discharges—that is, after repeated shallow discharges, the battery may be unable to discharge beyond the earlier points. It would seem that, under certain conditions, electrodes in the cell can develop a crystalline growth. This growth reduces the area of the electrode exposed to the electrolyte. This leads to a voltage reduction and a loss of performance.

Avoiding the memory effect is fairly simple. First, quick charge rather than trickle charge your NiCd batteries. Quick charging helps negate the effect of NiCd memory. Second, be sure to fully discharge your batteries to their 1-volt level, under a light load, on a regular basis.

Cell reversal is a condition that can occur with multiple NiCd cells connected in series, such as in a multiple-cell battery or a battery pack. Since not all cells are exactly the same, one cell in a chain may use up all of its charge before the others. As the pack continues to be used, a reverse charge is sent through the empty cell due to its charged neighbors. This reacts the water with the cathode, bonding the oxygen to the electrode and releasing hydrogen, which is then vented. This loss of water reduces the life of the cell.

To prevent cell reversal, don't perform a deep discharge on a battery pack. It is safe to cycle an individual cell to zero volts. In fact, timing the discharge cycle of a cell is one way of determining its exact capacity. With this information, a cell can be "matched" with other equivalent cells into a battery pack that is less prone to reversal.

Cell Voltage (nominal)	1.2
Cell Voltage (discharged)	0.9
Internal Resistance (Ri in Ohms)	0.04 – 0.10
Energy by Weight (Wh/Kg)	50
Cycle Life (recharge cycles)	500-5000
Shelf Life (to 80% capacity)	6-8 weeks

Table A-4 NiCd battery characteristics

NICKEL METAL HYDRIDE (NIMH)

The NiMH battery chemistry began its life in the 1970s, but it took more than ten years before its performance was good enough for commercial use. Modern batteries seem to be either incremental improvements on old technologies or inventions of large corporate research departments, so it's harder to name the inventors of this chemistry. Since the 1980s, the performance of the NiMH battery has been improved continuously by many companies, and it is now an excellent battery for portable applications.

The NiMH battery comes in the same sizes, with the same nominal voltage and same discharge curves as the NiCd battery. NiMh has a higher energy density and a lower internal resistance; the AA NiMH battery has a maximum continuous current draw rating of 10 amps, with 15-amp pulses. While the NiMH battery won't hold its charge as long as the NiCd, it can carry more power. The average AA NiMH battery has 1200Ah of capacity, compared to 800Ah for NiCd.

NiMh batteries are less prone than NiCds to the memory effect, but can still suffer from cell reversal problems when deeply discharged in packs.

Cell Voltage (nominal)	1.2
Cell Voltage (discharged)	0.9
Internal Resistance (Ri in Ohms)	0.03 – 0.04
Energy by Weight (Wh/Kg)	70
Cycle Life (recharge cycles)	500
Shelf Life (to 80% capacity)	2-4 weeks

Table A-5 NiMh battery characteristics

LITHIUM ION (LI-ION)

Continuing in the tradition of modern battery chemistries, the lithium ion battery has an increased energy density and can provide a respectable amount of current. High discharge rates don't significantly reduce its capacity, nor does it lose very much capacity after each cycle, still retaining 80% of its energy capacity after 500 recharge cycles.

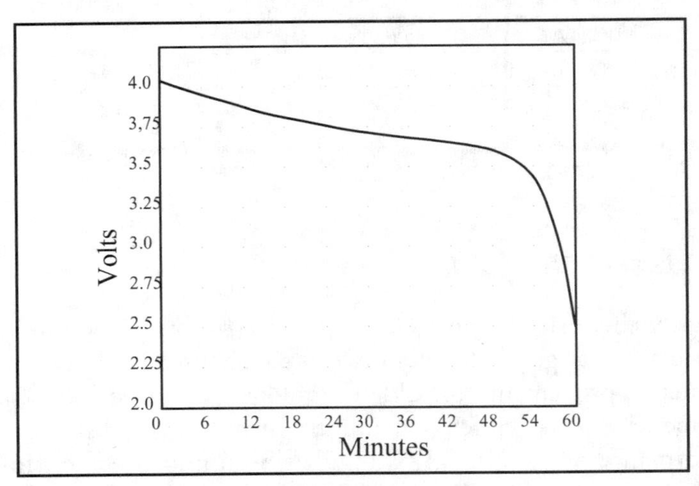

Figure A-3 AA Battery, 1600mA Load

This is a volatile technology—early versions were prone to exploding in the labs. It is the volatile nature of lithium that gives this battery its punch, though.

These benefits come with a price, of course (perhaps to pay for equipment damaged in the research?). As the technology matures, you should expect the price to drop.

Cell Voltage (nominal)	3.6
Cell Voltage (discharged)	3.0
Internal Resistance (Ri in Ohms)	0.06
Energy by Weight (Wh/Kg)	120
Cycle Life (recharge cycles)	500
Shelf Life (to 80% capacity)	2+ months

Table A-6 Lithium-ion battery characteristics

FUEL CELLS

Fuel cells aren't available for small, portable applications yet, but they are coming. The fuel cell isn't so much a battery as it is a catalytic chemical engine that creates electricity from fuel. The fuel is typically a variation of hydrogen, such as the hydrocarbon fuels methanol, natural gas, or even gasoline.

When these reach market you won't be recharging your batteries anymore, you will be refilling them.

APPENDIX B

BIBLIOGRAPHY

Allen, James. *Natural Language Understanding, 2nd edition.* Addison-Wesley, 1995.

Axelson, Janet Louise, and Jan Axelson. *Serial Port Complete.* Lakeview Research, 1998. www.lvr.com

Baddleley, Alan. "Working Memory: The Interface between Memory and Cognition." In *Cognitive Neuroscience: A Reader*, pages 292-304. Michael S. Gazzaniga, editor. Blackwell Publishers, 2000.

Briscoi, Garry. *Adaptive Behavioral Cognition.* Ph.D. Thesis, Curtin University of Technology, Australia, 1997.

Caelli, Terry, and Walter F. Bischof, editors. *Machine Learning and Image Interpretation.* Plenum Press, 1997.

Dowling, John E. *Neurons and Networks: An Introduction to Neuroscience.* The Belknap Press of Harvard University Press, 1992.

Dror, Ron O., David C. O'Carroll, and Simon B. Laughlin. *Accuracy of Velocity Estimation by Reichardt Correlators.* Journal of the Optical Society of America 18, 2001.

Fagg, Andrew H., et al. *Rapid Reinforcement Learning for Reactive Control Policy Design in Autonomous Robots.* Center of Neural Engineering, University of Southern California, 1994.

Fritzke, Bernd. *Growing Cell Structures - A Self-Organizing Network for Unsupervised and Supervised Learning.* TR-93-026, International Computer Science Institute, UC-Berkeley, 1993.

Gazzaniga, Michael S., editor. *Cognitive Neuroscience: A Reader.* Blackwell Publishers Ltd, 2000.

Goodale, Melvyn A., Lorna S. Jakobson, and Philip Servos. "The Visual Pathways Mediating Perception and Prehension." *Cognitive Neuroscience: A Reader*, pages 106-123. Michael S. Gazzaniga, editor. Blackwell Publishers Ltd, 2000.

Göppert, Josef, and Wolfgang Rosenstiel. *Dynamic Extensions of Self-Organizing Maps.* Proc. of ICANN'94, pages 330-333. Sorrento, Italy. Springer, 1994.

Hinton, Geoffrey E. *Training Products of Experts by Minimizing Contrastive Divergence.* TR 2000-004, Gatsby Computational Neuroscience Unit, 2000.

Hinton, Geoffrey, and Terrence J. Sejnowski, editors. *Unsupervised Learning: Foundations of Neural Computation.* MIT Press, 1999.

Hopfield, J. J., and Carlos D. Brody. *What is a Moment? Transient Synchrony as a Collective Mechanism for Spatiotemporal Integration.* Technical report, Princeton University and New York University. Also in *Procedures of the National Academy of Sciences USA 97,* pages 13919-13924.

Howe, Michael, and Risto Miikkulainen. *Hebbian Learning and Temporary Storage in the Convergence-Zone Model of Episodic Memory.* Technical report, Department of Computer Sciences, University of Texas at Austin, 2000.

James, Thurston. *The Prop Builder's Molding & Casting Handbook.* Betterway Books, 1989.

Kohonen, Teuvo. *The Self-Organizing Map (SOM).* Available at www.cis.hut.fi/projects/somtoolbox/theory/somalgorithm.shtml.

Kuipers, Benjamin, and Yung-Tai Byun. "A Robot Exploration and Mapping Strategy Based on a Semantic Hierarchy of Spatial Representations." In *Toward Learning Robots*, pages 47-64. Walter Van de Velde, editor. MIT Press, 1993.

Kuipers, Benjamin. *The Spatial Semantic Hierarchy.* Technical report, Computer Science Department, University of Texas at Austin, 1999.

Kuperstein, Michael. *Neural Network System for Adaptive Sensory-Motor Coordination of Multijoint Robots for Single Postures.* US Patent #4,884,216, 1989.

Lewis, M. Anthony, and Mark E. Nelson. *Look Before You Leap: Peering Behavior for Depth Perception.* Presented at the Simulation of Adaptive Behavior Conference, 1998.

Lewis, M. Anthony. "Visual Navigation in a Robot using Zig-Zag Behavior." In *Neural Information Processing Systems 10*, MIT Press, 1998.

Maass, Wolfgang, and Christopher M. Bishop, editors. *Pulsed Neural Networks.* A Bradford Book, MIT Press, 1999.

Marjanovic, Matthew J. *Learning Maps Between Sensorimotor Systems on a Humanoid Robot.* MSC Thesis, Massachusetts Institute of Technology, 1995.

Marjanovic, Matthew, Brian Scassellati, and Matthew Williamson. *Self-Taught Visually-Guided Pointing for a Humanoid Robot.* Technical report, Massachusetts Institute of Technology, 1996.

Marsland, Stephen, Ulrich Nehmzow, and Jonathan Shapiro. *A Real-Time Novelty Detector for a Mobil Robot.* European Advanced Robotics Systems Conference, Salford, 2000.

Martin, Paul, and Ulrich Nehmzow. *"Programming" by Teaching: Neural Network Control in the Manchester Mobile Robot.* Conference on Intelligent Autonomous Vehicles 1995, Helsinki, June 12-14, 1995.

McAuley, J. Devin. *Learning to Perceive and Produce Rhythmic Patterns in an Artificial Neural Network.* TR-371, Indiana University, 1993.

McClelland, James L., Bruce L. McNaughton, and Randall C. O'Reilly. *Why There are Complementary Learning Systems in the Hippocampus and Neocortex: Insights from the Successes and Failures of Connectionist Models of Learning and Memory*. PDP.CNS.94.1, Carnegie Mellon University and University of Arizona, March 1994. Also in *Psychological Review 102*, 1995: 419-457.

Miikkulainen, Risto. *Trace Feature Map: A Model of Episodic Associative Memory.* Technical report, University of Texas at Austin, 1991. Also in *Biological Cybernetics 66,* 1992: 273-282.

Mobus, George E. *Adaptrode-Based Neurons.* Technical report, Western Washington University.

Nehmzow, Ulrich. *An Episodic Mapping Algorithm for Mobile Robot Self-Localization: "Meaning" through Self-Organization.* Technical report, Manchester University, 1998. Also in *International Workshop on "Computation for Metaphors, Analogy and Agents",* Aizu Wakamatsu, 1998.

Pearman, Richard A. *Power Electronics: Solid State Motor Control.* Reston Publishing Company, 1980.

Platt, John. "A Resource-Allocating Network for Function Interpolation." In *Unsupervised Learning: Foundations of Neural Computation*, pages 341-354. Geoffrey Hinton and Terrence J. Sejnowski, editors. A Bradford Book, MIT Press, 1999.

Posner, Michael I., and Stanislas Dehaene. "Attentional Networks." In *Cognitive Neuroscience: A Reader*, pages 156-164. Michael S. Gazzaniga, editor. Blackwell Publishers, 2000.

Ross, J. N. *The Essence of Power Electronics.* Prentice-Hall Europe, 1997.

Saffiotti, Alessandro, Enrique H. Ruspini, and Kurt Konolige. *Blending Reactivity and Goal-Directedness in a Fuzzy Controller.* Procedures of the Second IEEE Conference on Fuzzy Systems, 1993.

Saffiotti, Alessandro. *Fuzzy Logic in Autonomous Robotics: Behavior Coordination.* Procedures of the 6th IEEE International Conference on Fuzzy Systems, 1997.

Sallans, Brian, and Geoffrey E. Hinton. *Using Free Energies to Represent Q-Values in a Multiagent Reinforcement Learning Task.* Technical report 19-2000, Department of Computer Science, University of Toronto, Canada, 2000.

Serling, Thomas Sterling, editor. *Beowulf Cluster Computing with Windows.* MIT Press, 2002.

Serling, Thomas, et al. *How to Build a Beowulf.* MIT Press, 1999.

Shephard, W., L.N. Hulley, and D.T.W. Liang. *Power Electronics and Motor Control.* Cambridge University Press, 1995.

Specter, David H.M. *Building Linux Clusters.* O'Reilly & Associates, 2000.

Squire, Larry R., and Eric R. Kendal. *Memory: From Mind to Molecules.* Scientific American Library, 2000.

Stevens, Roger L. *Serial PIC'n: PIC Microcontroller Serial Communications.* Square 1 Electronics, 1999.

Sun, Ron. *Introduction to Sequence Learning.* Technical report, CECS Department, University of Missouri, 1999. Also in *Sequence Learning: Paradigms, Algorithms, and Applications.* R. Sun and L. Giles, editors. Springer-Verlag, Berlin, 2000.

Suttin, Richard S., and Andrew G. Barto. *Reinforcement Learning: An Introduction.* MIT Press, 1998. Also available at www-anw.cs.umass.edu/~rich/book/the-book.html

Van de Velde, Walter, editor. *Toward Learning Robots.* MIT Press, 1993.

Whitledge, Jeff. *Pandamat: Controlling an Animat with Pandemonium.* Master's Thesis, University of Memphis, 1995. Also available at www.cswnet.com/~jwhitled/pandamat.htm

REFERENCES

4QD Electric Speed Controllers
www.4qd.co.uk
Abacom Technologies
www.abacom-tech.com

Acroname
5621 Arapahoe Ave, Suite C
Boulder, Colorado 80303
www.acroname.com

Ampro Computers
5215 Hellyer Avenue #110
San Jose, California 95138-1007
www.ampro.com

Cygnal
www.cygnal.com

Delphion, Inc.
3333 Warrenville Road
Suite 600
Lisle, Illinois 60532
www.delphion.com

DigiKey Corporation
701 Brooks Avenue South
Thief River Falls, Minnesota 56701
www.digikey.com

EMJ Embedded Systems
220 Chatham Business Drive
Pittsboro, North Carolina 27312
www.emjembedded.com

Globe Motors
www.globe-motors.com

Humanoid Robotics Group
Artificial Intelligence Laboratory
Massachusetts Institute of Technology
www.ai.mit.edu/projects/humanoid-robotics-group

HVW Technologies Inc.
3907 - 3A St. N.E. Unit 218
Calgary, Alberta T2E 6S7
Canada

ImageMagick
www.imagemagick.org

InnoMedia
90 Rio Robles
San Jose, California 95134
www.innomedia.com

Interlink Electronics
546 Flynn Road
Camarillo, California 93012
www.interlinkelec.com

Intel Performance Libraries
developer.intel.com/software/products/perflib

Intel Developer Site
developer.intel.com

International Rectifier
www.irf.com

Maxim/Dallas
Maxim Integrated Products, Inc.
120 San Gabriel Drive
Sunnyvale, California 94086
www.maxim-ic.com

Microsoft Research
One Microsoft Way
Redmond, Washington 98052
research.microsoft.com

Northern Tool & Equipment
www.northerntool.com

OOPic
Savage Innovations
www.oopic.com

Philips Semiconductor
1109 McKay Drive
San Jose, California 95131
www.philips.com
www.semiconductors.philips.com

Pittman Bulletin LCG, DC Gearmotors
www.pittmannet.com

Precision Navigation, Inc
5464 Skylane Blvd., Suite A
Santa Rosa, California 95403
www.precisionnav.com

Quality Kits
49 McMichael St
Kingston, Ontario K7M 1M8
Canada
www.qkits.com

Radiometrix
Hartcran House
Gibbs Couch
Carpenders Park
Hertfordshire
WD1 5EZ
England
www.radiometrix.co.uk

Robot Group, Austin Texas
www.robotgroup.org

Scott Edward's Electronics
www.seetron.com

Sharp Microelectronics Group
www.sharpsma.com

Small Parts Inc.
13980 N.W. 58th Court
P.O. Box 4650
Miami Lakes, Florida 33014-0650
www.smallparts.com
www.engineeringfindings.com

SRI International
Artificial Intelligence Center
333 Ravenswood Ave.
Menlo Park, California 94025
www.ai.sri.com

Team Delta
www.teamdelta.com

Tower Hobbies
PO Box 9078
Champaign, Illinois 61826-9078
www.towerhobbies.com

Vantec Remote Radio Control Systems
460 Casa Real Plz.
Nipomo, California 93444
www.vantec.com

Videre Design
www.videredesign.com

Vishay
www.vishay.com

Zeiner's Bass Shop
737 S. Washington #6
Wichita, Kansas 67211
www.zeiners.com

ZF MicroDevices
1052 Elwell Court
Palo Alto, California 94303
www.zfmicro.com

INDEX

C

D

E

F

G

H

I

J

K

L

M

N

O

P

Q

R

S

T

U

V

W